RECENT ADVANCES
IN
DEVELOPMENTAL BIOLOGY

Recent Advances in Developmental Biology

Vol. 6
Metamorphosis in Invertebrates

A.P. DIWAN
N.K. DHAKAD

ANMOL PUBLICATIONS PVT LTD
New Delhi-110 002

ANMOL PUBLICATIONS PVT LTD
4374/4B, Ansari Road
Daryaganj, New Delhi-110 002

Recent Advances in Developmental Biology

First Edition 1995
ISBN 81-7488-039-9

PRINTED IN INDIA

Published by J.L. Kumar for Anmol Publications Pvt. Ltd. and
Printed at Mehra Offset Press, Delhi.

Preface

The fundamentals of Developmental Biology are presented within the framework of scientific discovery. Researches in Embryology have made almost incredible strides in the past few decades. Consequently, the episting concepts of the developmental biology have expanded. There has been a revolution indeed in this derection.

The text integrates the desreptive, experimental and biochemical approches into a conceptual framework for the analysis of development. All important points are illustrated diagramatically. The title is not intended to be comprehensive, nor could it be at this length, but concentrates as putting across the basic principles of the subject as briefly and lucidly as possible. It does this with the aid of careful selected examples, some recent, and other classic of the field and with numerous illustrations. The aim is to enthuse the reader with this active and exciting area of research and to lay a solid foundation on which further study of its various facts may be based.

Authors

Contents

1

Metamorphosis in Crustaceans

There is general agreement that within the free-living Crustacea the sequence of larval stages and the accompanying changes that ultimately result in the juvenile animal bear little resemblance to the phenomenon of metamorphosis as it normally applies to development in the holometabolous Insecta. Most crustceans are anamorphic in their postembryonic growth, successively adding new segments and appendages in their progressive development through a number of larval stages to the adult. Some degree of metamorphosis, however, has been superimposed on these anamorphic stages of development. This varies from the relatively simple adaptation of appendages for swimming to a complete reorganization of the animal for a parasitic way of life. Snodgrass (1956) considers metamorphosis in a number of groups of Crustacea, giving examples of the progressive morphological changes, and suggests that from the variety of examples observed it would appear that metamorphosis in the Crustacea has developed separately in each order. From these examples it is not possible to define clearly a common, specific point during development when metamorphosis occurs in all crustaceans. However, if one accepts the more liberal definition of crustanean metamorphosis recommended by Passano (1961)—a change in form at a particular point in the animal's life—and includes the physiological and behavioural changes, it is then possible to consider the gradual and successive changes leading up to this transition, the external factors influencing these changes, the internal reorganization accompanying the more obvious morphological modifications, and the mechanisms controlling and regulating this period of development.

It is unfortunate that any consideration of crustacean metamorphosis is severely limited to a few examples, incomplete within themselves, by the paucity of information on most aspects of development. Within a few of the Cirripedia and Decapoda there have been limited efforts to describe the external and internal changes that occur during larval development and metamorphosis and the mechanisms that regulate these changes. In a consideration of these few examples a brief account of the normal pattern of larval development is necessary, including the point during development at which metamorphosis occurs. Inasmuch as this point is dependent upon definition it will be found that the literature frequently contains conflicting views on the exact stage of development of a particular species that represents metamorphosis.

Larval Development

In the present consideration, the larvae of the Cirripedia, Natantia, and Brachyura will serve as examples of three general types of metamorphosis in the free-living Crustacea. In the Cirripedia, development of the naupliar stages is a gradual progression that is interrupted by the extreme morphological modifications embodied in the cypris. In the Natantia, the gradual and progressive development of the zoeae is uninterrupted, and metamorphosis occurs without any comparable intermediate stage in the Brachyura, many of the morphological characters of the intermediate stage, the megalopa, appear during development of the later zoeal stages, and metamorphosis to the first crab is neither as abrupt as observed in the Cirripedia, nor has it the smooth transition observed in the Natantia.

The larval development of the Cirripedia has been described for a number of species, both from material in the plankton and from rearing the animals under laboratory conditions. There are six naupliar stages which, through successive molts, are distinguished by increased setation of the three pairs of appendages, changes in carapace shape, and the addition of carapace spines. The cypris larva, which does not feed, bears little resemblance to the nauplii, having modified antennules for attachment, six pair of thoracic appendages, and a completely different carapace shape. Following attachment to the substratum by a cementing substance produced by the cement glands, metamorphosis to the pinhead barnacle occurs and, although molting of the chitinous exoskeleton of the internal soma of the barnacle continues at regular intervals, the formation of external calcareous plates and their continuous growth marks a radical departure from either the nauplius or the cypris.

Larval development within the Natantia, which includes many of the shrimps, follows a variable number of zoeal stages. Through a series of molts, the thoracic appendages are added or increase in complexity, abdominal segments are added, and the appendages develop. In the development of *Palaemonetes kadiakensis,* metamorphosis to the first postlarva occurs after six zoeal stages without any intermediate stage comparable to the cypris of the Cirripedia.

Within the Brachyura, or true crabs, zoeal stages progress through a number of molts, at which time appendage setation becomes more complex, segmentation of the abdomen is completed, and abdominal appendages are added. The final zoeal molt gives rise to a megalopa in which the zoeal appendages are further modified, the thoracic appendages become functional, and which, in general morphology and behaviour, more closely resembles the juvenile crab. At the next molt the megalopa metamorphoses to the first crab stage.

The number of larval stages within any one species was once considered to be constant, but variability has been observed within the developmental stages of Crustacea representing a number of specific groups. Although there is general agreement on the occurrence of six naupliar stages prior to the cypris stage in development of *Balanus,* a number of authors have

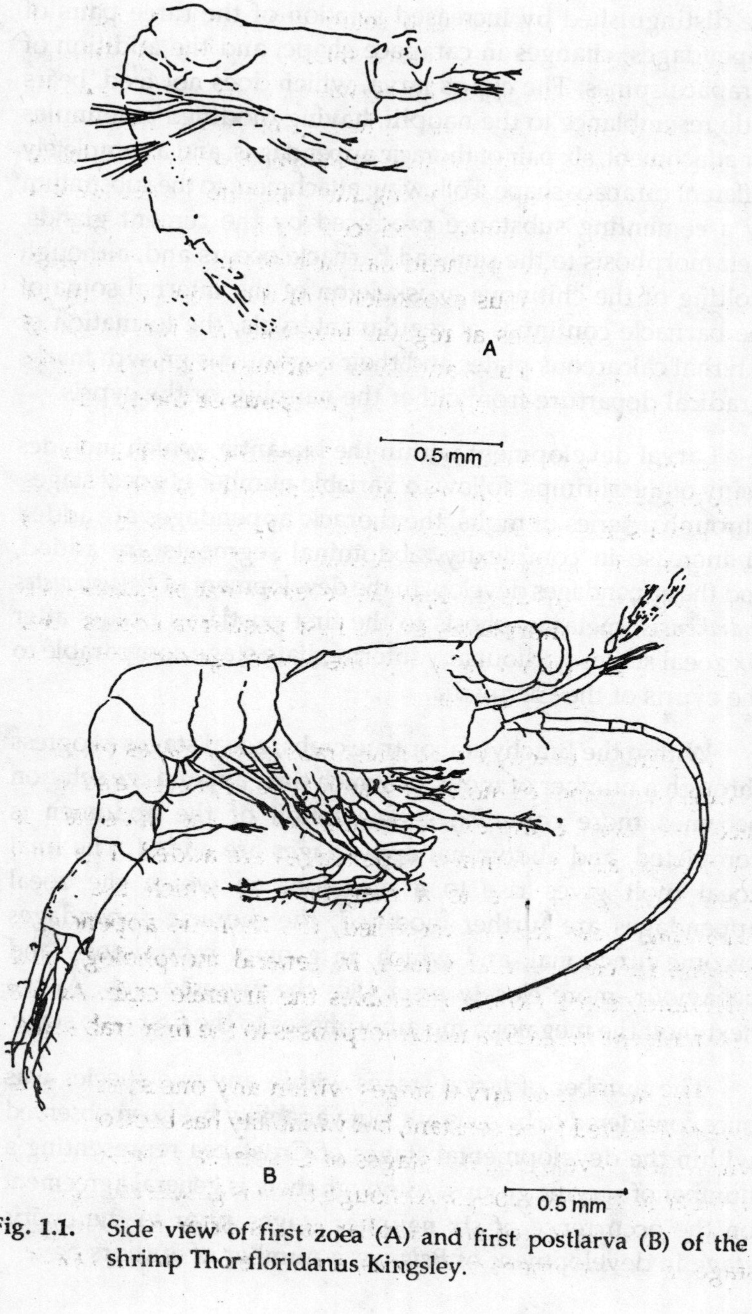

Fig. 1.1. Side view of first zoea (A) and first postlarva (B) of the shrimp Thor floridanus Kingsley.

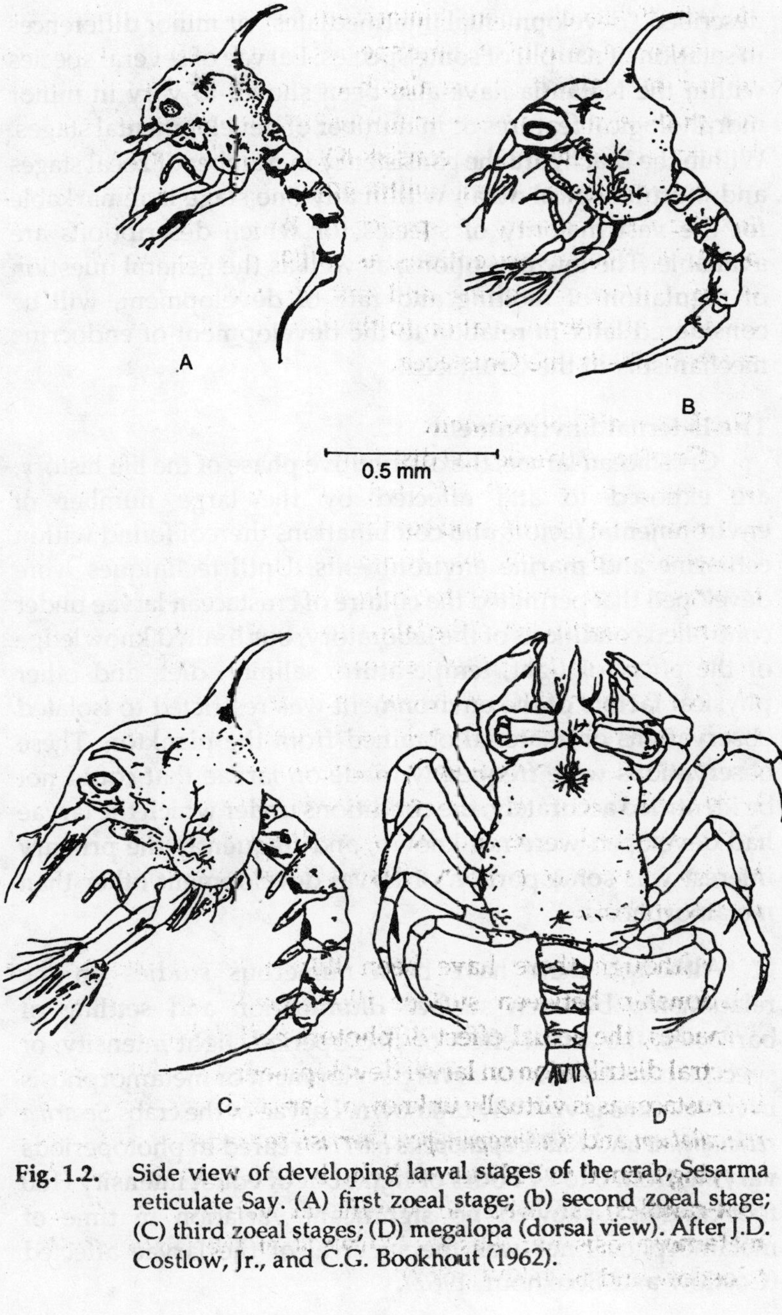

Fig. 1.2. Side view of developing larval stages of the crab, Sesarma reticulate Say. (A) first zoeal stage; (b) second zoeal stage; (C) third zoeal stages; (D) megalopa (dorsal view). After J.D. Costlow, Jr., and C.G. Bookhout (1962).

described "developmental intermediates" or minor differences in setation of nauplii of some species. Larvae of several species within the Natantia have also been shown to vary in minor morphological features or in number of developmental stages. Within the Brachyura the consistency in number of zoeal stages and morphological detail within any one stage is remarkable for the vast majority of species for which descriptions are available. The few exceptions, as well as the general question of regulation of molting and rate of development, will be considered later in relation to the development of endocrine mechanisms in the Crustacea.

The External Environment

Crustacean larvae, the distributive phase of the life history, are exposed to and affected by the large number of environmental factors and combinations thereof found within estuarine and marine environments. Until techniques were developed that permitted the culture of crustacean larvae under controlled conditions of the laboratory, our limited knowledge of the effect of light, temperature, salinity, diet, and other physical factors of the environment was restricted to isolated observations on material obtained from the plankton. These observations were frequently made on larvae that could not be identified accurately, the conditions under which the larvae had developed were not known, and frequently the primary interest was some portion of larval development other than metamorphosis.

Although there have been numerous studies on the relationship between surface illumination and settling of barnacles, the actual effect of photoperiod, light intensity, or spectral distribution on larval development or metamorphosis in crustaceans is virtually unknown. Larvae of the crabs *Sesarma reticulatum* and *Rhithropanopeus harrisii* reared in photoperiods varying from 0 to 24 hours of light, but of equal intensity (100 foot-candles), showed no significant variation in time of metamorphosis nor was the survival of the larvae affected (Costlow and Bookhout, 1962).

There have been a number of studies on the effect of temperature and salinity on survival and rate of larval development but, apart from noting and general effect on rate of development, few of these have considered specifically the problem of temperature effects on metamorphosis. Studies on the larvae of the Brachyura have shown that in some species the zoeal stages may develop normally in a wide range of combinations of salinity and temperature, whereas in other species mortality is extremely high at the time of the final zoeal molt to the megalops. The majority of the larvae that attain the megalops stage under these adverse conditions however, do, metamorphose to the first crab. Studies on development of larvae of the blue crab, *Callinectes sapidus*, have demonstrated the way in which the interaction of salinity and temperature can specifically affect the time of metamorphosis to the first crab (Costlow, 1967). As megalops maintained under a variety of temperature-salinity combinations showed a general delay in time of metamorphosis as the temperature decreased. In the, lower temperatures, however, this delay in the time of metamorphosis was further accentuated by an increase in salinity. For a series of megalops maintained at 20 and 30°C, the Q_{10} values for time of metamorphosis increased from 2.34 at 20 parts per thousand (ppt) salinity to 3.24 at 40 ppt salinity. For megalops maintained at 25 and 25°C, the values increased from 3.82 to 659 Q_{10}, respectively. From these values it is apparent that the delay in time of metamorphosis is not simply a function of reduced temperature; however, an explanation must await a more complete understanding of the physiological and biochemical changes that accompany metamorphosis.

Studies on the culture of Cirripedia in the laboratory have suggested that there is considerable variability in the nutritional requirements between different species. Thus the nauplii of some species will develop normally to the cypris on a number of different species of unicellular algae and diatoms and metamorphose to the pinhead barancle. With other species of

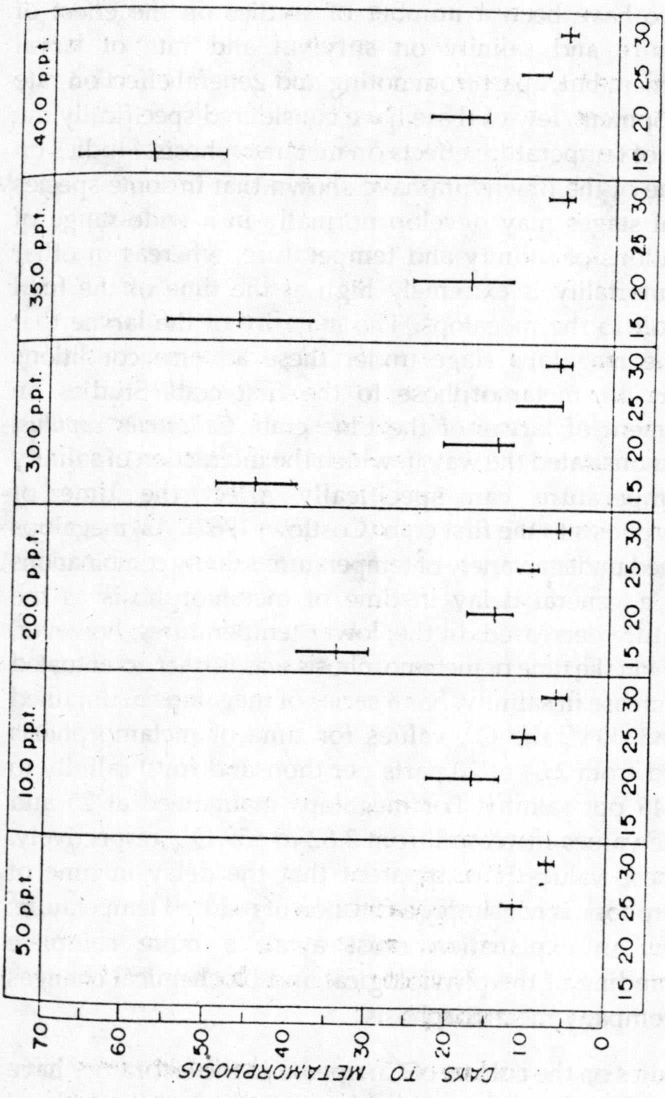

Fig. 1.3. Duration of megalops stage of Callinectes sapidus Rathbun maintained at 23 different combinations of salinity and temperature. The vertical line indicates the range, and the middle horizontal bar indicates the mean time to metamorphosis.

barnacles, however, there is no development beyond the naupliar stage on an algal diet, and only when food of an animal source as made available will the larvae metamorphose. Within the Natantia some, but not all, species of unicellular algae constitute an adequate diet for survival of the larvae and metamorphosis. Broad (1957a, b and c) has clearly demonstrated that the time required for development and metamorphosis is directly associated with the diet and that use of *Artemia* nauplii as a food source reduces the time required for metamorphosis and also increases survival at metamorphosis to the postlarval stage.

Survival of Brachyura larvae is extremely low on a diet of unicellular algae alone, and metamorphosis rarely occurs. While a mixed diet of algae and *Artemia* nauplii may increase survival, the time required for metamorphosis is sill considerably longer than that observed for zoeae and megalops maintained on either *Artemia* nauplii or on fertilized *Arbacia* eggs.

The Internal Environment

Although there have been detailed descriptions of the external morphological changes that occur at metamorphosis in a number of Crustacea, little is known about the internal reorgaization that accompanies them. Descriptions exist of the organ systems of some adult Crustacea, and the cytological changes associated with molting, sexual differentiation, and gonad development are known for representative examples within most of the major groups. Only a few of these have been complemented by studies on the physiological processes involved or the biochemical changes that occur. With very few exceptions these studies have been limited to the juvenile or adult Crustacea and have not considered the complicated changes that occur during larval development or those associated with metamorphosis.

Detailed descriptions of the histological changes and general reorgaization that accompany metamorphosis from the cypris to the pinhead barnacle are available (Doochin, 1951; Bernard and Lane, 1962). Walley (unpublished manuscript) has followed the cytological and histological changes during development of the six naupliar stages, the cypris, and the metamorphosed punhead. The sequence of changes she observed in different organ systems during development and metamorphosis of *Eialanus balanoides* is summarized. Although there are some differences in the observations of these three authors, it is apparent that metamorphosis in the Cirripedia involves a highly complex series of changes. These changes, as shown are not confined to the cypris stage. Some are innitiated during the naupliar stages, to be completed within the cypris stage, while other portions of the complex process of histolysis and reorganization take place during the cypris stage and carry over into the settle pinhead barnacle. These studies, plus the earlier observations by Walley (1964) that reorganization within the cypris stage is accompanied by a sequence of histolysis followed by phagocytosis, emphasizes that some similarities do exist between metamorphosis in barnacles and metamorphosis of holometabolous insects.

Although there have been isolated studies on particular tissues or cells within developing larvae of the Natantia and Brachyura, a complete description of the histological and cytological changes accompanying metamorphosis in these forms is not available.

Size, coupled with the difficulties in providing reasonable numbers of barnacle nauplii and cyprids of known age and stage, has discouraged studies on the physiology of development and metamorphosis in the Cirripedia. Clarke (1947) observed that barnacle larvae maintained a copper concentration that was higher than that found in sea water but also noted that this was not uncommon in ivertebrates in which hemocyanins are utilized as a respiratory pigment. Bernard and Lane (1961) used histochemical techniques in an effort to determine changes in the absorption and excretion of copper during settlement and attachment of cyprids of *B. amphitrite niveus*. In the planktonic cypris, copper was concentrated within the loosely organized tissues of the anterior one third of the animal as well as in the food mass of the gut. Although they did not identify the actual mechanisms involved, they suggest that absorption takes place through the respiratory surfaces and excess copper is excreted through the epithelial lining of the hind gut. They further suggest that the high concentration of copper may be indicative of a dynamic exchange and that these same cells may be involved in the concentration of other ions from sea water during the process of metamorphosis.

Although there have not been any studies on enzyme systems in the larvae of the Cirripedia, or possible changes in these systems associated with metamorphosis of the cypris, there is indirect evidence that certain changes do occur. Carbonic anhydrase, associated with respiration in many animals, has been identified in the shell-forming mantle tissue of *B. improvisus* adults (Costlow, 1959). Cypris larvae that had been reared from nauplii under laboratory conditions were maintained in sea water solutions of carbonic anhydrase inhibitors and, while metamorphosis to the pinhead was not delayed, development of the typical six-plate shell of the adult barnacle never proceeded normally. If the metamorphosed pinhead barnacles were removed from the inhibitor to the normal environment within three to six days normal shell development and growth were gradually resumed. If removal was delayed, however, to eight days following metamorphosis, normal shell development and growth did not occur. These results suggest that mechanisms may be present in the shell-forming tissues of the cypris and pinhead barnacle that determine, initiate, and direct the development of the typical six-plate shell within a "critical period" of four to six arms following metamorphosis. Four of these plates are define in the cuticle of recently settled pinheads, but if inhibition in shell growth occurs at this time the determinants may not function; normal development thus cannot be initiated, and the barnacle does not develop beyond the pinhead stage when the inhibitors are removed.

TABLE-1

Sequence of changes in different organ systems during metamorphosis in balanus balanoides.

Organ	*Naupliar Stages* *I II III IV V VI*	*Cypris* *Free-swimming*	*Cypris* *Settled*	*Adult*
Antennule	Simple 4 segment	4 Segmented, highly Specialized	Degenerate	Absent
Antennae	Biramous, swimming and feeding	Apparently	Absent	
Mandibles	Biramous, swimming and feeding	Rudimentary adult mandible derived		
Maxillae 1st and 2nd	Rudimentary	Incorporated into rudimentary oral cone	Achieve adult form	Functional, feeding Functional
Thoracic appendages	Rudiments develop in thoraces region	Function as swimming appendages	Internal histolysis, reorganation	functional, feeding appendages
Cuticle	Thin and flexible (1-2μ) except on mouth parts	Bivalve carapee. hard (7μ); im permeables thin (1-2μ) on mantle cavities and thorax		heavily ealeitied shell and eaticle over thorax and mantle
Epidermal and frontal glands	Scattered epidermal glands 2 gland cells present in each; frontal horn glands similar, become frontal gland in Cypris		Histolysis	Absent
Musculature	Naupliar muscles functional until nauplirs-cypris molt	Functional cypris muscles	Rudimentary adult	Adult
Nervors system	Pre-oral brain, 2 post-oral ganglia; developing post-mandibular ganglia	Morphological re-arrangement and limited histolysis	Extensive histolysis and regeneration	Adult

Naupliar eye	Single median eye of 3 components:				2 lateral pigmented photoreceptors "ophthalmie ganglion"
Compound eyes		Development	Functional	Lost	Absent
Frontal filaments	Attached to lateral lobes of brain		Incorporated with cmpound eyes	Lost	Absent
Gut	Cuticle-lined esophagus and hindgut; midgut divided by constriction, feeding phase		Esophagus and hindgut closed; some cells sloughed from epithelium and digested, nonfeeding phase		Feeding resumed
Digestive gland	Absent		Appear at junction of esophagus and midgut as outgrowth and enlarge		functional
Antennal glands	Present and functional	Degenerating	Absent		
Maxillary gland	Absent	Rudimentary	Presumed functional		Functional
Yellow cells	Absent	Appear in antennae	Migrate to posterior region of carapace after ecdysis	Disperse	Absent
Oil cells	Precursors around midgut constriction; in haemocoel; multi after ecdysis		Migrate to anterior region of carapace	Gradually	dueed and finally lost
Haemocytes	Present		Become Phagocytic	2nd phase phagocytosis	?
Cement glands		Rudimentary	Completely developed	Functiona	

* Adapted courtesy of Dr. L.J. Walley.

Although certain aspects of the physiology of adult Natantia and Brachyura have received considerable attention, comparable studies on the developing larvae have not been made. Of the few studies on the physiology of these larvae, most have either been made on planktonic forms, with little knowledge of the species or the particular stage of development (Zeuthen, 1947), or have been concerned primarily with changes other than those associated with metamorphosis (Vernberg and Costlow, 1966). Studies on the free amino acids in larvae of two species of Brachyura reared under controlled laboratory conditions have demonstrated a pamern in which the number of amino acids is sharply reduced during the megalops stage. This is followed by an increase immediately following metamorphosis to the first crab (Costlow and Sastry, 1966). In addition to the changes associated with metamorphosis, the relative abundance of a number of these amino acids is directly associated with the salinity and temperature of the environment in which development and metamorphosis occur.

Recent studies on osmoregulation in crab larvae suggest the changes that to occur are not directly associated with metamorphosis to the crab but actually take place late in zoeal development or at the time of the molt to the megalops (Kalber and Costlow, 1966). Working with larvae of *Rhithropanopeus harrisii* reared under controlled conditions in the laboratory it was found that osmoregulation in the final stage zoeae (stage IV) and the megalops is more comparable to that observed in the adult crab that the earlier zoeal stages. Subsequent studies (unpublished) have shown that the larvae of *Callinectes sapidus* show a gradual loss of hyperregulatory ability from the first to the seventh zoeal stage but rapidly regain it following the molt to the megalops stage.

Behaviour

Reese (1964) has recently reviewed the literature on the behaviour of marine animals, including the Crustacea. He comments on the extensive literature dealing with aggressive behaviour in marine decapod Crustacea, but most of these studies are limited to the adult animals and have not considered other types of behaviour. Research on the development of behaviour during the larval stages or changes in behaviour patterns associated with metamorphosis is very limited.

Most crustacean larvae are usually pelagic for a considerable portion of their development and, during this period, may be attracted to light. Toward the end of their pelagic phase they may either become negative to light or the response to light may be replaced by a positive response to gravity or the substratum. These differences have been observed in the field for many years, frequently with views on their adaptive value (Bousfield, 1954), and Knight-Jones and Morgan (1966) have recently reviewed the effects of pressure on marine animals, discussing in considerable detail the different types of response observed with nauplii and cypris larvae and with zoeae and megalops.

Within the Cirripedia the larval behaviour at the time of settling has received considerable attention, and a number of factors are thought to be involved. In discussing the orientation of barnacles at settlement, Crisp and Barnes (1954) point out that the cypris larvae explore the available surfaces extensively before actually settling, and a detailed account of this exploratory behaviour is given by Barnes (1955). The cypris settle in grooves and concavities, a tendency to which the term "rugophilic" has been applied (Crisp and Barnes, 1954). When settling in a groove, rugophilic settlement was usually characterized by rugotropism, the orientation of the anterior-posterior axis of the cypris to correspond with that of the groove, and was shown to be a tactile response to the contour of the surface. Established barnacle communities also affect settling, and the stimulus for this gregariousness is thought to be associated with a water insoluble chemical originating in the epicuticle of other barnacles.

Other environmental factors that have been shown to affect settling behaviour of barnacle cypris are the bacterial film of the substrate, light, water currents,. and surface angle. It should be noted, however, that settling precedes metamorphosis and that nothing is known of the physiological mechanisms underlieing the observed pattern of behaviour or the way in which these develop.

Thompson (1903), working with the final larval stage of the hermit crab *Pagurus longicarpus,* found that development of the nervous system was a continuous process but that the musoulature was further developed with each molt. At the time of the molt from the final zoeal stage of the glaucothoe, the musoulature was complete. From his observations on shell selection he concluded that the response of the glaucothoe to shells was identical to that observed in adult crabs and that they showed a preference for dextral shells. Coffin (1954) observed that the glaucothoe, although not normally occupying the shell until the end of the stage, did behave toward shells in a manner similar to the adult crabs. Reese (1962), working with the glaucothoe of three species of hermit crab reared under laboratory conditions, concurred with the earlier findings. The inception of behaviour of selecting shells occurs in the glaucothoe stage and is fully expressed when the animal encounters and enters its first shell. He further concluded that the preference for shells of a certain weight, as well as the ability of the glaucothoe to select between shells of different weights, appears to be innate.

The Mechanisms of Control

There has been very little interest in the mechanisms involved in the regulation of rate of larval development, molting in the larval stages, or metamorphosis. Although the endocrine systems and the processes they are thought to control in adult Crustacea have received considerable attention, little is known of the development of any endocrine system in crustacean larvae, the extent to which these mechanisms are altered during metamorphosis, or if osmoregulation, chromatophores, molting and metamorphosis in the larval stages are controlled by the same endocrine mechanisms that function in the adult animals.

The zoea of *R. harrisii* (Gould) have been shown to be isosmotic with seawater of 30 to 40 ppt salinity and to hyper-regulate in lower salinities, except during the first diecdysis, when they remain isosmotic (Kalber and Costlow, 1966). Zoeae in the last few hours of proecdysis also hyperregulate against 40 ppt salinity. Eyestalk extirpation during the second zoeal stage caused the larvae to lose the ability to osmoregulate against 10 to 30 ppt salinity within two hours after the operation. Preliminary experiments, however, have shown that the removal of both eyestalks from the final zoeal stage, stage IV, has very little effect on osmoregulation. Inasmuch as osmoregulation during the fourth stage zoea and the megalopa appear to be more comparable to that observed in the adult crabs, one must conclude that the mechanisms within the eyestalks that are associated with osmoregulation have become differentiated well in advance of metamorphosis to the first crab.

Numerous investigators have studied the chromatophores and their control in a number of adult Crustacea (Kleinholtz, 1961; Fingerman, 1965). Chromatophores have been described from the larvae of many species usually as an aid in identification, but little is known about their control or the manner in which this control may be affected by metamorphosis.

Larvae of the shrimp *Palaemonetes vulgaris* have been shown to respond to background colouration with changes in the chromatophores (Broch, 1960) and, although not specifying the source of the extracts, he further observed that the red chromatophore of an excised thoracic segment of a second zoea contracted on the addition of brain extract, expanded upon the addition of extract from the abdominal nerve cord, and contracted on the addition of eyestalk extract. From these results, Broch concluded that the endocrine control of chromatophores in larval *P. vulgaris* was comparable to that described for the adults.

Eyestalk extirpation in larvae of *P. pugio,* resulting in a response opposite to that reported for the adult animal, produced a pale condition that persisted through metamorphosis and indicated that there is some influence on the chromatophore system by the larval eyestalks (Hubschman, 1963). Further evidence of this influence has been found from injections of extracts of eggs and larval eyestalks of *Sesarma reticulatum* into eyestalkless fiddler crabs, *Uca pugilator* (Costlow and Sandeen, 1961). Chromatophorotropins were first identified within the eggs on the eleventh day of embryonic development. The activity of the chromatophorotropins increased gradually until the time of hatching, remained relatively constant throughout the nine days of zoeal development, and attained a higher level following development to the megalops stage. No appreciable change in chromatophorotropin level was noted following metamorphosis to the first crab, although extracts of adult *S. reticulatum* were found to have an activity approximately three times greater than the activity of the first crab stage.

Cycles of levels of activity of chromatophorotropins from eyestalk extracts have also been reported during the development of *S. reticulatum* larvae (Costlow, 1961). At the time of hatching the total activity was lowest, increased during the next 36 hours, and was reduced at the interval preceding the first larval molt. During the intermolt period of the second zoea the activity increased again but was reduced immediately prior to and following the second larval molt, increasing again during the intermolt period of the third zoeal stage. This same pattern was observed for extracts of eyestalks of the megalops. It was further noted that zoeae that were delayed in molting usually showed a reduction in total activity of the extracts at the time when molting would normally be expected to occur. If ecdysis did not follow, the total activity of the extracts again increased. On the basis of these results, several hypotheses are possible. First, the chromatophorotropin within the eyestalks of the larvae may fluctuate because of changes in titer associated with molting, but the hormone is not actually involved in the molting process. Although evidence has accumulated since these experiments which suggests that the X organ sinus gland complex of the larvae is not functional, one should not completely ignore the second hypothesis, i.e., that the cycle of activity also applies to the molt-inhibiting hormone. If the X organ sinus gland were functional and concentration of the molt-inhibiting hormone followed a cycle similar to that described for the chromatophorotropins, ecdysis would be preceded by a reduced titer, followed by an increase in concentration of the molt-inhibiting hormone and another larval intermolt period. The pattern of reduced activity or titer prior to molting and increasing titer following-molting would be repeated through all larval molts and carried over into the crab stages.

Shedding of the chitinous exoskeleton of the internal soma, the gills, and the mantle tissue occurs every three to five days in adult barnacles of temperate waters, but nothing is known about the mechanisms that control ecdysis. Histological studies of the body and mantle tissue have failed to disclose any tissues that might be associated with the control of molting. Within the Cirripedia, neurosecretory cells have been described from the central nervous system of the adult animals, but their function is not known, nor have comparable cells been described from any of the larval stages. The nauplii and adult animals have been shown to have chromatophorotropins that are similar in activity to those described from the eyestalks of adult and larval crabs, but the function is unknown. Removal of the paired eyes from recently metamorphosed pinhead barnacles, remnants of the naupliar eye that in the adult barnacles constitute the paired eyespots and the "ophthalmic ganglion" of Darwin, does not affect the molting frequency, but comparable experiments have not been successful with the late naupliar stages.

In adult malacostracans the efforts of numerous investigators have led to the concept that molting is controlled by two hormones: the molting hormone produced by the Y organ and the molt-inhibiting hormone produced by the X organ sinus gland complex of the eyestalks. Molting is initiated by the release of molting hormone from the Y organ, following the decrease in concentration of molt-inhibiting hormone from the X organ. If both the Y organs are removed, proecdysis is inhibited, but molting can be reinduced by implanting the glands into crabs from which they have been removed. Bilateral extirpation of the eyestalks, thus removing the X organ sinus gland complex, clim inates the molt-inhibiting hormone and molting is accelerated. If, however, both Y organs are absent, molting cannot be induced (Echalier, 1959; Passano, 1960).

Our knowledge of the development of these endocrine systems in the Macrura and Natantia is limited to three studies. Dahl (1957) described the embryology of the X organ in *Crangon allmamnni.* No figures or information on the activity of the cells were included, however, and his consideration of further development was limited to the observation that organization of the X organs and the sensory pore complex is essentially similar to that found in the adults. Dahl did comment on the absence of neurosecretory activity in these cells when appropriate strains were used. Pyle (1943), describing the X organ from eggs and larvae of the lobster *Homarus americanus,* posed questions on the development and function of the X organ in the larval stages but did not present any evidence about their function. Hubschman (1963), describing the detailed changes in the larval and early postlarval eyestalks of a number of species of *Palaemonetes.* found that the sinus gland could be recognized by the time of the fifth larval stage but not earlier. Even at the fifth larval stage the sinus gland was not a functional complex, but following metamorphosis it developed its definitive form. During larval life it was not possible to detect the accumulation of secretory material, but by the time of the third or fourth postlarval molt the sinus gland assumed functional importance. In addition, the ganglionic X organs found in the adult eyestalks were not functional during larval development, but secretory activity could be demonstrated in shrimp that completed several postlarval molts. Eyestalk extirpation did not produce any acceleration in the larval molting frequency or in the time of metamorphosis of the *Palaemonetes* larvae, Hubschman suggests that the molting cycle in the larvae, in the absence of a function X organ sinus gland complex. proceeds uninhibited. He further suggests that the rapid molting cycle in *Palamonetes larvae*, i.e., every two days, results from the production of a molting hormone by a larval molting gland, "if one exists," but is unable to offer any evidence to its location.

Within the larvae of the Brachyura, and the closely related Anomura, the limited studies on development of endocrine systems associated with larval molting and metamorphosis have also produced a poorly defined picture. Orlamunder (1942) found cells corresponding to the X organ of adults in eyes of developing embryos of *Birgus latro*, and Pyle (1943) described the same structure from eggs and larvae of the crab, *Pinnotheres maculatus*. Because of rearing problems, Pyle was not able to follow the development of the larvae through successive stages, but he did indicate that the sinus gland was not evident in late embryos or in the first zoeal stage. Neither Orlamunder 1942) nor Pyle (1953) considered the possible role of these structures in molting and metamorphosis of the larvae.

As had already been pointed out, the consistency that has normally been ascribed to the number of zoeal stages and molting frequency within any one species of the Brachyura was interpreted to mean that larval molting and rate of development were controlled by the same mechanism. Two exceptions to this consistency were the larvae of *Callinectes sapidus* and *Menippe mercenaria* in which an "extra" zoeal stage was first reported late in zoeal development under laboratory conditions. A zoea comparable to the "extra" eighth zoeal stage of *C. sapidus* subsequently has been described from the plankton. Detailed morphological studies on the larval development of *C. sapidus*, provided additional evidence that "extra" stages and "intermediate" stages were not unusual and gave some support to the hypothesis that larval molting and rate of development were not necessarily controlled by one mechanism (Costlow, 1965). The three general types of variability that were observed in the study of *C. sapidus* larvae suggested that at least two mechanism were involved: one that controlled molting and a second, normally synchronized with the first, that regulated the rate of morphological development in the larval stages. In the first type of variability, molting without any perceptible changes in morphology, the frequency of molting was maintained while the mechanism associated with regulation of development was inhibited. In the second type of variability, the "skipping" or elimination of a larval stage, the molting frequency was constant but the mechanism controlling development was accelerated. In the third type of variability, the molt resulted in a larva that combined morphological characters normally attributed to two zoeal stages. In these, the constant molting pattern would appear to have been preceeded by normal or accelerated development in the anterior portion of the larvae. In the posterior portion, however, morphological development was inhibited or normal. These observations suggested that the variability in morphology in associated with the malfunction of a regulatory mechanism. The fact that the variability is limited to the later larval stages was attributed to a gradual decline in the secretory activity of an endocrine mechanism rather than an abrupt cessation of activity.

In the absence of a complete description of the structure of the larval eye and any functional endocrine sites that may appear during development, it was necessary to begin with the hypothesis that the larval eyestalks did contain sites of endocrine activity comparable to those found in the adult crabs. If the X organ sinus gland complex of the larval eyestalks were functional, producing molt-inhibiting hormone and thus regulating the frequency of larval molts, removal of both larval eyestalks should result in accelerated molting within the zoeal and megalops stages. Eyestalk extirpation experiments on the larvae of three different species of Brachyura have suggested that the adult endocrine systems thought to control molting begin to function at different times in different species and do not function during zoeal development (Costlow, 1963, 1966a and b).

In *C. sapidus,* metamorphosis to the first crab stage is accelerated considerably by the extirpation of both megalops eyestalks within twelve hours following the final zoeal molt. If both eyestalks are removed on the first to fifth days following the final zoeal molt, the time of metamorphosis is not affected. Removal of one eyestalk on zero day of megalops life, the first twelve hours following the final zoeal molt, does not affect the time of metamorphosis, but extirpation on the first to fourth days resulted in a significant delay. That the acceleration observed is not a short-term effect is demonstrated by comparing the time of the first four postlarval molts for crabs that metamorphosed from megalops with both eyestalks removed on zero day, on the first to fifth days, and those that had only one eyestalk removed. Acceleration of molting continued only in those carbs from which both eyestalks were removed on zero day of megalops life, although the exaggerated increase in size was the same for all eyestalkless crabs. Observations on the metamorphosis of eyestalkless megalops of *C. sapidus* suggest that the X organ sinus gland complex is not functional within the first twelve hours of megalops life. If both eyes are removed during this period, i.e., before secretion of the molt-inhibiting hormone is initiated, metamorphosis is accelerated. If both eyestalks are removed after secretion has begun (first to fourth days). The time of metamorphosis is not accelerated. Unfortunately, zoeae from which both eyestalks were removed at any point in development did not survive long enough to provide for a comparison of zoeal molting frequency.

In an effort to determine the effect of eyestalk extirpation on molting frequency of the zoeae, similar experiments were conducted on the larvae of another crab, *Sesarma reticulatum.* In these experiments, however, extirpation was affected as early as stalked eyes developed in the zoeae, stage II. Removal of one eyestalk from the zoeae of *S. reticulation* during any

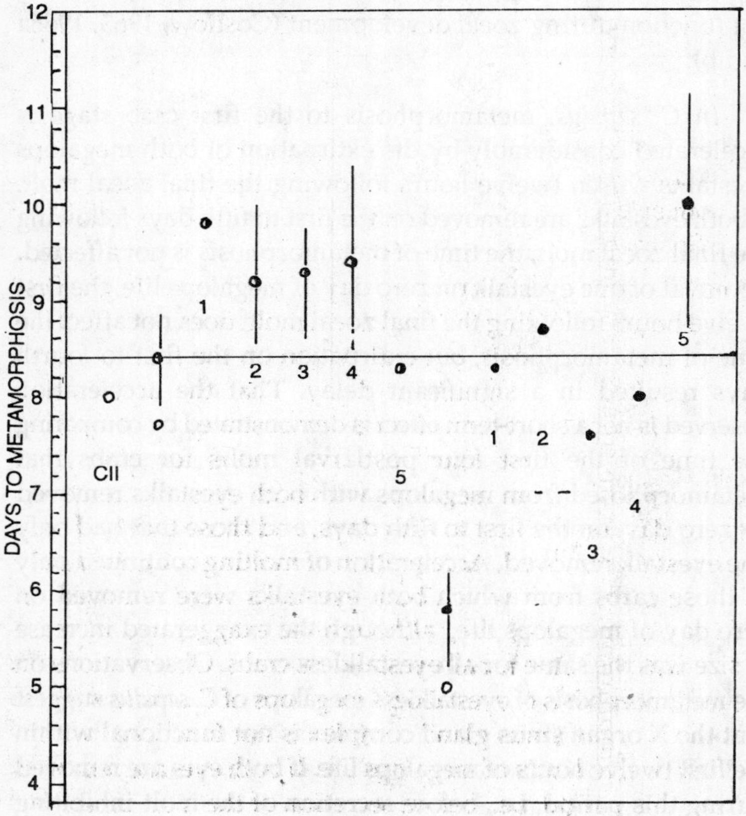

Fig. 1.4. Time required for metamorphosis of *Callinectes* sapidus Rathbun megalops of control series (horizontal line), megalops with centenna removed (CII), megalops with one eye removed on zero to fifth day 'following the final zoeal molt (O), and megalops with both eyes removed on zero to fifth day (●). Vertical line indicates the least significant difference of each series (LSD = SD. T_{o5}.

period of development did not affect the molting frequency or the number of larval stages. Removal of both eyestalks, simultaneously or individually at different times of development, resulted in the acceleration of metamorphosis of the megalops to the first crab, but only when extirpation

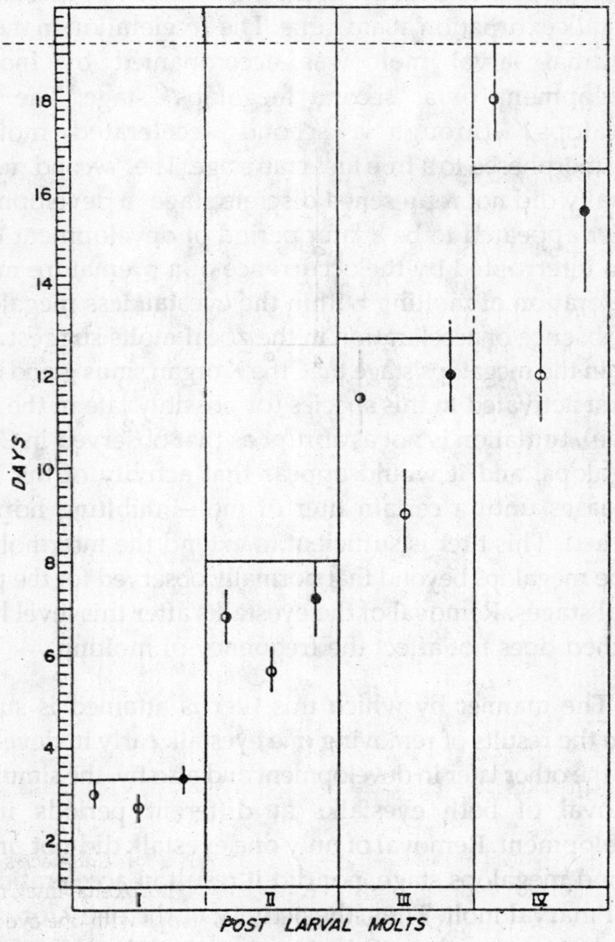

Fig. 1.5. Time of postlarval molts for crabs, *Collinectes* sapidus Rathbun, which metamorphosed from megalops of control series (horizontal line); megalops with one eye removed (O); megalops with both eyes removed on zero day following the final zoeal molt (O)′ and megalops with both eyes removed on first to fifth days following the final zoeal molt (●). Vertical line indicates the least significant difference of each series (LSD = SD X T_{o5}).

occurred prior to a critical period; fourth day of megalops development. The earlier zoeal molts were not accelerated by eyestalk extirpation at any time. The acceleration in the time of the final larval molt was accompanied by incomplete development, or a "second megalops" stage. The "second megalops," through a second accelerated molt, then metamorphosed to a true first-crab stage. The "second megalops" actually did not represent a discrete stage in development but rather appeared to be a later period of development that had been interrupted by the occurrence of a premature molt. The acceleration of molting within the eyestalkless megalops and the absence of acceleration in the zoeal molts suggest that it is within the megalops stage that the X organ sinus gland complex is first activated in this species (or possibly late in the stage III zoeae). Initiation is not as abrupt as that observed in *C. sapidus* megalops, and it would appear that activity of the X organ increases until a certain titer of molt-inhibiting hormone is reached. This titer is sufficient to extend the intermolt period of the megalops beyond that normally observed for the previous zoeal stages. Removal of the eyestalks after this level has been reached does not affect the frequency of molting.

The manner by which this titer is attained is suggested from the results of removing one eyestalk early in development and the other later in development and also by the simultaneous removal of both eyestalks at different periods in larval development. Removal of only one eyestalk did not produce a second megalops stage, nor did it result in acceleration of the final marval molt. Thus, the activity of the one X organ sinus gland as apparently sufficient to regulate the molting frequency of the megalops if it remains functional throughout larval life. However, if one eyestalk is removed during the final zoeal stage (stage III) and the remaining eyestalk is removed at different periods in megalops development, a sufficiently high level of activity of the one gland has apparently been reached by the fifth day of megalops development to regulate metamorphosis and prevent acceleration even if the remaining eyestalk is then removed. Also, if only one eyestalk remains up to the fourth day of megalops life and is then removed, a higher percent of second megalops result than if both eyestalks are intact through the fourth day and then removed. The degree of acceleration of metamorphosis is greater in megalops with only one functional eyestalk than in megalops that have intact eyestalks up to fourth day of megalops life. Thus, it would appear that the activity of the X organ sinus gland of one eyestalk is not eyestalk is not equal to the combined activity of two, and the resultant titer of moltinhibiting hormone at any point prior to the fourth day is not sufficient in one-eyed larvae to completely prevent accelerating of metamorphosis and or the resultant second megalops stage if the second eyestalk in removed on the fourth day.

While removal of both eyestalks early in zoeal development of *S. reticulation* did not result in acceleration of zoeal molts, it did result in a small number of zoeae which, at the time of the molt to the megalops stage, molted instead to an "extra" or supernumary zoeal stage. This occurred only in some zoeae from which both eyestalks were removed prior to the first day of the stage III zoeae, whereas others molted directly to the megalops stage. The stage IV zoeae, while continuing to develop toward the more advanced morphological condition of the megalops stage, retained many of the characteristics normally associated with the zoeae. This extra stage cannot be attributed to a change in the molting frequency of the larvae but rather suggests that early in larval development a second endocrine site or storage organ is developed within the zoeal eyestalks. The product of this site would appear to regulate the rate of morphological changes associated with a consistent number of zoeal stages, the transition from the final zoeal stage to the megalops stage, and possibly the changes involved in the metamorphosis of the megalops to the first crab stage. It is suggested that the activity of this second site begins during the final zoeal stage (stage III) and, if both eyestalks are removed prior to this point in development, the normal transition from the final zoeal stage to the megalops stage is not possible. The mechanisms controlling molting frequency in the larvae, however are not affected, and since morphological development has been retarded, an additional zoeal stage results rather than the megalops.

Inasmuch as *S. reticulatum* normally has only three zoeal stages during its complete development, one may question if there was a sufficiently long period of development to permit detection of any significant reduction in length of intermolt period that might result from removal of both eyestalks. That is, development of the three zoeal stages of *S. reticulatum* is normally completed in approximately 10 days at 25° C. The first zoeal stage does not have stalked eyes and occupies approximately three days of this total period. Only in the remaining two zoeal stages, when the eyes are stalked, can extirpation be conducted. If both eyes were removed on day one of stage II and one day was allowed for recovery of the larvae, a total of approximately five days of zoeal development remain in which acceleration of the additional two molts could occur.

Further evidence for acceleration of molting during zoeal development and the mechanisms that control rate of development and metamorphosis was sought in the development of *Rhithropanopeus harrisii*, a species of mud crab that normally has four zoeal stages and one megalops.

Removal of both eyestalks as early as the first day of the second zoeal stage did not result in a consistent, significant acceleration of any of the larval molts, further substantiating the hypothesis that the X organ sinus gland complex does not contribute to the control of molting in the zoeal stages. Also, there was no acceleration of the time of metamorphosis to the first crab, and the first indication of any decrease in duration of the intermolt period for crabs that metamorphosed from eyestalkless larvae was at the time of the third postlarval molt. Thus, it would appear that the X organ sinus gland complex of

R. harrisii is not functional until sometime after metamorphosis to the first crab and further demonstrates the extent of variability in mode of action and time of endocrine control of molting within larval stages of the Brachyura.

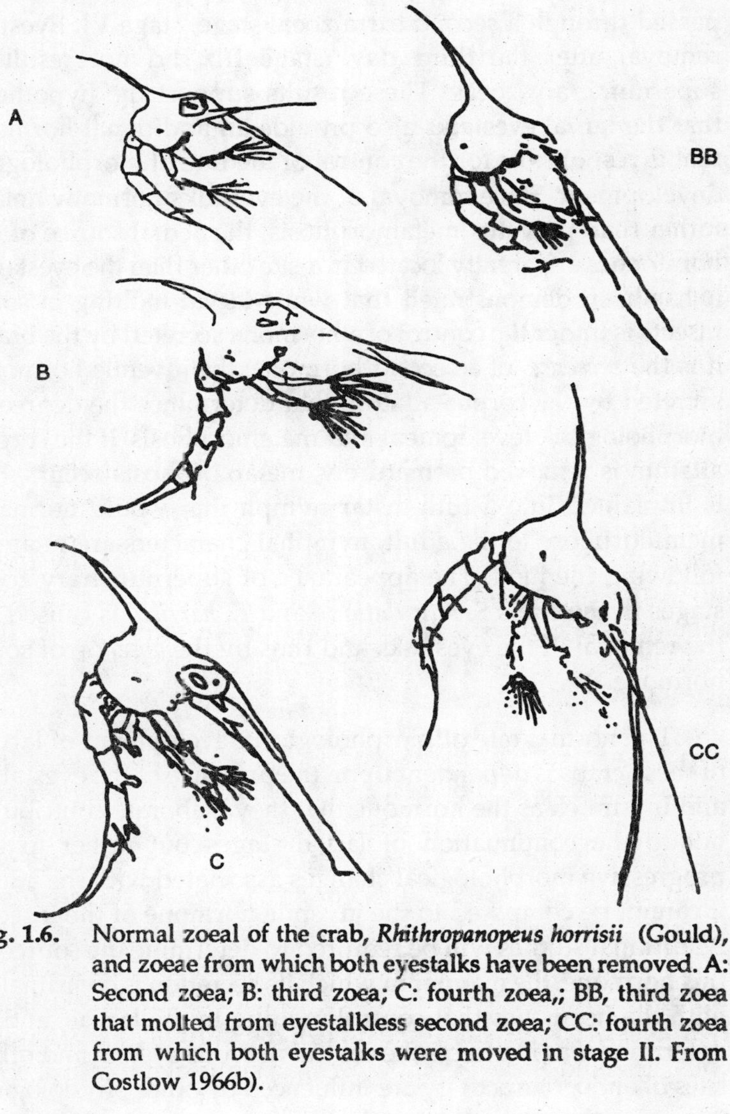

Fig. 1.6. Normal zoeal of the crab, *Rhithropanopeus horrisii* (Gould), and zoeae from which both eyestalks have been removed. A: Second zoea; B: third zoea; C: fourth zoea,; BB, third zoea that molted from eyestalkless second zoea; CC: fourth zoea from which both eyestalks were moved in stage II. From Costlow 1966b).

Removal of both eyestalks from zoeae of *R. harrisii* prior to the third day, stage III resulted in 45 to 70 per cent of the larvae passing through an extra or supernumerary zoeal stage before molting to the megalops. From 1 to 10 per cent of the destalked zoeae, depending upon the time of eyestalk extirpation, passed through a second extra zoeal stage, stage VI. Eyestalk removal after the third day (stage III) did not result in supernumerary zoeae. These results support the hypothesis that the larval eyestalks also provide an additional hormone that is responsible for the control of the rate of morphological development. Since removal of the eyestalks normally delays rather than prevents metamorphosis, the actual source of the hormone is apparently located in a site other than the eyestalks. It has been demonstrated that while larval molting in some insects is under the control of a hormone secreted by the brain, it is the *presence* of a second hormone, the juvenile hormone, secreted by the corpus allatum that determines the degree of morphological development and metamorphosis. If the corpus allatum is removed prematurely, metamorphosis occurs. If it is implanted into a fifth-instar nymph that would normally metamorphose to the adult, nymphal characters are retained following the molt. The appearance of supernumerary zoeal stages in the crabs *S. reticulation,* and *R. harrisii* is caused by the removal of the eyestalks and thus by the *absence* of some hormone.

The normal rate of morphological development of larvae of these crabs is dependent upon the presence of the eyestalks, and inasmuch as the hormone that they alaborate contributes not to the continuation of larval stages but rather to the progressive morphological changes associated with the adult, it cannot be compared to the juvenile hormone of the insects. Additional studies will be required to determine the source of this hormone, the manner in which its secretion is coordinated with the hormones that control molting in the larvae, and to determine if as in insects the central nervous system and other sites of endocrine activity are influenced by other physiological processes during larval development.

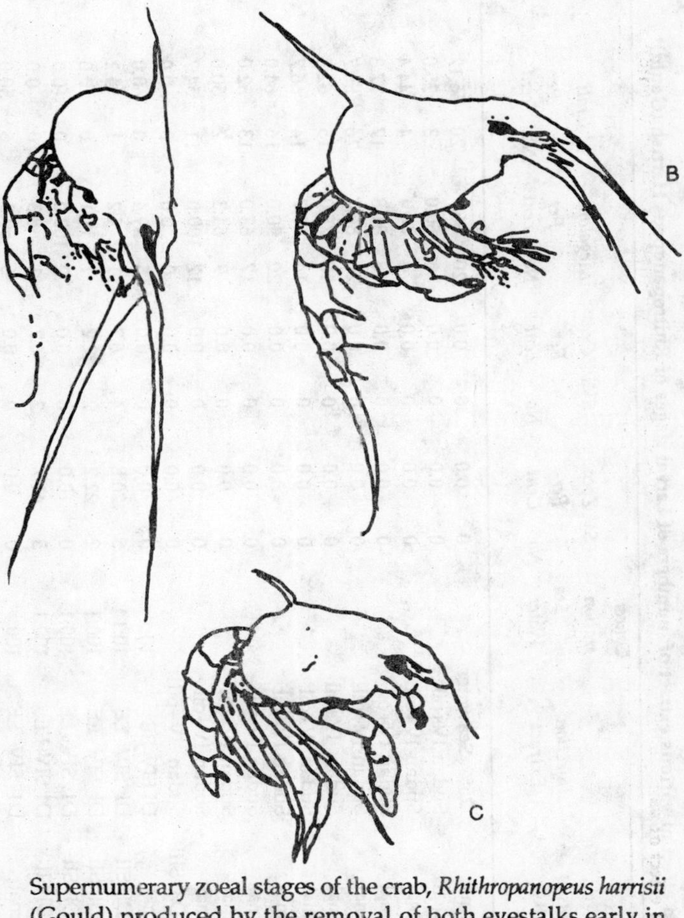

Fig. 1.7. Supernumerary zoeal stages of the crab, *Rhithropanopeus harrisii* (Gould) produced by the removal of both eyestalks early in larval development. A First "extra" zoea (stage V); B: second "extra" zoea (stage VI); C: megalops that molted from eyestalkless stage VI zoea. From Costlow (1966b).

Within the larvae evidence for the existence of the Y organ that produces the molting hormone in the adult is fragmentary and very indirect. It has not been described in the larvae of any of the Natantia or Brachyura. In adult crabs, the removal of the Y organ results in the inhibition of premolt growth of regenerating appendages, and the implantation of this gland stimulates limb regeneration (Echalier, 1956; Jyssum and Passano

TABLE 2

Number of Zoea Injected and effect of various extract of numbers of Larval stages of Rhitropanopeus Harrish (Gould)*

Injection No. Inje-cted	Zoea Sur. Zoea No.	Zoea Sur. Per-cent	Time of Injection	Conditions of Eyestalks	Injection Extract	Concen-tration in Sea Water	SV Zoea No.	SV Zoea Per-Cent	SVI ZOea No.	SVI ZOea Per cent	Megalops No.	Megalops Per cent	Crali No.	Crali Per cent
60	30	50.0	DS2II	Normal	25% Seawater		0	0.0	0	0.0	20	66.7	20	66.17
60	12	20.0	D2 SI	Normal	Sudan IV+oil		0	0.0	0	0.0	6	50.0	3	25.0
18	9	50.0	D3 SI	Normal	Sudan IV+oil		0	0.0	0	0.0	4	44.4	4	44.4
60	22	36.7	D1 SII	Normal	Sudan IV+oil		0	0.0	0	0.0	18	81.8	17	77.3
50	30	60.0	D2 SII	Normal	Sudan IV+oil		0	0.0	0	0.0	26	86.7	26	86.7
50	24	48.0	D1 SIII	Normal	Sudan IV+oil		0	0.0	0	0.0	18	75.0	15	62.5
50	21	42.0	D2 SIII	Normal	Sudan IV+oil		0	0.0	0	0.0	14	66.7	14	66.7
50	25	50.0	D3 SIII	Normal	Sudan IV+oil		0	0.0	0	0.0	20	90.0	16	64.0
30	25	83.3	D1 SIV	Normal	Sudan IV+oil		0	0.0	0	0.0	17	68.0	13	52.0
30	14	46.7	D2 SIV	Normal	Sudan IV+oil		0	0.0	0	0.0	9	64.3	7	50.0
20	17	85.0	D1 Meg	Normal	Sudan IV+oil		0	0.0	0	0.0	17	100.0	11	54.7
35	10	28.6	D3 SIII	2off D3 SIII	Sudan IV+oil		0	0.0	0	0.0	0	0.0	0	0.0
56	17	30.4	D2 SII	Normal	DI SIV EE	10:1	0	0.0	0	0.0	5	29.4	0	0.0
90	24	26.7	D2 SII	2off DI SII	DI SIV EE	10:1	5	20.8	4	6.7	1	4.2	1	4.2
37	9	24.3	D2 SII	2off DI SII	DI SIV EE	100:1	2	22.2	2	22.2	1	11.1	0	0.0
20	2	10.0	D2 SII	2off DI SII	D1 SIV EE	100:1	0	0.0	0	0.0	2	100.0	0	0.0
35	12	34.3	Mixed SII	2off DI SII	D1 SIV EE	100:1	5	41.7	2	6.7	1	8.3	0	0.0
20	12	60.0	Mixed SII	Normal	D1 SIV EE	100:a	0	0.0	0	0.0	6	50.0	6	50.0
44	25	56.8	D3 SII	Normal	D2 SIV EE	10:1	0	0.0	0	0.0	18	72.0	8	32.0
10	4	40.0	D3 SII	2off D2 SII	D2 SIV EE	10:1	4	100.0	4	100.0	0	0.0	0	0.0

47	4	8.5	D3 SII	2off D2 SII	Mixed SIV EE	100:1	3	75.0	1	25.0	2	50.0	2	50.0
38	17	44.7	D3 SII	Normal	Mixed SIV EE	100:1	0	0.0	0	0.0	9	52,9	5	29.4
50	19	38.0	D2 SII	Normal	Mixed SIV EE	400:1	16	84.2	4	21.1	7	36.8	5	29.4
150	87	58.0	D1 SII	Normal	D1 SIII CNS	100:1	0	0.0	0	0.0	59	67.8	44	50.6
120	31	25.8	D2 SII	Normal	D1 SIII CNS	10:1	0	0.0	0	0.0	20	64.5	16	51.6
150	84	56.0	D1 SII	Normal	D2 SIII CNS	100:1	0	0.0	0	0.0	62	73.8	47	56.0
160	54	33.8	D1 SII	Normal	D2 SIII CNS	10:1	0	0.0	0	0.0	37	68.5	33	61.1
80	67	83.8	D1 SII	Normal	D3 SIII CNS	10:1	0	0.0	0	0.0	39	58.2	20	29.9
40	4	10.0	D2 SI	Normal	D1 SIV CNS	10:1	0	0.0	0	0.0	3	75.0	3	75.0
40	8	20.0	D2 SI	Normal	D1 SIV CNS	100:1	0	0.0	0	0.0	3	37.5	3	37.5
30	29	96.7	D1 SII	Normal	D1 SIV CNS	10:1	0	0.0	0	0.0	20	69.0	20	69.0
40	29	72.5	D1 SII	Normal	D1 SIV CNS	100:1	0	0.0	0	0.0	21	72.4	21	72.4
97	39	40.2	D2 SII	Normal	D1 SIV CNS	10:1	0	0.0	0	0.0	10	25.6	10	25.6
27	9	33.3	D2 SII	2off D1 SII	D1 SIV CNS	10:1	3	33.3	2	22.2	1	11.1	0	0.0
45	25	55.6	D2 SII	Normal	D1 SIV CNS	100:1	0	0.0	0	0.0	12	48.0	5	20.0
34	13	38.2	Mixed SII	2off DI SII	D1 SIV CNS	100:1	3	23.1	3	23.1	1	7.7	1	7.7
20	3	15.0	Mixed SII	Normal	D1 SIV CNS	100:1	0	0.0	0	0.0	1	33.3	1	33.3
130	113	86.9	D1 SIII	Normal	D1 SIV CNS	10:1	0	0.0	0	0.0	74	65.5	44	38.9
45	33	73.3	D2 SII	Normal	D2 SIV CNS	10:1	0	0.0	0	0.0	28	84.8	24	72.7
50	18	36.0	D2 SII	Normal	D2 SIV CNS	100:1	0	0.0	0	0.0	15	83.3	15	83.3
20	8	40.0	D3 SII	Normal	D2 SIV CNS	10:1	0	0.0	0	0.0	3	37.5	2	25.0
14	10	71.4	D3 SII	2off D2 SII	D2 SIV CNS	10:1	3	30.0	2	20.0	1	10.0	0	0.0
15	9	60.0	D3 SII	Normal	Mixed DIV CNS	100:1	0	0.0	0	0.0	4	44.4	4	44.4
20	10	50.0	D3 SII	2off D2 SII	Mixed SIV CNS	100:1	3	30.0	1	10.0	1	10.0	0	0.0
140	73	52.1	D1 SII	Normal	Adult optic nerve	50:1	1	1.4	0	0.0	59	80.8	55	75.3
100	55	55.0	D2 SIII	Normal	Adult optic nerve	85:1	2	3.6	0	0.0	25	45.4	13	23.6
100	53	53.0	D2 SIII	Normal	Adult optic nerve	85:1	0	0.0	0	0.0	18	34.0	17	32.1

TABLE 2 (continued)

No. Zoea Injected	Zoea Sur. Injection No.	Zoea Sur. Injection Per-cent	Time of Injection	Condition of Eyestalks	Injection Extract	Concentration in Sea Water[b]	SV Zoea No.	SV Zoea Per-cent	SVI Zoea No.	SVI Zoea Per-cent	Megalops No.	Megalops Per-cent	Crab No.	Crab Per-cent
61	46	75.4	D1 SIV	Normal	Adult optic nerve	120:1	0	0.0	0	0.0	40	47.0	33	71.7
100	72	72.0	D2 SIV	Normal	Adult optic nerve	120:1	0	0.0	0	0.0	49	68.1	39	54.2
60	45	75.0	D2 SII	Normal	Adult "Y" organ	160:1	0	0.0	0	0.0	21	46.7	16	35.6
110	8	0.7	Mixed SII	Normal	Adult "Y" organ	100:1	0	0.0	0	0.0	5	62.5	3	37.5
45	29	64.4	D1 SIII	Normal	Adult "Y" organ	100:1	0	0.0	0	0.0	17	53.6	13	44.8
60	38	63.3	D2 SIII	Normal	Adult "Y" organ	160:1	0	0.0	0	0.0	27	71.0	17	44.7
40	19	47.5	D1 SIV	Normal	Adult "Y" organ	100:1	0	0.0	0	0.0	6	31.6	5	26.3
25	17	68.0	D2 SIV	Normal	Adult "Y" organ	160:1	0	0.0	0	0.0	14	82.3	11	64.7
200	95	47.5	D1 SII	Normal	Juvenile hormone		0	0.0	0	0.0	11	13.6	0	0.0
200	105	52.5	D2 SII	Normal	Juvenile hormone		0	0.0	0	0.0	6	5.7	0	0.0

[a] From Costlow (1966b). D day; S zoeal stage; EE eyestalk extract; CNS central nervous system

[b] Number of eyestalks, central nervous systems, optic nerves, or Y organs per cc sea water.

1957). In some insects that normally lose the ability to regenerate appendages after metamorphosis to the adult, the implantation of corpora allata from larval forms (source of juvenile hormone) restores the capacity for regeneration (Pflungfelder. 1939). Studies on autotomy and regeneration of the chela of megalops of *C. sapidus* have shown that the complete and functional claw was regenerated at the time of metamorphosis if autotomy were induced prior to the fourth day following the final zoeal molt (Costlow, 1963b). If autotomy of the claw occurred later than the fourth day of megalops life, the regenerating appendage first appeared as a limb but at metamorphosis and was not normally functional until after the first or second postlarval molt. One interpretation of these results is that prior to the megalops stage, only the Y organ is present and functional in the zoeae, providing a molting hormone and maintaining a molting frequency of three to four days throughout zoeal life. Regeneration of the complete claw, if autotomy occurs on or before the third day of megalops life, is possible because either the Y organ is sufficiently active or enough molting hormone is present to initiate regeneration and also stimulate metamorphosis to the first crab on the eighth day. If autotomy occurs after the third day, the titer of the molt-inhibiting hormone produced by the now-functional X organs of the eyestalks has increased to a point where the comparable amount of molting hormone cannot initiate regeneration of the complete claw and also induce molting or.

If removal of the source of a hormone that controls molting or rate of morphological development in the larvae interferes with these processes, in theory the replacement of the hormone in sufficiently high concentration at an appropriate period of development should restore the proper endocrine balance, and development should proceed in the normal manner. As the injection of extracts from a variety of larval and adult tissues did not reduce the number of supernumerary zoeae of *R. harrisii* that resulted when both eyestalks were removed. Also, the pattern of development of normal zoeae was not altered by the injection of extracts from larval and adult tissues. The injection of insect juvenile hormone extract of unknown titer into normal, eyed larvae did not after the pattern of morphological development or the frequency of molting. It should be noted, however, that in spite of the fact that approximately 50 per cent of the zoeae survived the injection and completed the normal four zoeal stages, an unusually small number molted to the megalops stage. None of these metamorphosed to the first crab. If the juvenile hormone extract had no effect on the developing crab larvae one might expect survival to the first crab to be comparable to that observed when extracts of sea water or Sudan IV plus oil were used. With the latter extracts, 25.0 to 86.17 per cent of the injected larvae did successfully metamorphose to the first crab. Thus, while the injections of insect juvenile homone did not discernibly affect the morphological development of the larvae, it is possible that the inability to metamorphose to the first crab was directly associated with the extract.

Conclusions

Earlier studies on metamorphosis in the Crustacea were confined largely to observations on material obtained from the plankton, for efforts to culture the larvae in the laboratory had been unsuccessful. Within recent years, however, techniques have been developed that permit the rearing of larvae of most of the major groups within the Crustacea. Development, from early embryology through metamorphosis and into the postlarval stages, can be followed under, controlled environmental conditions and, with some species, survival as high as 80 per cent has been attained consistently. Thus, it is now possible to consider many aspects of metamorphosis on a level that was not previously feasible with planktonic material.

For purposes of discussion it is convenient to consider the external environment, the internal environment, behaviour, and the mechanisms of control, but it is obvious that a study of "metamorphosis" cannot be restricted to any one of these any more than it can be defined as a single point in time. For a more complete understanding of the complex morphological, physiological biochemical, and behavioural changes that together constitute metamorphosis, it is imperative that we have a clearer picture of the cytological and histological changes as development progresses. This should extend well into the period of embryonic development, since some systems—e.g., the chromarphores—have been shown to develop well before hatching of the larvae. An understanding of the internal morphology of all developmental stages is also essential because, as has been described for the cypris and the megalops, the internal reorganization is not necessarily confined to a particular stage but may represent a gradual progression through a number of earlier stages. Evidence that the X organ sinus gland complex of *R. harrisii* does not begin to function until well after metamorphosis to the first crab suggests that a more thorough examination of the early postlarval stages is also required.

Advances in our understanding of morphological changes must be complemented by comparable efforts toward the elucidation of the underlying physiological and biochemical processes. Using the techniques that are available, or by modifying them to the microlevel, it should be possible to determine the enzyme systems involved during development, the time at which they appear, the role they play in metamorphosis, and changes that may occur during the transition from the pelagic existence of the larvae to the more benthic life of many of the adults. Problems dealing with the synthesis, elaboration, and utilization of proteins, carbohydrates, and lipids during the development of most crustaceans have been neglected, and little is known of the qualitative or quantitative aspects of biosynthesis during embryonic and larval development. At what point in larval development are specific proteins first synthesized, and what tissues are associated with their elaboration? When does heterogeneity first become apparent, and in what way are the larval patterns modified by metamorphosis, early postlarval development, and sexual differentiation? Is the reduction in number of free amino acids during zoeal development followed by an increase at the time of metamorphosis, associated with osmoregulation, or does it represent the utilization of these units in same phase of "metamorphosis" that extends throughout most of the zoeal life?

The interrelationship between the "external" environment and the "internal" environment is an equally important facet to put understanding of metamorphosis. Why, in larvae of some estuarine species, do different optimum conditions of salinity and temperature exist for each stage while survival of larvae of other species is not dependent upon narrow limits of these two environmental factors? In what ways are larval development, is well as metamorphosis, affected by pressure, light of different intensities and wavelength, O_2 and CO_2 tensions, organic and inorganic constituents of sea water, and differences in quality and quantity of available food? In what way is development affected by fluctuations is any one or all of these factors, and is acclimation of the larvae to any one or all of these factors affected by changes in other parameters of the environment?

The complete picture of metamorphosis will also depend upon a clearer understanding of the various mechanisms controlling the rate and sequence of larval development, larval molting, osmotic balance, sexual differentiation, and metamorphosis itself. Here the studies on internal morphology should contribute to the picture of where and when these systems develop, the time at which they become functional, and the changes that may occur during development through the juvenile animal. This should involve a clear delineation of the sites or mechanisms that are present within the larval eyestalks and other portions of the developing central nervous system. What activates the elaboration of neurosecretory products or hormones at predetermined intervals within the regular pattern of larval development? In what way is the loss or disruption of these mechanisms by extirpattion compensated for by other portions of the nervous system? Is there an association between the results obtained from experiments on eyestalk extirpation of larval stages—i.e., the production of large supernumerary larvae—and the observations on the natural occurrence of gigantism and neotony in larvae of some crustaceans?" In what ways are these systems co-ordinated and, at the cellular level, what is the nature of the response? Studies on the chemical structure of hormones in larval Crustacea should provide a more basic understanding of the origin and evolution of groups within the Arthropoda and provide for further research on similarities in the synthesis of the basic chemical components associated with early embryonic and larval development in many of the invertebrates.

As information on the larval mechanisms becomes available we should constantly reexamine the existing concepts of endocrine control in the adult Crustacea. Eyestalk extirpation during a specific time in the megalops of *C. sapidus* has resulted in crabs that continue to follow an accelerated molting pattern. Other crabs, metamorphosing from megalops that had lost both eyestalks 12 hours later, do not display an acceleration of the molting frequency. If molting in adult crabs is dependent upon the activation of the Y organ by a reduced titer of molt-inhibiting hormone from the X-organ sinus gland complex, has the Y organ of megalops of the first series remained activated? If so, what source of molt-inhibiting hormone, in the absence of both eyestalks, serves to regulate the normal frequency of molting in megalops of the second series? If an alternate source of molt inhibiting hormone exists within the central nervous system of the megalops, how is it activated in the megalops of the second series, and why is it not activated in megalops in which both eyestalks were removed only 12 hours earlier? Crabs that metamorphosed from megalops of both series show a much greater increase in carapace size at molting than crabs that metamorphosed from normal megalops. Thus, while molting initially may be controlled by a hormone produced and stored by endocrine sites within the eyestalks, the regulation of size increase, or water uptake, cannot be under the control of this same hormone. If separate hormones are involved, what is their chemical difference and in what way is their action co-ordinated in normal adults?

Metamorphosis in the Crustacea remains "a problem in developmental biology," The solution to the problem does not lie in isolated studies on the cellular reorganization, the physiological changes, the chemical structure of the hormones, or in the injection of insect juvenile hormone into larval stages, but rather through an integrated approach that recognizes that development involves the animal in all of its parts.

2

Metamorphic Changes in Insects

The aim of this chapter is to set the scene for the detailed accounts that will be given in the following three chapters. Metamorphosis has long been recognized as a characteristic of insects. Even the earliest systems of classification distinguished between those insects that metamorphosed and those that did not. From paleontological records the oldest insects determined with certainty are winged forms. However, it is fairly firmly established that the origins of the insects lie with the symphylan group of the Myriapoda, and that ancestral primitive insects were wingless and therefore would have lacked a metamorphosis.

What exactly do we understand by "metamorphosis"? A dictionary definition may be as follows: "A change in form structure, or function, as a result of development, specifically the physical transformation during postembryonic development, as of the larva of an insect to the pupa or the pupa to an adult or the tadpole to the frog." We speak, within the insects, of the absence of metamorphosis in the groups Collembola, Diplura, and Thysanura, which are described as ametabola since no dramatic changes in appearance occur during development. These insects are wingless and therefore are apterygotes.

The rest of the insect orders are primitively winged or pterygote. Of these, the exopterygote insects—the bugs (Hemiptera), the may flies (Ephemeroptera), the stone flies (Plecoptera), the termites (Isoptera), the dragonflies (Odonata), cockroaches, and locusts (Orthoptera), to name a few, have their wings developing externally, hence their name Exopterygota, These are characterized by a series of larval stages or nymphal stages hat lack wings and have immature reproductive systems. They show "incomplete" metamorphosis, i.e., a gradual transition to the winged, reproducing adult. Each nymphal stage molts into the next stage, which may have slightly more developed wing buds and reproductive system. However, even here the greatest "jump" or change is generally seen at the last nymphal-adult molt. At this time the most dramatic growth in wings and reproductive systems occurs. The second group of pterygote insects are endopterygotes with wings developing internally and include the scorpion flies (Mecoptera), the alder flies and lace wings (Neuroptera), the caddis flies (Trichoptera), beetles (Coleoptera), moths and butterflies (Lepidoptera), bees and wasps (Hymenoptera), and flies (Diptera). In these insect orders metamorphosis is described as "complete". i.e., the larval stages completely lack any external evidence of wings and reproductive system, and the transformation into the adult is a two-step process, involving first a most from the larva into a pupa, and then a molt from a pupa into an adult. Often, as in the more specialized forms of the Lepidoptera, Hymenoptera or Diptera, the larva is completely unlike the adult in appearance. Thus, in the Lepidoptera we have the larva or caterpillar metamorphosing into the chrysalis or pupa from which the very different adult moth or butterfly emerges. In this last category there are degrees of change. At the least dramatic, we have cases like *Sialis* the alder fly, or the lacewing or ground beetles, where the larva, pupa, and adult still show strong resemblances, particularly in the abdominal segments and the pupa is capable of movement. Larva and adult differ mainly in the development of compound eyes, wings, and external genitalia.

Significance of the Pupal Stage

Much has been written of the "pupal stage" and its significance in, the evolution of the insects. One of the more recent discussions is that by Hinton (1963), who compares some of the older theories of Berlese and Poyarkoff. Hinton favours the argument that a pupal stage has been necessitated by the development between larva and adult of flight muscles that require a "mold." He homologizes the various nymphal stages of exopterygotes with the larval or nymphal stages of the endopterygotes, the last nymphal stage of exo-with the pupa of endo-, and the adults of both. Berlese was of the opinion that the endopterygote larval stages corresponded to embryonic forms and that all of the exopterygote nymphal stages were concentrated into the endopterygote pupa. Poyarkoff suggested that the adult exopterygote corresponded to the pupal endopterygote and that the adult endopterygote was an *extra* adult stage.

Hinton draws attention to examples of "pupal" forms among the exopterygote orders—Thysanoptera (thrips). Coccidae (mealy bugs), and Aleurodidae (greenhouse white flies). In these, the general structure of the feeding larva or nymph has departed very widely from that of the adult; the structural differences are bridged in he last two nymphal stages. As the differences between the stages have become greater, this has involved a greater degree of structural reorganization to bridge the gap of the internal tissues. As a result, the last two nymphal stages. instars have become more and more quiescent and have eventually ceased to feed, becoming closely analogous to endopterygote pupae. The structural reorganization required to bridge the gap, in some cases, may be greater than is found in some primitive endopterygotes.

General Changes Involved During Metamorphosis

Besides the development of wings, compound eyes, and of external and internal genitalia, one of the other main changes that occurs in metamorphosing insects may be in their feeding habits. Thus, a caterpillar and a moth or butterfly, or a fly maggot and the adult fly, may have completely different habits, and these are reflected in structural changes in the mouth parts, gut, salivary glands, and gut muscles. In some frogs, a change from the larval herbivorous diet to the adult carnivorous one is accompanied by a shortening of the gut at metamorphosis. In a fly, the gut shortens also, and the larval crop, larval caecae, and larval salivary glands all disappear the very much shorter adult gut develops a crop and an elaborate rectum with characteristic rectal papillae. Clearly, changes in mouth parts accompany changes in feeding habits, as can be seen in a chewing caterpillar and a sucking moth or butterfly. Also, changes in feeding habits and mouth parts are accompanied by changes in gut musculature.

General body musculature may change considerably, as is the case with the higher flies where all larval muscles are replaced, or these changes may be far less extreme, involving only the musculature to the newly developing wings. For example, insects such as the alder fly (Neuroptera), scorpion fly (Mecoptera), or even the generalized members of some of the more specialized families such as Lepidoptera and Diptera, retain essentially larval musculature in the abdomen. The motile pupa of a mosquito or of a chironomid nonbiting midge is very different from the immobile pupa of a higher fly in which all of the larval muscles are broken down.

In the case of an amphibian such as the frog, metamorphosis involves the migration from water to land, and correlated changes in the respiratory system occur. In the pterygote or winged insect there are changes in the respiratory system of the terrestrial larval form during metamorphosis to the flying adult form. The elaborate adult thoracic musculature is heavily tracheated, often with an equally elaborate system of air sacs. A simpler system of air sacs is also found in aquatic forms, where one function would be to provide buoyancy. Insects appear to have evolved from terrestrial ancestors with well-developed tracheal systems, paralleled today in the Myriapoda. Aquatic forms therefore are secondary and are found with normal tracheal systems that may have closed spiracles; these become open and funçtional in the pupal and adult stages. aquatic larvae may have tracheal gills or a superficial dense tracheation beneath the thin body wall, but these are lost during metamorphosis to the terrestrial flying adult form. The tracheal system is intimately involved in the metamorphic changes undergone by the other organ systems. For instance, it has been found within the Diptera that the degree of breakdown of the respiratory system follows closely the degree of histolysis of the abdominal musculature (Whitten, 1960).

The most specialized examples among the Endopterygota will be considered in most detail in the present general survey, for it is here among the specialized lepidopteran, hymenopteran, or dipteran that one finds the extremes in metamorphic change as one proceeds from larva to adult. Not only are reproductive systems, wings, and compound eyes present in adult and absent in younger larval stages, but with changes in the feeding habits extensive changes in mouth parts and gut may occur, and muscle requirements may be drastically different in larva and adult. With the acquisition of a flying habit the sensory environment is completely different (this occurs to a lesser or greater extent in various exopterygotes also) so that larval sensillae are broken down and new adult ones appear. Thus, drastic internal changes are reflected in changes in the respiratory and nervous systems, and recent evidence shows equally elaborate changes occurring in the hemocytes (blood cells), associated with changes in other tissues. In fact, there would appear, in the specialized endopterygote, to be no system that remains unaffected by the metamorphic changes occurring during the pupal stages.

The specialized endopterygote is however best considered first in relation to its generalized relatives, and then to other endopterygotes. Since all of the endopterygote orders cannot be given equal emphasis, the Diptera will be treated in some detail in the present chapter. Within the Diptera, the specialized problems found in the Diptera Cyclorrhapha (higher flies) are placed in better perspective when compared with those of the more generalized forms within the order. The mosquito is a member of the more generalized Diptera Nematocera and it is significant that the extremes of larval tissue histolysis and reorganization are absent here. Clements (1963) has stated for the mosquito that "metamorphosis starts early in larval life with slow development of imaginal buds, compound eyes and gonads, processes accelerated in the fourth instar, and largely completed in the pupa. It is clear that important changes continue to take place in the first days of adult life that only in the older adult is metamorphosis completed." Pigmentation of the adult eye ommatidia apparently occurs in *Culex pipiens*, by the second larval instar!

The evolution of extremes in metamorphosis is therefore seen in many cases to have occurred within the higher endopterygote orders, just as there has been a general progressive evolution in the insects as a whole form the a-metamorphic apterygotes, through the exopterygotes to endopterygote forms. It should be noted that apterygote, ametamorphic insects have a complement of endocrine organs exactly comparable with those of the "higher" pterygotes (Watson, 1963). It would appear that the juvenile hormone has the same relative role—that of blocking the factors responsible for differentiation of tissues into adult forms, and a second role, that of stimulating yolk deposition in adult females. In all insects ecdysone has the same function also, that of inducing molting.

Changes in hormone balance influence the *progress* of metamorphosis as will be seen in this and succeeding chapters, but the endocrine organs themselves do not *cause* the differences between a non-metamorphosing and a metamorphosing insect.

The question of why the cockroach begins life as a wingless individual or why a fly has evolved into a form that begins life as wingless, legless, eyeless maggot remain unresolved. It is possible that, as Berlese conjectured, the latter represents an embryonic form, hatching prematurely. In the vertebrates the old concept of "ontogeny recapitulating phylogeny," where embryonic features tend to be retained while the adult form is more susceptible to change, is seen for example in the transient development of gill arches and circulation in the embryos and early stages of development of higher vertebrates. In an amphibian such as the frog we find the aquatic tadpole, with its fishlike respiratory and corresponding circulatory systems, "metamorphosing" into the adult, terrestrial amphibian with lungs, no gills, and modified gill-arch circulation. If the ancestral insect is a symphylan-diplura type, and the pterygote insect larva is prematurely hatched in the ancestral wingless form, then the acquisition of wings at metamorphosis closely parallels the amphibian condition, with its acquisition of legs at metamorphosis. These are academic questions, the answers highly speculative, but the facts of metamorphosis remain significant to the biologist, with the insect and amphibian remaining striking examples with which to study the process.

Problems Posed by Insect Metamorphosing Tissues

In a metamorphosing pupa one finds various larval tissues degenerating at specific times. What initiates cells death? Can this predetermined event be experimentally altered, delayed, or permanently prevented? How far back in this is the process determined? Also, in the developing pupa we have tissues that seemingly regress or dedifferentiate, and new adult tissues redifferentiate from them. Do we know what happens to individual cells? Do cells that are differentiated for a particular function in the larval stage dedifferentiate and then redifferentiate into cells performing like or dissimilar functions in the adult stage? Are some adult cells developed from undifferentiated "reserve" larval cells? In the metamorphosing pupa we may have tissues that appear as new structures not represented previously. Some of the insect flight and leg muscles come into this category. Since muscles are characteristically mesodermal, can the leg muscles develop from epidermal, ectodermal elements of the imaginal discs or are mesodermal myoblasts present in the discs? It has been shown that radioactive proteins in the cuticle of larval insects reappeár in the developing adult cuticle. In this case, how are these materials removed from one and transferred to the other in space and time? The mass of larval muscle in a fly is relatively enormous.

At the time of muscle breakdown are the materials lost to the body or is there conservation and use in the build-up of adult tissues? Are the blood cells involved to any great extent in tissue breakdown? Are the cells destined to cell death autocytolyzed or are living cells attacked by phagocytic blood cells? Are there ultracellular components comparable to focal degradation points in other animals, and how general are lysosomes? How does a tracheal system laid down at a previous instar solve the problems of providing respiratory gases, firstly to larval tissues destined to die, secondly to larval tissues to be retained, and thirdly to adult tissues not yet differentiated...all requirements within the same pupal instar? A vital and fundamental problem involves the role of the central nervous system. How much metamorphosis occurs in this system? Are we any further forward now than we were in 1952 when Power, in one of a remarkable series of papers on the *Drosophila* nervous system stated, "The morphogenesis of the insect's nervous system is still largely unexplored in spite of a large bibliography." Among the problems that remained unanswered he named the basic sequence of events in the origin and differentiation of the component parts of the central nervous system, the fate of the peripheral nervous system at the time of metamorphosis, the development of the innervation of the imaginal organs, and the changes that occur in the nervous system correlated with the changes that occur in the various end organs. An understanding of the events occurring in the nervous system is basic to any understanding of metamorphosis of individual organs, yet work on the central nervous system during metamorphosis is far behind that on the endocrine system. Even here it has been shown possible to change, enhance, or retard metamorphic events by altering the hormonal environment, yet we are still left with the basic problem of control of the endocrine organs themselves and, ultimately, of the brain neurosecretory cells. What, e.g., are the controlling mechanisms behind the circadian rhythms of development in the metamorphosing *Drosophila* pupa, shown so elegantly by Harker (1965a, 1965b)?

THE SIGNIFICANCE OF METAMORPHOSIS TO GENERAL BIOLOGICAL PROBLEMS

Embryology

In the more extreme cases of metamorphic change there is a virtual return of most tissues to an embryonic state with the constituent cells in a undifferentiated form. As metamorphosis proceeds, differentiation of individual cells of the various tissues occurs. Such is the case in a higher fly or bee where the muscular system is entirely broken down and reorganized, and there are correspondingly elaborate changes in the gut, general epidermis and tracheal system, with correlated reorganization of the nervous system. One of the most valuable aspects of insect metamorphosis is as subject material for study of embryological and general developmental problems. Problems of cell and tissue differentiation, of pattern formation, of cell determination and of cell death can all be studied in the metamorphosing insect. Of primary significance also is the fact that insects are amenable to experimental techniques such as those of implantation (e.g., Bodenstein 1950 and Hadron 1951).

Genetics

Studies using the fly *Drosophila,* and to a lesser extent, the hymenopteran *Habrobracon,* have played a large part in our understanding of genetical problems; these are essentially studies of metamorphic events in these insects. The various eye mutants or wing mutants or mutants showing abnormal bristle pattern result from the action of various mutant genes. However, an understanding of the steps which intervene between the formation of the initial protein and the manifestation of the visible effect, is in essence an understanding of the events of differentiation within the metamorphosing pupa, at least where the effects are on the adult phenotype. There may be a large number of developmental sequences between the activity of the mutant gene and the expression of for example, "vestigial wing" in *Drosophila*. Studies on the formation of biological "patterns" are often also metamorphic problems, as in the case of bristle patterns in *Drosophila* (e.g., Sondhi, 1963).

Chromosome Structure and Gene Action

Insects have another tremendous advantage over other organisms in possessing giant chromosomes. It is unfortunate that they are known only from the order Diptera among the insects, but at the same time it makes this order of greater interest. These giant polytene chromosomes arise by chromosomal DNA replication in the absence of corresponding cell division. They are best known from larval salivary glands; in *Drosophila* has been possible to "map" the different bands of the salivary cell chromosomes, so that specific regions represent the positions of particular genes whose action has been determined by genetical means. Giant chromosomes are characterized by regions of intense "puffing" activity. These regions represent areas of active RNA synthesis. The RNA is thought to be messenger RNA which passes into the cytoplasm, where, in association with the ribosomes, it organizes the amino-acids into particular sequences to form specific proteins. In this way the origin of a specific protein could be traced back to the activity of a specific puff at a definite position on the giant chromosome.

Cells other than salivary gland cells may contain polytene chromosomes, although they are relatively smaller. These include cells of Malpighian tubules, male seminal vesicles, nurse cells of the ovary, heart cells, pericardial cells, and even fat body cells. Recently, cells having chromosomes as large as salivary gland chromosomes have been recorded for epidermal cells of the fly foot (Whitten, 1964a). They are responsible for secretion of the various layers of cuticle forming the dorsal surface of the adult foot-pad of the higher flies. These cells are thus intimately involved in general adult "development", whereas salivary gland cells have the disadvantage of undergoing cell death at the beginning of the pupal stage. Changes in salivary chromosome puffing, brought about by changing the hormonal milieu, have been correlated with effects on "general development." Titles such as "Regulation of gene action in insect development" (Kroeger and Lezzi, 1966), and "Chromosome changes associated with differentiation" (Clever, 1965), indicate the extent to which studies of salivary gland chromosome puffing activity have attempted to relate events at the chromosomal level to events at the level of the whole organism, in this case the metamorphosing pupa. Giant epidermal cells, such as those of the foot-pads in the higher flies, should prove of particular interest in pursuing these problems.

Questions of growth by cell division, as opposed to growth by cell size, have also been raised as a result of studies involving metamorphosing insects, particularly the mosquito (e.g., Trager, 1937 and discussed in Clements, 1963). It has been found that, as a rule, cells which increase in size (for example of salivary glands, Malpighian tubules, rectum, abdominal muscles, trichogen cells, larval oenocytes) are incapable of cell division, and except for Malpignian tubules are incapable of transforming into adult organs, so that these then are the cells which histolyse in the metamorphosing pupa. During growth, replication of chromosomes proceeds without separation and without cell division, resulting in the characteristic polytene chromosome. On the other hand cells that multiply by cell division remain small, and are retained and pass through into the adult; presumably they are undifferentiated and therefore capable of subsequent differentiation. A seemingly unique condition is found in the various mosquitoes—that of *endopolyploidy* with somatic reduction. Here, in the small intestine, and to lesser extent in other parts of the gut, epidermis, tracheae and neurilemma, cells become polytene but apparently not differentiated. In *Aedes* the cells may be 16 N with 48 instead of the normal 6 chromosomes, with the homologues closely associated in a polytene chromosomes. However, at the onset of pupation the polytene chromosome splits up and cell division occurs, without chromosome replication. In this way there are produced many small diploid cells which differentiate to reconstitute the small intestine and rectum. An answer to the question of why this should happen is not at present available.

Gross Morphological Changes at Metamorphosis

One of the most striking aspects of insect metamorphosis, particularly to the casual observer, is the dramatic change from larva to pupa at the larval-pupal ecdysis. This is all the more remarkable when one realizes that the pupal "mold" produced at the larval-pupal molt follows the future adult shape, particularly with respect to the large muscle-filled thorax. Yet at the time of its formation none of these adult structures may be developed in some instances, as is the case with the higher flies. The mechanical aspects have been discussed at length recently in a review article by Cottrell (1964): The external shape of the insect is largely limited by the inherent form of the different parts and more particularly by the extent to which the outermost layer of cuticle—the epicuticle—can be unfolded. Then again, the unfolding of the epicuticle and the stretching of the underlying endocuticle is brought about by hydrostatic pressures produced by two activities: swallowing of air or water and the contraction of muscles. The relative importance of these two may differ in different insects. Final form is also attained by the expansion of soft, not yet expanded parts, but not of any prehardened areas. There is a very exact relationship between the two. Kohler has shown that the eversion of the wing buds in *Ephestia*, the clothes moth, is brought about by hydrostatic pressures and by muscles opening the mouths of the peripodial cavities. Fraenkel showed that in *Calliphora* eversion of the pupal head is brought about by peristaltic abdominal contractions. Cottrell discusses other examples also. Personal observations with *Drosophila* and *Sarcophaga* have shown the presence of a series of larval suspensory muscles that seem to play a role in attaining pupal form. Typically, the cyclorrhaphan nephrocytes are shown as slung from the two lobes of the salivary glands. In fact, their attachment in both *Drosophila* and *Sarcophaga* (and probably other Cyclorrhapha also) is by way of a muscle that runs along the proximal third of each lobe of the salivary glands, with firm attachment between gland, muscle, and nephrocytes being maintained by the continuous connective tissue basement membranes (micrographs in Whitten, 1964c). These muscles are attached at one end near the spiracles and at the other on the gut. Two other muscles, with origins in the prothoracic region and insertions asymmetrically on the midgut, pass through the body cavity to their points of attachment on the midgut. These muscles show rhythmical contractions and resemble the alary muscles, being clearly striated under phase optics. In the fly, a "cryptocephalic pupa" has its head still invaginated, but leg and wing discs are everted though flaccid and unexpanded; the abdomen is larval in shape being disproportionately long, and larval tracheae are still within the body. Simultaneously with eversion of the head, there is extension of the wings and legs and contraction of the abdomen. Contraction of the abdomen must reduce the internal volume and help force blood into the legs and wings and head. A large air bubble present before, and absent after eversion of the head, also probably plays some part. Observations of internal changes reveal a dramatic overall contraction of the entire system of longitudinal visceral muscles and unpaired suspensory muscles. This simultaneous contraction of the muscles effects the shortening of the larval gut to the dimensions of the much shorter adult gut. At the same time, contraction of the longitudinally running visceral muscle through the midgut caecae reduces these to four small knobs on the gut. Their cells later histolyze, and these structures are not represented in the adult gut. At the same time, the contraction of the "slinging" or suspensory muscles plays a role in the form of the pupa and future adult. By anchoring the gut and salivary glands they seem to help prevent their being everted into the head along with the brain. This is probably also helped by muscles such as the alary muscles, which also attach to tracheae. At the same time the muscles, whose origins are on the body wall behind the prothoracic spiracles, may help during their contraction to draw in the "neck" region between head and thorax. Nothing is fortuitous, and the establishment of shape in the pupa is a very precise process.

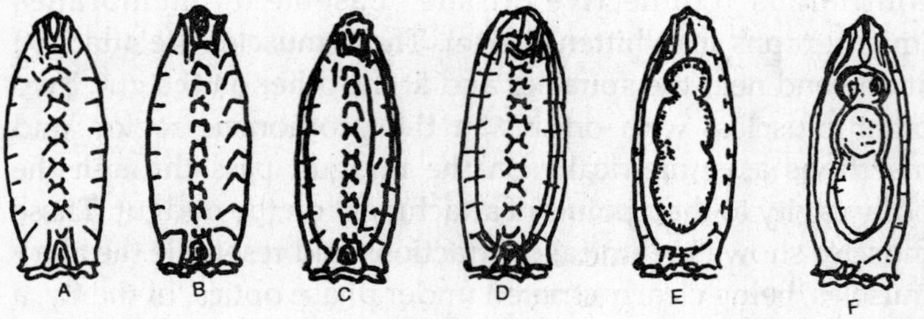

Fig. 2.1. General sequences in the transformation of the larva to a pupa in the fly, in dorsal view showing cuticular layers and outlines of the tracheal system. A: larva within larval cuticle; B: larva after beginnings of puparium formation, with epidermis of posterior end retracted away from larval cuticle; C: secretion of the so-called "prepupal cuticle" (may be ecdysial membrane) by the larval epidermis; D: a cryptocephalic pupa with pupal cuticle formed and with legs and wings partically everted, but with head still invaginated; E: pupa, lying within the hardened larval cuticle (puparium), "prepupal membrane," and its own pupal cuticle; F: later pupa with newly developed adult cuticle and tracheal system, enclosed within the earlier cuticles. Size $^{1}/_{2}$ in. long.

Attainment of adult shape is equally intriguing. Clearly, it is genetically determined that the slightly bilobed saclike pupal leg appendage of the higher fly should develop by epidermal cell retraction, cell division, and differential growth into an adult leg, yet little is known of the factors that determine this shape. There is a suggestion in the development of the fly foot that the ecdysial membrane (c.f. Taylor and Richards, 1965), the *first* secretion of the adult epidermal cells, may play a mechanical role in maintaining shape in the delicate, otherwise naked, cytoplasmic extensions of the various epidermal cells that form hairs of various sizes. Once the adult structures are laid down, it can be seen from Cottrell's account that the adult shape is attained by a dual process—by the prehardening of structures that will *not* expand after ecdysis (the various leg bristles and the terminal pretarsal structures shown in the fly foot) and the postecdysial expansion of soft structures that subsequently harden, e.g., the wings and general abdominal cuticle. Evidently, timing is delicate and there are elaborate control mechanisms involved. Again it is with the flies that much work has been done. Fraenkel and Hsiao (1962) and Cottrell have separately found a "darkening factor," named by former "bursicon." More detailed discussion is given in Cottrell's review. It is sufficient here to note that the factor is released after the flies have ceased "digging" their way through the soil and that darkening can be produced in these "digging" flies by injection of blood from air swallowing, darkening flies. The cessation of digging and the onset of air swallowing and darkening are closely interrelated; " shut off" of digging movements is permanent. It cannot be related to the hardening process and, according to Cottrell, the mechanism of "shut off" is probably nervous. Push through the ground is accompanied by and partially brought about by alternate eversion and retraction of the blood-filled sac at the anterior end of the head—the ptilinum. The muscles producing the movement breakdown within three days of emergence, demonstrating the remarkable co-ordination of events. The nervous stimulation for cessation of digging may well result from signals received (or no longer received) by the tiny tactile hairs dispersed over the surface of the ptilinum (Whitten, 1963c;). Each hair possesses a sensory nerve fiber, and avoiding reactions with soil particles are seen to result from stimuli received by these hairs. When the fly appears above ground, no resistance is felt by the ptilinum, and signals cease to be received by the sensory nerve fibers. This in itself could *initiate* the chain reaction which involves retraction of the ptilinum, air swallowing, darkening, and finally breakdown of the ptilinal muscles. Particularly puzzling is that if the darkening factor is a hormone whose release is necessary for the final synchronous darkening of all areas to be hardened and darkened following expansion, what is responsible for the progressive predarkening of areas such as the legs before ecdysis? This occurs from thorax forward and backward and proximodistally along the legs. The predarkening process has been used as a very reliable aging device, e.g., for general development aging in *Drosophila*

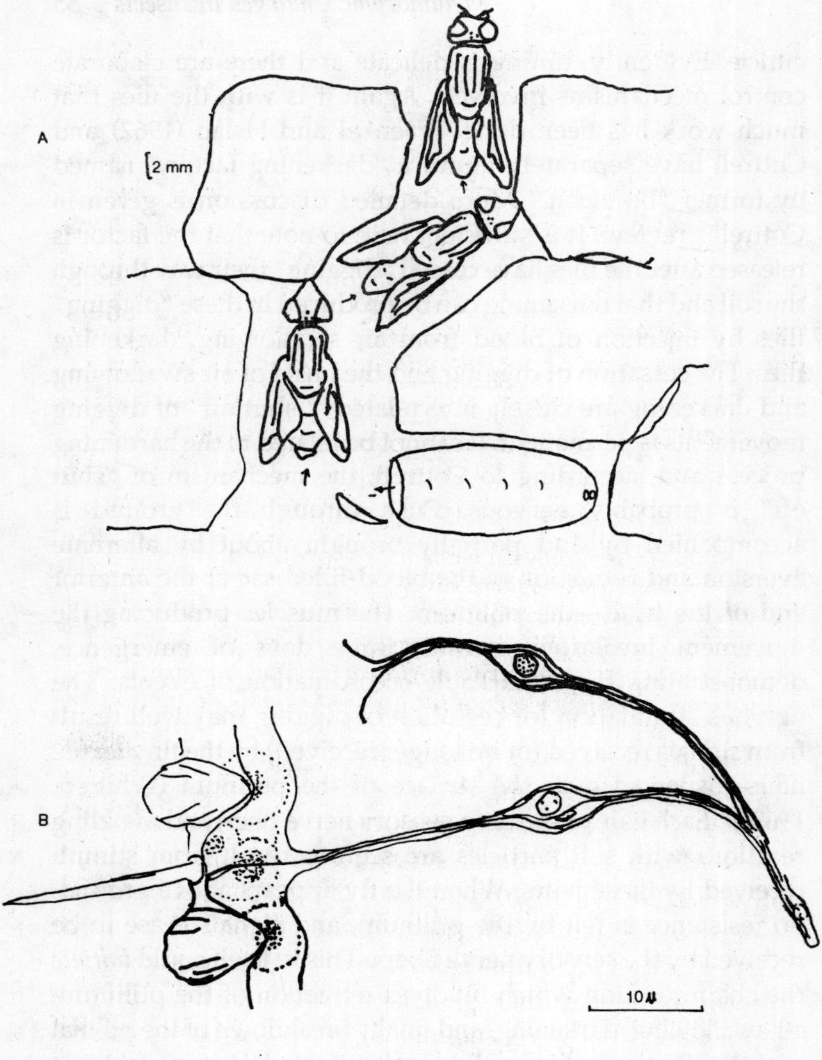

Fig. 2.2. A: Diagrammatic representation of a newly emerged fly pushing its way through the soil by alternate inflation and deflation of the ptilinum", B: one of the sensory hairs occurring between scales on the ptilinum, which may function in triggering off the complex sequence of events leading to the eventual darkening of the adult, cessation of digging, withdrawal of ptilinum, and histolysis of the ptilinal muscles. Sensory nerve fibers, sheath cells, and innervated hair.

(Harker, 1968a, 1965b) and in the development of the fly leg (Whitten, 1968). Are the preecdysial and the postecdysial hardening and darkening processes really controlled and effected by different agents?

Developmental Rate in Metamorphosing Pupae: Internal Clocks

A very different but equally intriguing aspect of the metamorphic process occurring between larval and adult stages can be followed in two papers by Harker in a study of developmental rates and their controlling mechanisms. Working with the fruit fly *Drosophila melanogaster*, Harker (1965a) first studied the course of development in two strains in an attempt to establish the factors affecting the time taken to complete three stages of: head eversion to yellow eye, yellow eye to wing pigmentation; wing pigmentation to eclosion. Time measurements were taken for pupae entering each stage at each particular hour of the day when insects were kept at 12-hour light : 12-hour darkness; 12-hour bright light: 12-hour dim light; or in continuous darkness. She found that the duration of each stage in both strains is affected by the time of day relative to the *light cycle at which the stage is entered*. The duration of each stage for pupae kept in continuous darkness is affected by the time of day at which the stage is entered, relative to the light cycle to which they had been exposed as larvae. She further found that the time interval curves for all three stages of any one strain take the same form. Consequently, because of the very wide range of development rates, dependent on the time within the light cycle at which each stage begins, a population in which the larvae all pupate within a 24-hour period will continue to produce adult flies over several days. The eclosion rhythm is a population effect and does not reflect the phasing of individuals at a dawn eclosion; the majority of individuals emerge at dawn because of the summation effect of circadian rhythms of development at earlier stages.

In a second series of experiments, Harker (1965b) studied the time taken for *D. melanogaster* to complete the same three stages of development for pupae entering each stage at each particular hour of the day in cycles of 12-hour light: 12-hour darkness; 4-hour light: 20-hour darkness; 18-hour light: 6-hour darkness. Similar experiments were carried out on insects in which the larvae were subjected to the light cycles but the pupae remained in constant darkness. It was found that the duration of each stage is affected by the time of day, relative to the light cycle, at which the stage is entered. When pupae are in constant darkness, the development rate of each stage is affected by the time of day at which the stage is entered relative to the *particular light cycle* to which the larva had been exposed. It was further shown that the eclosion rhythm, arising as a "summation effect" of rhythms of development at earlier stages, may become bimodal in light cycles with suitable photofractions. It was further found that the rate of development of a pupa entering a stage during the light period is related to the time interval since the light "on" signal; the preceding dark period has no effect.

Finally, the development rate of a pupa entering a stage during the dark period is affected by the time interval elapsing since the light "off" signal, but may also be affected by the previous light "on" signal, although there is no simple relationship between them.

Harker concluded that as the development rates are maintained in constant darkness, the rate is affected by factors following a *diurnal rhythm*. The *form* of the rhythm is determined by both light "on" and light "off" signals, but the *timing* of the rhythm is determined by the two signals acting independently of each other.

Metamorphic Changes in Individual Systems

In order to follow the types of problems presented by individual tissues one needs to look at each system in turn, for each presents its own characteristic problems. Cell differentiation and pattern formation may be best studied in various epidermal cell types, cell death perhaps in muscle systems, along with considerations of blood cell relationships, cell determination in the various imaginal rudiments, and controlling mechanisms in endocrine and nerve-tissue relationships. As general a picture as possible will be given in the space available, but slightly more emphasis will be given, in tracing sequences in individual systems, to members of the Diptera Cyclorrhapha, relating them as far as possible to events occurring in the lower Diptera Nematocera and other insects. For each system the same question will be asked. "In what way does the mosquito (Diptera Nematocera) differ from the fly (Diptera: Cyclorrhapha), and how then do other insects compare with both?"

Metamorphic Changes in the Alimentary Canal

Changes occurring in the alimentary canal vary from one insect type to the next and cannot be covered comprehensively here. The cyclorrhaphan system again will be taken as an extreme example of the type of changes that occur. The feeding apparatus in larva and adult are quit different: It has been mentioned earlier that the larval gut is longer, possesses a larval "crop" and four midgut caecae and is otherwise more or less a long, straight tube. The adult gut lacks the crop and midgut caecae, and is characterized instead by the development of an adult crop and of a rectum containing four "rectal papillae" of characteristic appearance but until recently of little known function.

The mouth parts of the larva consist of a pair of strong mouth hooks; the adult mouth appendages—labial, maxillary, and mandibular—are highly specialized and arise from the median portion of the frontal sac of the larva, an imaginal rudiment lying between the brain lobes. Changes in the gut during metamorphosis involve degeneration of the midgut epithelium, disappearance of the midgut caecae, disappearance of the labial "crop" and the development of adult crop and rectum from two "imaginal disc" areas located at the junctions of foregut and midgut and midgut and hindgut, respectively. Also, the larval salivary glands break down, and adult salivary glands develop from an "imaginal disc" area located between the base of the larval gland and the duct. From the undifferentiated cells of this area the adult glands arise by cell mitosis, growth and differentiation. Other imaginal areas may include labial and a ring of peri-anal cells.

Co-ordinated longitudinal contraction of the gut muscles and those of the midgut caecae bring about shortening of the gut and reduction of the caecae to stumps. This contraction throws the basement membrane into elaborate folds. Slightly later, the basement membrane breaks down as does that the other tissues in the pups (Whitten, 1962a), and the gut lining cells degenerate, adult epithelium being formed from small interstitial cells in the gut wall. The age of a midpupa can be fairly accurately followed by determining the length and degree of development of the adult crop. In the cyclorrhaphan pupa the Malpighian tubules are retained throughout pupation and evidently function at least in the late pupa, where they are filled with a milky material. This is passed into, the accumulates in, the rectum and later partially contributes to the gut content or "meconium," which is eliminated from the gut at the time of emergence.

Virtually nothing is known of muscle reorganization. Do the adult visceral muscles arise from dedifferentiated gut muscle cells or do some or all arise from myoblasts, as has been suggested in older literature and again recently by Crossley (1965) for certain of the abdominal skeletal muscle cells in *Calliphora*, and by Daly (1964) for a large proportion of the thoracic muscles in *Apis*, the bee? In the case of the fly gut, the muscles apparently disappear, with adult muscles appearing in the late pupa. However, it is very probable that the muscles dedifferentiate, to redifferentiate later in a fashion similar to the cyclical muscle changes shown in *Rhodmus* by Wigglesworth (1956). Here the muscle fibers disappear between molts, with only the sheath and nuclei remaining; then at the onset of each molt the muscles are completely differentiated again.

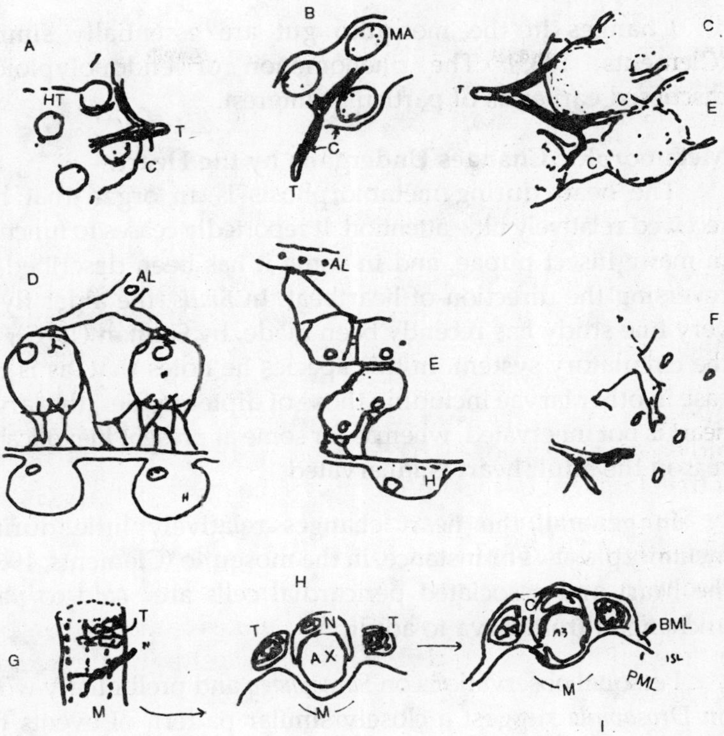

Fig. 2.3. "Connective tissue" relationships in various fly tissues; larval basement membranes characteristically break down in the early pupal stage to allow growth of the developing tissues; new adult basement membranes (connective tissue) develop in the late pupa when growth and differentiation are completed. Note continuity of the membranes (c). A: larval epidermis (hypodermis) and trachea; B: larval Malpighian tubule and trachea; C: larval fat body and trachea-these cells are released in the early pupa by breakdown of the basement membranes; D and E: alary muscle-pericardial cell-heart connections; membranes are retained during the pupal stage; F: muscle with nerve and trachea; G: developing adult muscle with developing nerve and tracheal air sac; H: cross section of developing adult muscle, nerve with axon sheath cell, and trachea; I: same as H, a little later with adult basement membranes (BML) laid down; PML: plasma membrane, ISL: interspace. Partially reproduced with permission form Quart. J. Micr. Sci., 103, 1962.

Changes in the mosquito gut are essentially similar (Clements, 1963). The phenomenon of endopolyploidy, discussed earlier, is of particular interest.

Metamorphic Changes Undergone by the Heart

The heart during metamorphosis is an organ that has received relatively little attention. It reportedly ceases to function in many insect pupae, and in some it has been described as reversing the direction of heartbeat. In *Sialis,* the alder fly, a very fine study has recently been made, by Selman (1965), on the circulatory system. In this species he notes that, as is the case in other larvae including those of dipteran flies, the larval heart is not innervated, whereas for some at present inexplicable reason the adult heart is innervated.

In general, the heart changes relatively little during metamorphosis. For instance, in the mosquito (Clements, 1963) the heart and associated pericardial cells are said to pass unchanged from larva to adult.

Personal observations on *Sarcophaga* and preliminary work on *Drosophila* suggest a closely similar pattern of events for these cyclorrhaphans. The heart is seen to pass without dissolution through to the adult stage, but in the midpupa metamorphic changes do occur. Also the larval posterior tracheal plexus is absent. The heart wall itself becomes somewhat thickened, and this thickening does not extend to the aorta. Correlated with the increase in wall material, the intrinsic heart cells become active and display well-developed polytene chromosomes (Whitten,1963a). The smaller anterior and the large posterior pericardial cells undergo changes in the midpupal stage, and the larger cells also develop well defined polytene chromosomes. The thickening of the heart but not of the aortal region produces a humped effect at the junction of the two regions, and this comes to lie in the raised abdominal region immediately posterior to the constriction between adult thorax and abdomen. In a midpupa the heart muscles have degenerated and degeneration is also seen in the alary muscles. In these latter, there would appear to be events occurring that are comparable with the cyclical muscle dedifferentiation and redifferentiation in the abdomen of *Rhodnius*. In a fly midpupa, heart pumping ceases and there is virtually no circulation of blood. Developing adult nerves can be followed migrating over the heart surface of a mid to late pupa. An attempt to review events recorded for other insects will not be made; the scarcity of information concerning metamorphic changes can be seen from the recent review of Jones (1964). The absence of circulation during the midpupal period would seem to be pertinent to questions of hemocyte relationships in the pupal stages, and these will be discussed later in more detail.

There are many other peripheral issues in connection with the developing adult circulation during metamorphosis, including the development of pulsatile organs, dealt with in current textbooks on entomology. As with all the other systems, the more one looks into the problem the more one realizes how little, relatively, is known of the subject. Few insects have their circulatory systems as thoroughly investigated as has the alder fly *Sialis* in Selman's recent study.

Metamorphic Changes in the Fat Body, Oenocytes and Nephrocytes

The fate of the fat body varies in different insects. For instance within the Dipters, in a generalized dipteran such as the mosquito, it is said to pass from larva to newly emerged adult unchanged and little if at all diminished in size (Clements, 1963). Changes in the higher flies are however somewhat more complex. It was formerly considered that the larval fat body of the fly broke down in the early pupa and thereafter ceased to function (e.g.,Bodenstein, 1950). However, what happens at pupation is that the connective tissue sheath (basement membrane) that binds the cells into a composite organ breaks down and release the component cells (Whitten-1962a;). The breakdown, as with all the events in the metamorphosing pupa, is perfectly timed and co-ordinated. Breakdown of thoracic segments of the fat body occurs in the cryptocephalic pupa while the abdominal portion breaks down later, in the early pupa. Although appearing diffuse, the fat body arises segmentally and functions as such in later life. A branch of a middorsal nerve has been traced to the fat body in *Sarcophaga*, and the basement membranes of the nerve and fat body are confluent. The fat body in this fly would therefore seem to be innervated.

The cyclorrhaphan larval fat body cells function right into the adult stage. They have been the subject of much recent research, particularly in *Drosophila*, where they seem to play an important part in intermediary metabolism in the developing pupa. Changes in the fat body are characteristic of certain *Drosophila* mutants.

In many insects including flies and moths a separate adult fat body originates within the pupal stage. Older workers have suggested that adult fat body cells may have originated from hemocytes. The whole subject is by no means clear. Evidence in the Diptera Cyclorrhapha recently has suggested that this is not as unlikely a possibility as might at first be supposed (Whitten, 1964b). Phagocytic hemocytes, which at the time of pupation ingest larval fragments, later undergo a second cycle of activity. They persist right into the adult stage, and at this time appear very like the adult fat body cells. The mosquito is seen not to have a distinct adult fat body, and the larval fat body is said not to histolyse in the early adult (Clements, 1963). Certainly phagocytic hemocytes would be relatively much less active in the mosquito where tissue histolysis is less extensive.

Oenocytes are cells which are thought to be involved in intermediary metabolism, in growth and in reproduction (Wigglesworth,1965). In many insects including the mosquito and the fly there are two generations—larval and adult. The larval oenocytes in the fly are large cells with polytene chromosomes, located segmentally in groups along the side of the body. Adult oenocytes are smaller, appear in the pupal stage and are often associated with the adult fat body cells. In the mosquito the larval cells grow in size, develop polytene chromosomes, and breakdown in the pups. The many smaller adult oenocytes appear in the abdomen during the fourth larval instar (Clements, 1963).

In the fly, *ventral nephrocytes* thought to be excretory in function and related to pericardial cells, are slung as a garland between the lobes of the salivary glands (attached by muscles and basement membrane material). They break down in the early pupa, at the time of the disappearance of the salivary glands. Comparable cells are present in the mosquito where in some species they also form a chain between the salivary lobes.

Metamorphosis of the Muscular System

Among the Endopterygota there is a considerable range in the degree to which muscles are retained from larva to adult. In some of the more primitive orders, but also in some of the more primitive members of the more specialized orders including the Diptera, the majority of the larval muscles are carried over into the adult stage. The muscles, particularly of the abdomen, are carried over and changes occur mainly in the muscles of the thorax that are responsible for flight. In the mosquito for example, larval abdominal muscles are used in the pupa for swimming and these degenerate early in adult life. The adult abdominal muscles arise from myoblasts in the pupal stages. However, among the specialized from larva to adult. In the thorax of the honeybee, a very comprehensive survey made recently by Daly (1964) should that no larval muscles pass unchanged into the adult, that 85 per cent of larval muscles are associated with the development of adult muscles, and that 15 per cent of the larval muscles degenerate and have no relationship to adult muscles. Some 61 per cent of adult muscles arise from aggregations of myoblasts, including those of appendages and direct flight muscles. Crossley (1965), in a recent study on the blowfly *Calliphora erythrocephala*, similarly showed the origin of certain of the abdominal muscles from myoblasts, an observation substantiating the opinions of several of the older workers at the turn of the century (e.g., Perez, 1910). The description by nuclei that subsequently differentiate into adult muscles in the same location as the larval is difficult to conceive but is all the more fascinating if actually the case.

In recent years interest has centered largely on the breakdown of muscles and the cause of the breakdown. This remains one of the most intriguing problems of insect metamorphosis. Of particular interest is the fact that not all larval muscles degenerate at the same time. Some are retained while their neighbours degenerate. Initial work on this subject was in the Lepidoptera (see e.g., Finlayson, 1956). More recently, an extremely interesting series of papers have appeared by Lockshin and Williams (1965), also on the Lepidoptera. Whereas the earlier workers concentrated on muscles that were carried over from the larva into the early pupa and then degenerated. Lockshin and Williams studied muscles that passed into the adult and were involved in emergence; these degenerated three days later under normal conditions. Similar differential breakdown occurs in other insects: e.g., in the cyclorrhaphous Diptera certain larval muscles do not degenerate with the remainder after puparium formation (Cottrell,1964), but they do so some 48 hours later. Likewise, there are muscles that degenerate in the early adult, including those of the ptilinum and other muscles expressly involved in the process of eclosion. The ptilinal muscles degenerate following the sequence of events that starts with cessation of digging. Similar controlling mechanisms to those shown by Lockshin and Williams for the lepidopteran may apply in these and other cases. They have concluded that the initiation of cell death in the muscle cells is brought about by the coordination of several factors-humoral, nervous, and an ability of the cell to respond to the factors. Thus, the breakdown of certain intersegmental muscles in the adult of first potentiated by exposure to the molting hormone ecdysone during the first few days of adult development. Three weeks later, when development is completed, actual breakdown is triggered by a neural mechanism. The muscle cells attain the capacity to respond to this during the three days before emergence.

Johnson (1959) has shown that where degeneration of flight muscles normally occurred in adult aphids this was prevented by decapitation. Further, the thoracic muscles of decapitated aphids broke down when they were joined in parabiosis with intact aphids. The complete story may well be similar to that shown for the postadult muscle breakdown in the Lepidoptera. Much has been done and much remains to be done on this subject. It is an important biological problem because the whole question of "programmed cell death" is involved.

The problems of muscle dedifferentiation and subsequent redifferentiation are equally interesting. Why should the muscles of *Rhodnius* (Wigglesworth, 1956) dedifferentiate between molts, and shortly before molting redifferentiate contractile muscle fibers? The same problem is posed in the metamorphosing pupa by the larval muscles that the transformed into adult muscles, constituting the 39 per cent of the total thoracic musculature in the adult honeybee (Daly,1964).

Reorganization of the alary muscles in the fly pupa presents a similar problem. Another problem is posed by muscles that develop in the pupal stage and degenerate before emergence. The functions of muscles of this type are discussed at length by Daly for the honeybee, and their possible involvement in determining shape is also considered.

Hinton (1961) has described how the disposition of the indirect flight muscles of insects can be brought about at the time of the larval-pupal molt by activity of the epidermal cells. He demonstrates this with *Simulium,* which as a generalized nematoceran member of the Diptera has (unlike the more specialized Cyclorrhapha) its epidermal cells carried over from larva to adult. The indirect flight muscles at the time of the larval-pupal molt are undifferentiated strands, attached to the epidermis but in relative positions quite unlike the relative positions of the corresponding future adult muscle. He concludes that the definitive shape of the adult mesothorax and the final position and orientation of the skeletal muscles are determined by the way the epidermis grows between the time of secretion of pupal cuticle (larval-pupal molt) and the time that the larval cuticle is shed (larval-pupal ecdysis).

The Controlling Centers of Metamorphic Change: Endocrine Glands and the Nervous System

A brief look wili be taken of the hormone system that controls the metamorphic changes in the various organs. As, was seen earlier, the endocrine system is essentially similar in all of the groups of insects. The corpus allatum secretes juvenile hormone, the prothoracic (ecdysial) gland cells (and the ventral gland cells, their homologues in the apterygote thysanurans) secrete ecdysone, the molting hormone, while the corpus cardiacum, which has been assigned various functions, contains distinct glandular and storage regions. The cephalic endocrine system in the Thysanura Apterygota is seen to resemble closely that of the generalized pterygote (Watson, 1963). It is currently thought that the brain hormone, secreted by neurosecretory cells of the brain, stimulates the prothoracic glands as the principal target organs. The secretion of ecdysone in the presence of juveline hormone in the winged pterygote insects produces a juvenile from; the secretion of ecdysone in the presence of a reduced titer of juvenile hormone initiates metamorphosis, so that at the time of the last larval molt in an endopterygote insect, a pupa is formed instead of a further larval stage. With relatively few exceptions it has been shown that after emergence the corpus allatum again becomes active, and the hormone secreted stimulates the deposition of yolk in the female.

Metamorphosis of the Nervous System

Along with the endocrine system, the nervous system is the controlling center of the body's activities, so that changes in the nervous system are fundamentally concerned with metamorphic changes in other systems. It has already been seen for example, that the nervous system plays a role in the process of differential muscle breakdown. Also, the neurosecretory cells of the brain are virtually the controlling agents of the endocrine system: through regulation of the release of brain hormone molting is controlled; through regulation of the release of juvenile hormone maturation is controlled (Schneiderman and Gilbert, 1964). What then happens to the nervous system itself during metamorphosis? As was noted earlier in reference to Power's work, surprisingly little is known about metamorphosis of the insect nervous system at the individual cellular level. Neurometamerphosis has been the subject of study recently by several workers, e.g., pipa and Woolever (1965) and Ashhurst and Richards (1964a and b), both for Lepidoptera. Many older workers considered the fate of individual nerves; evidence is extensive, although often contradictory, and Wigglesworth (1965) should be consulted for major references on the subject.

For *Drosophila* along (Bodenstein,1950) there is no detailed account of nervous system changes. What happens to the paired lateral nerves of the larva? Do these disappear in the pupal stage? Are the adult nerves all new developments? How is the relatively longer and narrower ventral thoracico-abdominal nerve mass of the larva converted into the shorter, broader thoracic mass of the adult where the whole abdominal center is reduced to the small posterior region? An answer to some of these questions will be attempted below in discussing metamorphic changes in the cyclorrhaphan nervous system.

Power (1952) drew a graph for the *Drosophila* nervous system development in the pupa, plotting volume increases in the components of the central nerve mass, of cells relative to fibers. He concluded that there was essentially little cell breakdown, but that there was a great increase in neuropile. Clearly, where larval organs break- down their nerve supply is no longer required. Conversely, where new adult organs are formed, a new nerve supply also has to be available. Do the old nerve fibers degenerate? Are the old nerve fibers taken over by other tissues? Do new neurons come into being in the pupa? Do the larval neurons dedifferentiate, their nerve processes degenerating, and later redifferentiate and form new nerve fibers? These are problems touched on but not completely solved for any of the insects in which changes are extreme. The field lies open for investigation at the light and electron microscope levels. Of course, the problem is far more extreme in a case like the Cyclorrhapha than in, say, the Diptera Nematocera such as mosquitoes or crane flies, where most tissues and organs are carried over more or less unchanged from larva to adult. In the mosquito, it is said that the central nervous system grows by cell multiplication in the larva, with very slight increase in cell size. Only slight histolysis occurs in the pupa, and in the adult there is again cell multiplication (Trager, 1937). However, even here we will have the large flight muscles developed. How is the new innervation to these developed? How do nerve fibers grow out of and into (sensory) a central nerve mass that is invested in a thick neural lamella? Breakdown of the connective tissue sheath would seem to be one prerequisite for this (Whitten,1962a, Ashhurst and Richards, 1964; Pipa and Woolever, 1965).

General Metamorphic Changes in the Cyclorrhaphan Nervous System

Instead of complete breakdown and reorganization of the peripheral nervous system there is actually continuity in the *basic* framework from larva to adult (Whitten, 1965). At the time of puparium formation there are numerous mitoses among the sheath cells of the lateral abdominal nerves. Similar mitoses are present in the recurrent nerve from hypocerebral ganglion to the midgut junction, and the mitoses are paralleled in the string of cells (mesodermal gonad duct rudiment) arising posteriorly from the testis and ovary. This burst of mitotic activity is followed by the breakdown of the sheaths, thus allowing cell growth and migration to occur. In the ventral nerve mass, growth occurs along the region and the net result

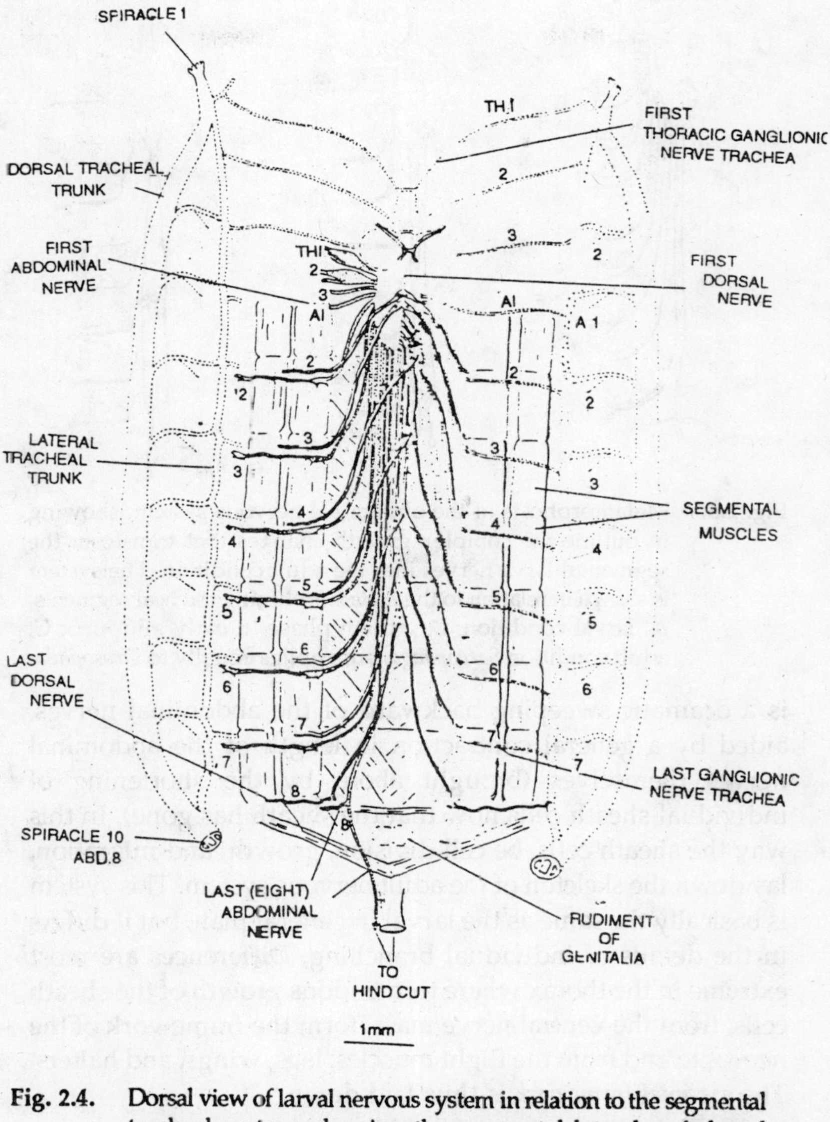

Fig. 2.4. Dorsal view of larval nervous system in relation to the segmental tracheal system, showing the segmental lateral and dorsal nerves, together with some of the larval muscles, the first and second leg imaginal discs, and the genital disc. *Sarcophaga* and other Cyclorrpha.

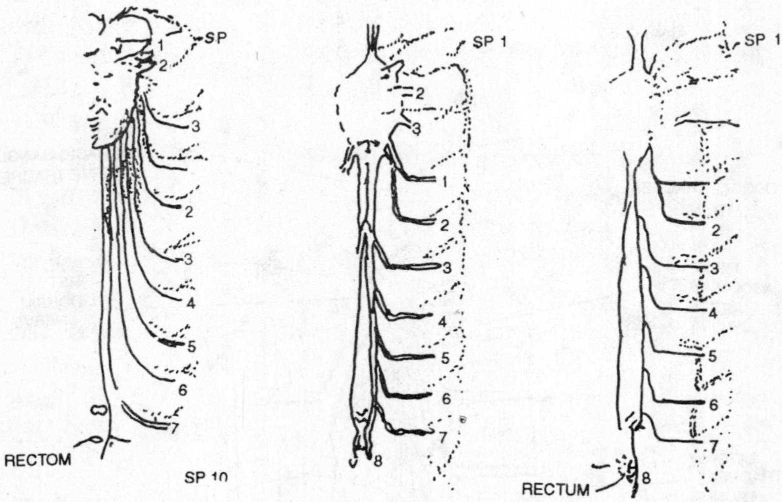

Fig. 2.5. Metamorphosis of the obdominal nervous system, showing in outline the complex growth changes that transform the segmental larva nerves into the adult condition. The system is shown in relation to the metamorphosing tracheal segments. A: larval condition; B: growth phase in early midpupa; C: adult condition *Sarcophaga*, but applies equally to *Drosophila*.

is a dramatic sweeping backward of the abdominal nerves, aided by a general contraction in length of the abdominal nerves themselves (brought about by the shortening of individual sheath cells now that the sheath has gone). In this way the sheath cells, be cell division, growth, and migration, lay down the skeleton of the adult nervous system. This system is basically the same as the larval in general plan, but it differs in the details of individual branching. Differences are most extreme in the thorax where tremendous growth of the sheath cells, from the ventral nerve mass, form the framework of the nerves to and from the flight muscles, legs, wings, and halters. The *external framework* is thus laid down.

Internal to this, individual nerve fibers may degenerate and new fibers regenerate. Evidence so far suggests this and also that there may be dedifferentiation and redifferentiation of fibers of individual neurons. Pipa and Woolever, using electron microscopy, observed little evidence for "axoplasmic degradation," but the situation in the lepidopteran nerve cord may well be less extreme than in the cyclorrhaphan system. Muscle breakdown and tracheal histolysis are certainly less extreme in the former group of insects. Early in pupation, at the time of breakdown of the larval neural lamella, there is a concurrent breakdown in dorsal nerves and in the larval abdominal ganglionic tracheae—the reduction in volume of the abdominal center is such that these tracheae could not even be accommodated in the adult nerve mass. One of the most spectacular observations, which will be discussed in connection with the tracheal system, is the fact that these abdominal ganglionic tracheae *before puparum formation* occurs do not produce pupal cuticular linings. Their fate is already determined at this time. They are empty tracheal tubes, their cells destined to die. Why these? If apparently not innervated, why do they die and other tracheae in the same hormonal milieu survive through to the adult? What changes occur in the abdominal nerve mass that can cause such drastic reductions in respiratory requirements? Late in the pupal stage, when the processes of growth and differentiation are completed, the new adult nerve sheath is laid down. In the meantime, new adult tracheae have been formed, including the air sacs. Their sheaths are formed at the same time as, and are confluent with, the adult neural lamella. These events have been followed in *Sarcophaga bullata, Phormia regina* and *Drosophila melanogaster* and are essentially similar in all three. It may be mentioned that staining for neurosecretory material in sections of the nervous system of *Sarcophaga* has revealed the presence of dorsal neurosecretory cells throughout pupal life. They are always gorged with PAF staining material, which does not imply that they are active but does show that they pass through unchanged from larva to adult.

Nerves and tracheae to organs destined to disappear in the pupa, breakdown; new nerve branches grow out at a time when the innervated organs themselves are still developing. What stimulates the sheath cells of the abdominal nerves to undergo mitosis? The number and location of these mitoses determines the future pattern of the abdominal nervous system. How do the nerves at this stage know where future adult structures will develop and require innervation? This brings us into the realm of cell differentiation and pattern formation, problems that will be considered in more detail later. The co-ordination of the different metamorphic processes is amazing, as is the realization that all these events occur within, in response to, and at speeds determined by light darkness cycles such as those shown by Harker for *Drosophila* (1965a and b).

Hemocyte Activity in Relation to Metamorphosing Tissues

Older works on the hemocytes were largely concerned with the role of blood cells in relation to developing tissues. More recently, work has centered more on hemocyte identification, origins, and their relative numbers at various stages, together with their role in coagulation. The vast amount of work up to 1962 is reviewed by Jones (1962). Since that time renewed interest has arisen in the part that blood cells may play in 1962, described briefly the fate of blood cells during the life history of the alder fly, *Sialis lutaria*. Here, phagocytes were found to ingest all particles freed in the hemolymph as a result of the cytolysis of the cells of the fat body, and in the gills the epidermal cells are ingested by the blood cells also only after cytolysis. The phagocytes or morula cells, as they are later called, were said to disintegrate immediately before darkening of the cuticle. Ashhurst and Richards (1964b) described adipohemocytes in the vicinity of the nerve cord at the time of breakdown of the sheath in the lepidopteran *Galleria Mellonella*. The participation of these cells in the breakdown of the neural lamella was subsantiated by autoradiographic studies by Shrivastava and Richards (1965) and by electron microscopic observations by Pipa and Woolever (1965) on the same species. The latter demonstrated the presence within the adipohemocyte of neural sheath fibrils. These blood cells were not found to be

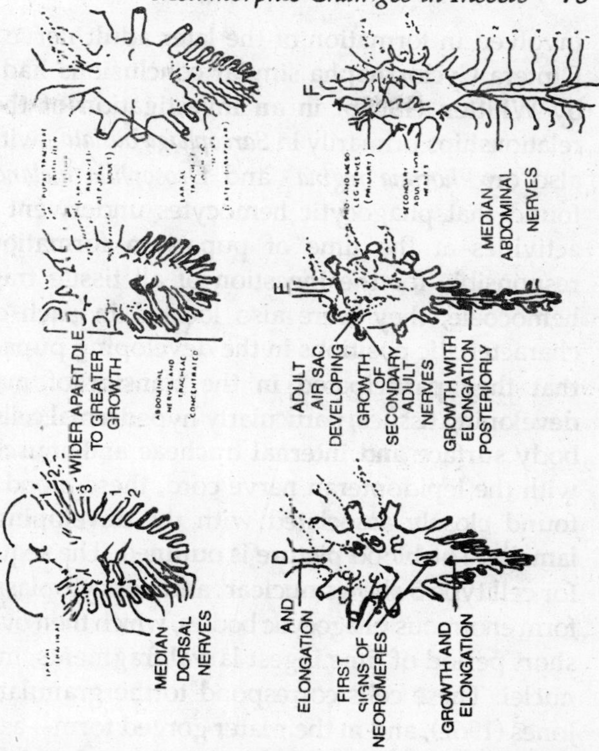

Fig. 2.6. Metamorphosis of the thoracico-abdominal nerve mass showing general changes in shape due to growth of sheath cells and the formation of adult "neuropile" in the thoracic, and general reduction in the abdominal region. A: larval system in dorsal view, with lateral and dorsal nerves and segmental tracheae (dated); B: cryptocephalic pupa showing general elongation of the thoracico-abdominal mass; C: first-day pupa showing beginnings of relative enlargement of thoracic and reduction of abdominal centers; D: second-day pupa, showing disappearance of abdominal tracheae and apparently also of the segmental dorsal nerves with neural lamella broken down; growth and elongation occurring in the abdominal region; F: midpupa showing growth and elongation of the abdominal nerves, growth of secondary thoracic nerves and the development of adult tracheal epithelium; F: adult system completed, with lining of air sac secreted and new adult" connective sheaths" formed continuously around nervous system and tracheae. *Sarcophaga*, but events are similar in *Drosophila*.

involved in formation of the later adult nerve sheath. In the Diptera Cyclorrhapha similar conclusions had been reached by Whitten (1964b) in an investigation of the blood-tissue relationships primarily in *Sarcophaga bullata,* with observations also on *Phormia regina* and *Drosophila melanogaster.* It was found that phagocytic hemocytes underwent very complex activities at the time of puparium formation. They were responsible for the ingestion of all tissue fragments in the hemocoele; they were also located in such consistent and characteristic positions in the developing pupae as to suggest that they participated in the transfer of materials to the developing tissues, particularly hypodermal cells of the general body surface and internal tracheae and muscle tendons. As with the lepidopteran nerve cord, these blood cells were not found closely associated with the developing adult neural lamella. The blood picture is outlined. The sequence of events for cell type 5 shows nuclear and not cytoplasmic division to form enormous phagocytic bodies, which then over an extremely short period of time ingest larval fragments, including whole nuclei. These cells correspond to the granular hemocytes of Jones (1962), and in their later gorged form—as "spherules"—they correspond to the "spherule cells" and "spheres of granules" of earlier workers. These cells have recently been followed at light and electron microscope levels (Whitten, 1968) through the pupal to the adult stage. Their continuing close relationship with the developing adult tissues has convincingly shown the importance of the cells not only in removing larval fragments but also in processing and passing on materials to the developing adult tissues. By the time the adult emerges the blood cells contents are depleted. These cells undoubtedly correspond to the type F hemocytes described by Crossley for *Calliphora* (1964). In a later paper (1965), Crossley described breakdown of type F cells in the midpupa, whereas it will be shown below in a study of co-ordinated development in the fly foot that the corresponding cells in *Sarcophaga* pass through into the adult stage, functioning even at the time of emergence, in the ingestion of selected epidermal cells. We have now reached the stage in

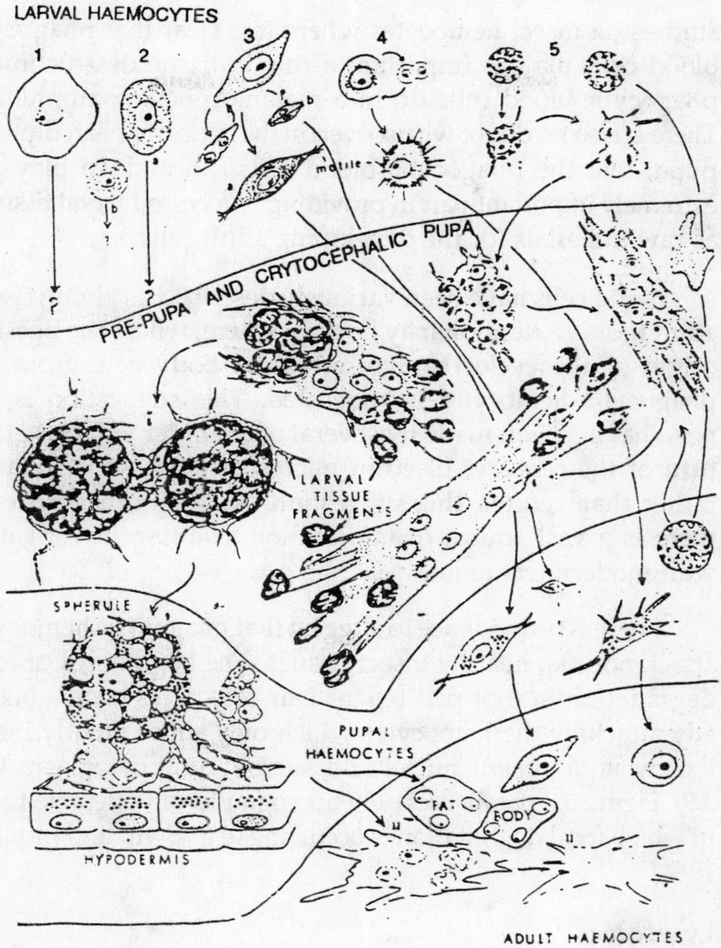

Fig. 2.7. Diagrammatic representation of the changes occurring in the hemocytes, accompanying metamorphosis in *Sarcophaga bullata*. 1: hyaline cell; 2 to 4: plasmatocytes, 2-aval, 3-fusiform, 4-round, 5: granular hemocytes that by complex processes undergo nuclear division, forming multinucleate masses that phagocytose larval fragments and become the pupal "spherules," these later appear uninucleate but may or may not retain cytoplasmic bridges throughout pupation. They appear again later in the pupal stage to be multinucleate and enter a second phase of phagocytosis, selectively ingesting adult hypodermal (epidermal) cells at very specific times.

studies on insect hemocytes where it is clear that phagocytic blood cells play as important a role in tissue destruction as phagocytic blood cells do in a metamorphosing amphibian. There is also no doubt whatsoever, in the cyclorrhaphan dipteran pupa, that the phagocytic blood cells can and do play an extremely important part in providing "processed larval tissues" as raw materials for the developing adult cells.

Blood cells have been variously described as giving rise to adult tissue. Attention may be drawn here, where the possible origin of the cyclorrhaphan adult fat body cells from the phagocytic hemocytes is suggested. This suggestion is not new, having been made by several of the older workers at the turn of the century. Electron microscopy has substantiated, rather than refuted this suggestion. As with other systems, there is a vast source of information awaiting investigation with modern techniques.

There is no evidence to suggest that phagocytic hemocytes attack normal, healthy insect tissues. The beginnings of focal degradation in foot pad tenant hair cells. This occurs before any attack by the hemocytes which only subsequently ingest the pycnotic tenent nuclei and surrounding cytoplasm (Fig 3.9). Figure 3.8 parallels closely micrographs of vertebrate tissue in which focal degradation is occurring (e.g., Swift and Hruban, 1964).

Lysosomes

The cyclorrhaphan phagocytic hemocyte provides ideal material for study of lysosome origin and fate and function. DeDuve's article on the lysosome (1963) describes lysosome activity with respect to the tail resorption and tissue breakdown in the metamorphosing frog. A similar story can be given for lysosome activity in the cyclorrhaphan phagocytic hemocyte. Figure 3.10 shows a typical picture of different stages of degradation of *larval* fragments in the vicinity of the nucleus in a day four (at 25°C pupa. Acid phosphates has been demonstrated in these cells and in their apparent homogoues in *Calliphora crythrocephala* (Crossley, 1964). This is particularly good material for such studies of lysosomes in that within a time span of 12 days at 25°C these blood cells can be followed at all stages of their various phagocytic activities.

The Reproductive System

Since the reproductive system matures in all insects regardless of whether they undergo metamorphosis or not, considerations of general development of the reproductive system are not directly relevant to the present discussion. However, certain aspects bear on the subject. In the apterygote ametamorphic insects, the reproductive system is laid down at the time of hatching from the egg, and maturation occurs later, although even in the Apterygota, genitalia may be undeveloped at the time of hatching and develop in later instars. In insects with complete metamorphosis, the larvae may hatch with the reproductive system in an embryonic state. Gonads are present but the ducts are limited to the mesodermal elements that extend from the gonad distally, to end posteriorly in a coelomic sac or ampulla, located in different segments in male and female. During metamorphosis, development of the reproductive system is completed when the genital imaginal rudiment develops both the external genitalia and the ectodermal elements of the internal reproductive system, including median ducts, accessory glands and spermathecae. When growth is completed and ectodermal and mesodermal ducts have met and fused the breakdown of cells and continuity of the ducts to the outside is established. In one of the best-known insects, *Drosophila*, onset of metamorphosis in characterized by immediate and rapid growth of the imaginal discs, including the genital disc. In this same example an anomaly exists in that the ectodermal component is considered to give rise to the entire duct system right to the gonad. This anomaly has recently been doubted in the higher fly *Sarcophaga*, where the mesodermal ducts have been shown to be present in the larva in a typical embryonic state. The ducts themselves represent true examples of "imaginal rudiments" (Whitten, 1965). The onset of papation is characterized by extensive mitoses in this strand of cells. Subsequent dissolution of the connective tissue "basement membrane" and shortening of the strand (closely similar to events in the peripheral nerves) produces the proximal mesodermal genital duct in both male and female. This meets up and fuses with the ectodermal components derived from the genital imaginal disc. Thus *Sarcophaga* falls into line with other endopterygote insects in which metamorphosis involves passing from the embryonic to the adult condition. It would be surprising if *Drosophila* is really exceptional in this respect.

During metamorphosis, extensive changes occur within the reproductive system, the Degree of maturation attained by the time of adult emergence varies from male and female and from species to species. In many cases gametes are fully developed at the time of emergence of the adult insect. Review articles of King (1964) and Telfer (1965) provide interesting and instructive reading on this subject.

The Epidermis in the Metamorphosing Pupa

The epidermis and its derivatives have been left for consideration until last because in many respects this is the most important and interesting cell layer to the insect developmental biologist. Included here are the imaginal discs which are essentially epidermal sacs containing nerve and tracheal endings along with the hemolymph and contained blood cells. Strictly speaking, the ectodermal regions of the reproductive system and of the gut also fall into this category, but they have already been considered briefly along with the rest of these systems. The tracheal system is an invaginated extension of the general epidermis and will be considered first.

The Tracheal System

The tracheal system is composed of two cell types: the tracheal cell and the tracheolar cell. Essentially, the difference between a trachea and a tracheole is that the former is formed as a layer of cuticle by the tracheal epithelium, while the tracheole is terminal and formed intracellularly by a single cell. At one time it was thought that tracheae possessed the spiral or annular thickenings, whereas the tracheoles did not. However, electron microscopy has shown that tracheoles do also. A physiological distinction made between tracheae and tracheoles is that little gaseous exchange occurs through the tracheal wall, exchange being mainly by way of the tracheoles at the tissue level. It will be seen below that the metamorphosing pupa may be an exception to this generalization. The pupa has respiratory problems that are unique in that a single system laid down at the larval-pupal molt has to tracheate tissues that are destined to disappear during the pupal stage, tissues that persist through from larva to adult, and more significantly, yet other tissue that develop during the pupal stage and are not in existence at the time that the tracheal system of the pupa is laid down. These demands on the tracheal system will be taken up following a general survey of the development of the system from first instar to adult.

Details can be obtained from Buck (1962) and Miller (1964) on the large amount of work that has been done on the tracheal system; however, problems of development are relatively unexplored. In general, it has been thought that at each molt the tracheal lining is replaced but that the tracheolar lining is retained throughout the life of the insect. However, the situation is much more complex than this, particularly in the metamorphosing insect where drastic changes are occurring in other systems. A schematic drawing of the types of development that can occur in individual tracheae. In general, the insect tracheal system is divided into 10 pairs of tracheal metameres, occurring in segmentally repeating order and, in the pterygote insects, joined laterally and longitudinally at junction points of *tracheal nodes.* The original segmental pattern can be seen at each molt when the continuous tracheal tube linings "break" at the tracheal nodes, and each segment is removed through its respective spiracle. This occurs even in

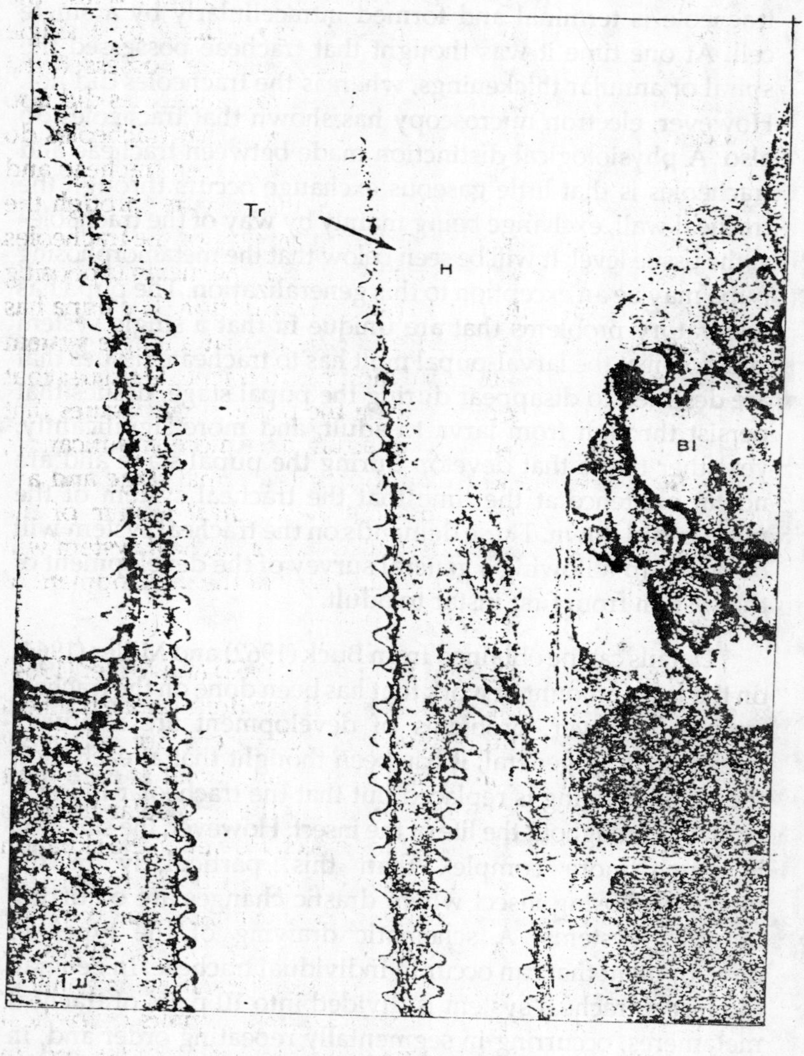

Fig. 2.8. Electron micrograph of a pupal tracheole from the leg of *Sacophaga;* note taenidia the very thin layer of cytoplasm between tracheal; lining and hemocoele, and the virtual nonexistence of basement membrane. It is suggested that any gaseous exchange is between tracheal lumen and hemocoele. Bl: adjacent blood cell; H; hemocole; Tr: tracheole.

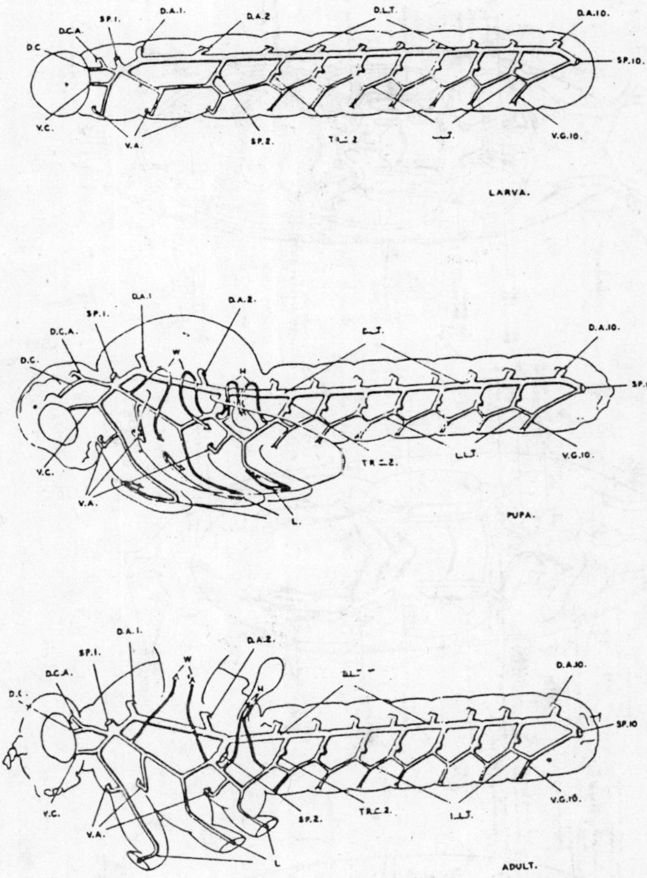

Fig. 2.9. Diagrammatic representation of the development of the tracheal system in a generalized member of the Diptera. A: larval system is left lateral view, B: pupal system; C: adult system. At each molt the tracheoles are retained, new tracheae developing around those of the preceding stage, with new branches also being formed, particularly in the thorax. Abbreviations: D.C. dorsal cervical; V.C.: ventral cervical D.C.A.: dorsal cervical anastomosis; V.A,: ventral tracheal anastomosis; SP,: spiracle; D.A. : dorsal anastomosis; T.R.C. transverse connective, V.G.: ventral ganglionic, D.L.T. : dorsal longitudinal trunk ; L.L. T: lateral longitudinal trunk; W: wing trachea; H: halter trachea; L: leg trachea.

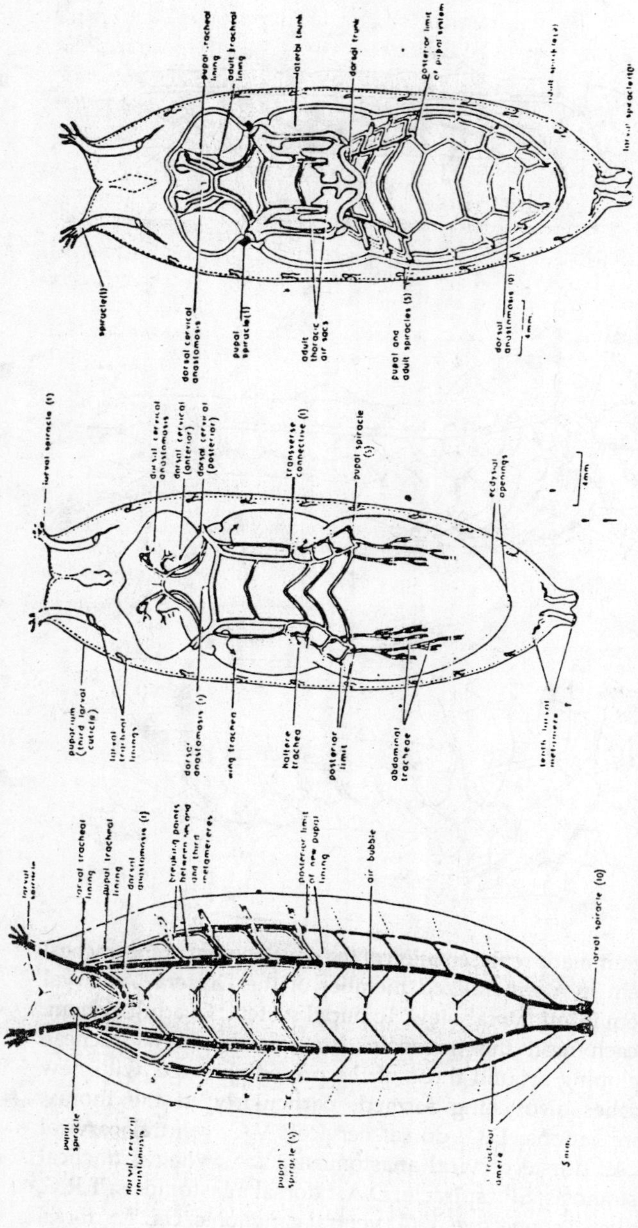
pupal tracheal lining
adult tracheal lining
lateral trunk
dorsal trunk
posterior limit of pupal system
adult spiracle (2)
larval spiracle (10)
spiracle (1)
dorsal cervical anastomosis
pupal spiracle (1)
adult thoracic air sacs
pupal and adult spiracles (5)
dorsal anastomosis (10)
4 mm
larval spiracle (1)
dorsal cervical anastomosis
dorsal cervical (anterior)
dorsal cervical (posterior)
transverse connective (1)
pupal spiracle (3)
ecdysial openings
4 mm
puparium (third larval cuticle)
larval tracheal linings
dorsal anastomosis (1)
wing trachea
haltere trachea
posterior limit
abdominal tracheae
tenth larval metamere
larval spiracle
larval tracheal lining
pupal tracheal lining
dorsal anastomosis (1)
breaking points between second and third metameres
posterior limit of new pupal lining
air bubble
larval spiracle (10)
pupal spiracle
dorsal cervical anastomosis
pupal spiracle (3)
5 mm

Fig. 2.10. Diagrammatic representation of the development of the tracheal system in a specialized member of the Diptera — *Drosophila melanogaster*; here development is highly complex. The posterior half of the larval system is histolyzed, being reformed when the adult system is laid down, although the last tracheal metamere is never replaced. The adult thoracic system develops around the preexisting pupal system and by elaborate branching of the epithelium of the basic system. A: prepupal stage with pupal system being formed around the larval; B: functional pupal system; C: adult system being formed around the pupal system.

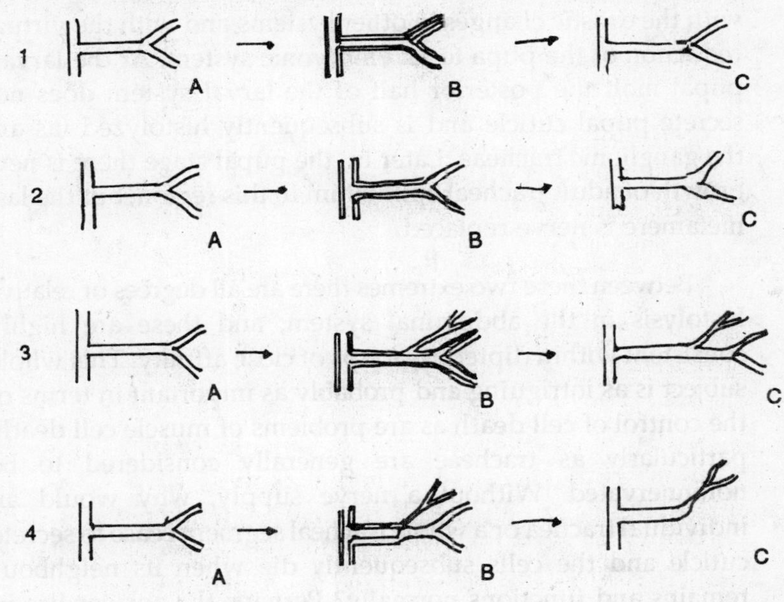

Fig. 2.11. Diagram showing the different types of fate that an individual trachea can undergo during development. Details can be found in Whitten (1962b).

forms in which the spiracles are physiological closed and not functional as far as respiration is concerned. The degree of change or growth in he tracheal system from instar to instar is determined by the physiological requirements, so that as far as the metamorphosing pupa is concerned, there is a direct correlation between the degree of metamorphic change in, say, the muscular system and that in the tracheal system. The pattern of development in a generalized dipteran, and members of the nematoceran Bibionidae would follow this closely. Here the larval basic pattern is taken right through from larva to adult, and the only drastic change is in the thorax where the tracheation to the new flight muscles, wings, halters, and legs is superimposed on the larval pattern by growth of new tracheal branches at the end of the last larval instar. In contrast, the sequence of events in *Drosophila melanogaster* where there is considerable change from larva to adult, correlated closely with the drastic changes in other systems and with the virtual reduction of the pupa to an embryonic system. At the larval-pupal molt the posterior half of the larval system does not secrete pupal cuticle and is subsequently histolyzed (as are the ganglionic tracheae. Later in the pupal stage there is new growth of adult tracheal epithelium in this region, but the last metamere is nerve replaced.

Between these two extremes there are all degrees of relative histolysis in the abdominal system, and these are highly consistent within dipteran groups of close affinity. This whole subject is as intriguing and probably as important in terms of the control of cell death as are problems of muscle cell death, particularly as tracheae are generally considered to be noninnervated. Without a nerve supply, why would an individual trachea or a whole tracheal segment cease to secrete cuticle and the cells subsequently die when its neighbour remains and functions normally? Perhaps the answer lies in physiological changes in the organ that is tracheated and destined to die. However, the converse cannot apply to the development of new tracheal branches to tissues that are not yet developed in the pupa.

It has been said that the elaborate tracheal development of the thoracic region of adult systems bears little relationship to that of the earlier larval system. This is not so. From a survey by the author of the development of the adult system within some 20 families of Diptera, it is evident that the very elaborate system of tracheae or air sacs represents branching of the tracheal epithelium of the earlier basic larval pattern. Superficially, they look very unlike each other and their corresponding larval systems. However, close inspection shows the same basic pattern of transverse and longitudinal elements, with very constant positions of origin of leg, wing, and halter tracheae. No comparable survey is apparently available for other endopterygote orders.

Some Unique Characteristic of Pupal Tracheal Systems

The pupal system of the cyclorrhaphous Diptera, including forms like *Drosophila*, *Sarcophaga* and *Calliphora*, is characterized by grapelike bunches of tracheoles suspended in the hemocoele. These arise by mitosis of cells of the tracheal system, and after the larval-pupal molt there is often extension of the tracheal cell cytoplasm toward various organs, including the developing epidermis. The pupal stage has been seen to be characterized by the absence of basement membranes: there is no penetration of individual tissues by the pupal tracheoles which lie essentially within the hemocoele. Some of these tracheoles. They are particularly interesting in that respiratory exchange is apparently occurring between the tracheole and the hemolymph. The lining of cytoplasm, particularly on one side, is infinitesimally thin and so is any basement membrane. These pupal tracheoles are seen to move in later pupal stages to come in closer proximity to developing adult organs such as the reproductive system. For example, an early pupal testis or ovary has no tracheation, whereas later pupal gonads have functional pupal tracheoles associated with them, as well as the as yet nonfunctional adult tracheae and tracheoles. The condition in the Diptera Cyclorrhapha is extreme, but the same condition prevails in at least the thorax of more generalized forms where there is extensive development of muscles in the later pupal stages. The situation diagrammatically represents in an organism such as *Drosophila*. In the present case it is the tracheal cell cytoplasm that extends towards the tissues, whereas in the bug *Rhodnius*, Wigglesworth (1959a) described the drawing of tracheoles, by

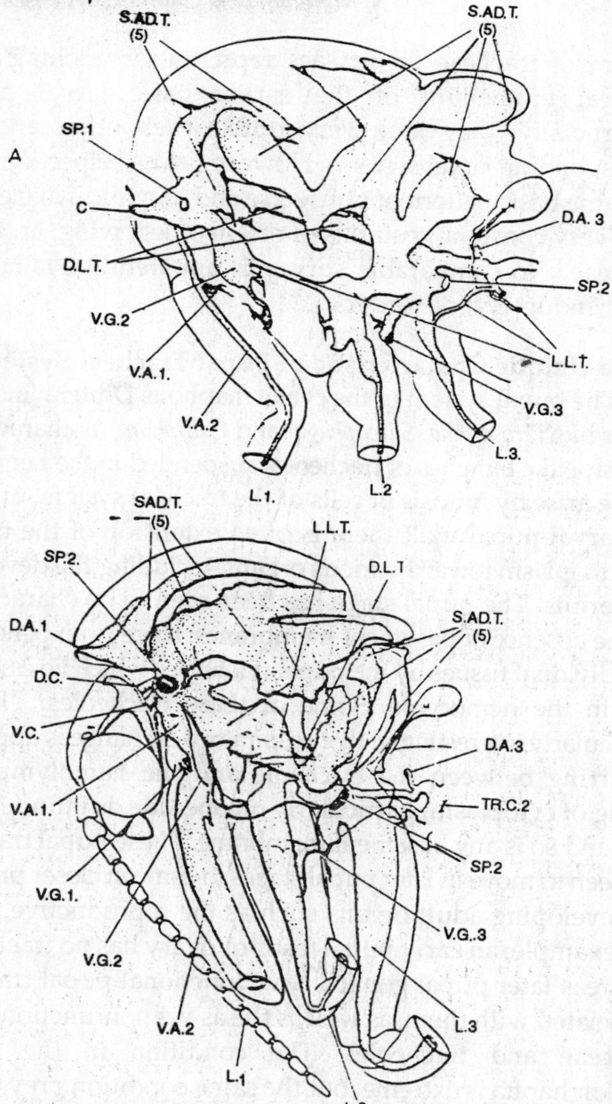

Fig. 2.12. Diagrammatic representation of two adult tracheal systems that superficially look very dissimilar, yet both arise by growth of the larval and pupal tracheal epithelium around the preexisting pattern. Secondary adult trachede arise by branching from the basic pattern. A: *Drosophila*, B: *Mycetophila* (Mycetophilidae, Diptera Nematocera. S.AD.T.(s) : secondary adult tracheae in the form of air sacs.

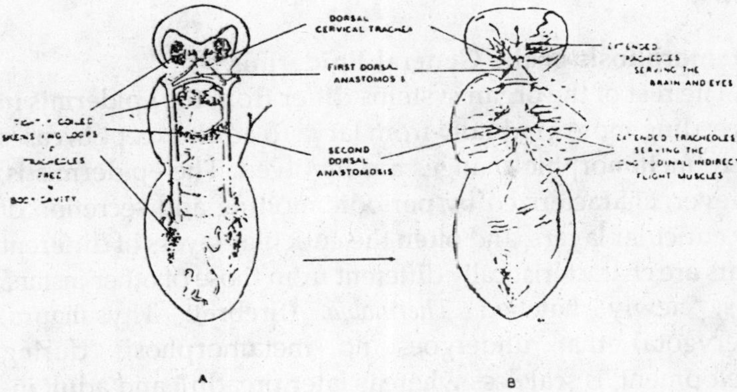

Fig. 2.13 Diagrams to illustrate the unique condition of the pupal tracheal system in a specialized pupa such as in *Drasophila* or *Sarcophaga*. Except for a few tracheae, such as those to the thoracic nerve mass, pupal tracheation consists of bunches of tracheoles lying free in the hemocoele. Later, many of these tracheoles migrate to become attached to developing adult organs such as gonads, muscles, etc. They are removed and not replaced at the pupal-adult acdysis. A: early pupa; B: late pupa.

epidermal cell extensions, to areas requiring oxygen. Epidermal cells at quite considerable distances away would send out processes towards tracheoles. "capture them," and draw them to the deprived area.

By comparison with other epidermal cell types the tracheal cell shows remarkable uniformity, the only distinction being between tracheal and tracheolar. Whether the cells at the tracheal nodes are structurally different seems never to have been investigated. That they are important physiologically in connection with gradients has been shown by Locke (1964). Cells of the spiracular region are not included in this tracheal system since they are most easily thought of as belonging to the general body surface. It would be of considerable interest to understanding of cell differentiation to know whether the same tracheal cell ever produces specific larval, as opposed to pupal and pupal as opposed to adult, tracheal cuticle. Whether the air sac cuticle is specifically different in an adult is not known.

Metamorphosis of the General Epidermis

The rest of the organ systems differ from the epidermis in proceeding more gradually from larva to adult except in cases where metamorphic changes are involved. The epidermis is, however, characterized by periodic molting and secretion of new cuticular layers, and often the cuticular layers of different instars are characteristically different from those of other instars. Thus, newly hatched *Thermobia* (firebrat: Thysanaura, Apterygota) that undergoes no metamorphosis during development, is scaleless whereas later preadult and adult instars have scales; a fly pupal cuticle is smooth, while that of the adult consists mainly of small and large bristles and their sockets. There has been a tendency in recent years to talk of the cuticles of metamorphosing endopterygote insects as "larval," "pupal," or "adult" and to relate the appearance of the specific type to the hormone balance between juvenile hormone and molting hormone at a particular period of development. Thus, the larval cuticle has been said to be produced in the presence of high levels of juvenile hormone and low ecdysone; the pupal cuticle is produced in the presence of high levels of ecdysone and lower juvenile hormone; and finally the adult in high ecdysone and virtually no juvenile hormone. In the case of the non-metamorphosing *Thermobia*, a piece of scaleless cuticle from a newly hatched larva, when implanted into an adult, is found to molt and prematurely produce scaled cuticle. The larva has a high and the adult a low concentration of juvenile hormone (Watson, 1963). While there are in most Endopterygota considerable differences between larval, pupal, and adult cuticles, one cannot say, for instance, that bristles are confined to adult cuticles. A particular insect type will have characteristic cuticles at different stages, but whether a larval cuticle is ultrastructurally diagnostic has not been shown for any insect. Certainly the tracheal cuticle appears to be essentially similar at all stages, and the epidermal cells secreting this are every bit as exposed to hormone fluctuations as are the cells of the general hypodermis. That changes are not simple responses to different hormone titers has recently been shown by Krishnakumaran and Schneiderman (1664). They questioned whether perhaps the cells of the lepidopteran adult differ from cells of the pupa in response to juvenile hormone, since it was seen in experiments that adult abdomens could be induced to molt again when joined to pupae, but that the second adult cuticle, although distinctly adult, seldom produced scales. A much older adult did produce normal scaled adult cuticle. A high level of juvenile hormone did not seem to influence the development of the adult cuticle since experimental results were similar with abdomens containing high levels of juvenile hormone (e.g., *cecropia, cynthia)* and others containing low levels *(polyphemus).* Krishnakumaran and Schneiderman quote a comparable phenomenon from the work of Piepho and his students, who showed that when fragments of adult cuticle were implanted into larvae they did not secrete larval cuticle immediately but required several molts before doing so. It may be of interest to observe in this context that the cells of the imaginal discs in the fly (which will be seen below to be determined but not differentiated), secrete first pupal cuticle and *subsequently* undergo extensive mitoses followed later by differentiation into adult cuticle secreting cells. These disc cells in an adult host continue mitoses indefinitely, but implanted into a metamorphosing larva they first secrete pupal cuticle, then further divide, and finally differentiate and produce adult cuticle.

Metamorphosis of the Epidermal Cells

The degree of change occurring in individual epidermal cells varies from insect to insect. In the more generalized endopterygotes or even the more generalized members of the specialized orders, e.g., the Nematocera of the Diptera, the epidermis passes more or less unchanged from larva to adult. On the other hand, in the higher flies, the Cyclorrhapha, there is said to be a complete replacement of larval epidermis by adult epidermal cells that originate from areas of undifferentiated imaginal cells—those of leg, wing, halter, genital, labial, antennal and eye discs, and paired groups of cells segmentally arranged on the abdominal segments. The pupal cuticle is secreted by the larval epidermal cells of the abdomen and small areas of the thorax, and the remainder is secreted by the undifferentiated cells of the different imaginal discs. Only in the cyclorrhaphan tracheal system might it be possible to follow individual epidermal cells through from larva to adult, whereas in the Nematocera this should be possible for general epidermal cells also.

The metamorphosis of the epidermis may best be looked at now from the viewpoint of "imaginal disc" development, since a tremendous amount of thought-provoking and fascinating work has been done on this subject. Following this, the fate of a few individual cells can be traced through from the time of the larval-pupal molt to the time of adult emergence. This will demonstrate the complexities on the one hand and the great potential on the other that the metamorphosing insect epidermal cells have in helping to solve some of the most puzzling of biology's and life's problems—those of nuclear-cytoplasmic relationships and of cell and tissue differentiation. The insect epidermal cell has been the fascinating subject of severai articles. including that of Stern (1954) and of Wigglesworth (1959b). There have been new developments involving the epidermal cell, that include gradient studies (Locke 1964) and the discovery of giant polytene chromosomes in the dipteran epidermis: these, along with the fact that the epidermis is undoubtedly the main "target organ" of the molting hormone ecdysone, give the epidermal cell greater importance than ever.

Metamorphosis of the Imaginal Discs

In the more generalized Dip.era the imaginal discs of wings and legs lie beneath the epidermis and are not withdrawn into the body as are those of the specialized Cyclorrhapha. Also, the cavities connecting with the general epidermis are never closed and the eyes do not develop from invaginated buds. Developing pigmented eyes can often be seen, as in the mosquito, through the semitransparent head capsule of a late larva.

The imaginal discs of a cyclorrhaphan fly include genital and leg discs, genital and first and second leg discs), wing, halter, labial, eye and antennal. The groups of cells between duct and larval salivary gland, those at the junctions of foregut and midgut and midgut and hind gut, the mesodermal genital duct rudiment of *Sarcophaga,* and the hypodermal discs of the abdominal segments, are also imaginal rudiments. It is difficult to draw a hard and fast line of distinction between some of these cell groups and those associated with individual systems, such as the tracheal cells that will suddenly at puparium formation undergo intense mitosis to produce the bunches of specialized pupal tracheoles, or the sheath cells of the peripheral nerves, which presumably functioned in the embryo to secrete the larval nerve sheath and now at puparium formation are stimulated into intense mitotic activity to give rise to the cells that form the skeleton of the adult peripheral nerves. Studies on the leg or wing or genital imaginal discs, are essentially studies on epidermal cells lines. Embryos of endopterygotes that undergo complete metamorphosis, such as *Drosophila,* possess two cell types. Firstly, there are those cells that are determined to form the larval body and, in these, differentiation immediately sets in. Secondly, there are those cells that are determined but are set apart and remain in an embryonic state until the time of pupation (Hadorn, 1965). Cells of the imaginal discs come into this category. They do divide and the discs grow considerably in size from their earliest appearance to the end of the larval stage, but they remain undifferentiated. Due largely to the classic earlier works such as those of Bodenstein using transplantation techniques and more recently the fascinating works of Hadorn and his collaborators, some amazing phenomena have come to light with respect to potentialities of imaginal disc cells in the metamorphosing insect. It would seem that the cells of these discs are already determined for "leg" or "wing" or "eye" early in embryonic life, as early as the first to fifth hour, at a time when the imaginal discs are not yet visibly set apart within the still apparently homogeneous blastoderm. There is even determination for particular areas of these discs. There is *determination* at this time but not *differentiation.* This occurs only at metamorphosis. As stated by Hadorn, the change from undifferentiated to differentiated state in holometabolous insects is controlled by hormones. The juvenile hormone somehow in the larval instars blocks any differentiation; a drop in the level of juvenile hormone at the onset of metamorphosis results in a "deblocking" of readiness for differentiation present in the discs throughout larval life. *Perhaps the juvenile hormone acts as a repressor of the sets of genes needed in differentiation* but here Hadorn warns that much more information is needed before the happenings that clear the path for differentiation in a determined cell can be interpreted in molecular terms. Until details are clearer of the responses by particular cells of known fate and history, it would be good to extend such reservations to work on hormone action, particularly with respect to the influence of the balance between juvenile hormone and ecdysone on the production of specific "larval," "pupal," or "adult" cuticle. The same reservations may be extended to the effects of hormones on chromosome puffing sequences. The results with larval salivary glands cannot necessarily explain hormone action on tissues that will undergo metamorphosis.

The exciting results of Hadorn and co-workers involve the maintenance of imaginal discs within adult hosts, where they increase in size by extensive mitosis but do not differentiate, Removal from the adult and transplantation into a metamorphosing larva brings about cessation of cell division and differentiation into adult structures. Of particular interest is the process of "transdetermination" whereby after some eight or nine transplants (from one adult host to another), the disc when allowed to determination" whereby after some eight or nine transplants (from one adult host to another), the disc when allowed to differentiate in a metamorphosing larval host, differentiates into an adult structure different from the original implant. Thus, a genital disc after many transplants may give rise to leg or wing or thorax. Transdeterminations occur in the probability order of genitalia, antenna, leg, palpus, wing, and thorax the latter occurring in the oldest transfer generations. Reversals were also found. No explanation can as yet be given for these amazing results, but it is suggested by Hadorn that possibly a degree of dilution reached by cell cultures might have a feedback action on gene activation. Instead of the genes that formerly determined the original organ, other genes come into play, resulting in the transdetermination. Some of the transdetermination effects are reminiscent of some Drosophila mutants such as aristapedia were a tarsus is formed on the antenna. However, most of the transdetermination effects are not known as mutants.

Metamorphosis of an Imaginal Disc: The Leg Disc

The type of development undergone by individual elements of an imaginal disc can be seen in the case of the leg. Details have been followed for *Sarcophaga bullata,* the flesh fly, but development is essentially similar in other Cyclorrhapha, including *Drosophila*. Event followed at light and electron microscope levels demonstrate the complex changes and co-ordinated development of the different components: various epidermal cells, tracheoles, and phagocytic blood cells. Some of these events have already been discussed in relation to hemocyte activity during metamorphosis. At this point the main concern will be with the development of individual epidermal cells.

The mass of thousands of determined yet undifferentiated epidermal cells (stage 1) first secrete the pupal cuticle of the leg disc. Retraction of the epidermal cells from the pupal cuticle follows and cell division continues (stage 2). Five cells forming the future end of the claw and four future giant cells become recognizable at stage 3 while cell growth and continued

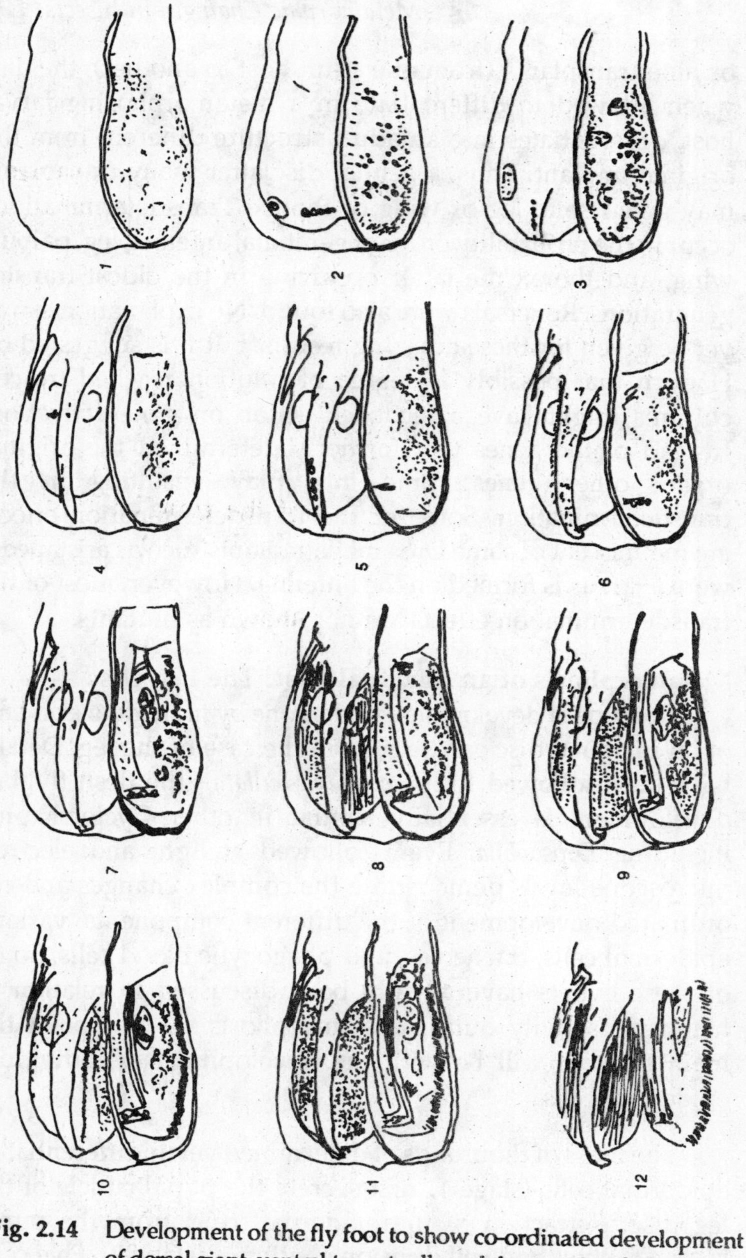

Fig. 2.14 Development of the fly foot to show co-ordinated development of dorsal giant, ventral tenent cells and the associated phagocytic blood cells. For details, see text. *Sarcophaga*.

mitoses occur among "mother" tenent hair cells whose descendants will occupy the future ventral surface of the foot. Growth and cell division thrust the five claw cells into the tip of the forming claw, and the foot pads take shape by progressive growth and median indentation. Dorsally, the two giant cells extend over each foot pad; their nuclei are characterized by giant polytene chromosomes. The tenent cells give out very fine cytoplasmic processes, while the dorsal cells grow in width as broad, flat cells. The tarsal hair cells at this sametime from, like the tenent cells, cytoplasmic processes, varying in size according to whether a large or small hair is to be produced. Stage 4 marks the end of growth and differentiation, yet at this time *no* cuticle is as yet deposited. Thus, the future shape of a hair is laid down by cell growth (c.f. Lees and Picken, 1945) and is completed by the time cuticle secretion begins. This is of interest in questions of abnormal bristle development in mutant *Drosophila*.

Cessation of growth and attainment-of-cell maturity is followed by a well-spaced sequence of cuticle deposition accompanied in the giant cell by constant and characteristic patterns of chromosome puffing activity. First is secreted an *ecdysial membrane* (1), followed by a cuticulin layer (2), a layer of dense exocuticle (3), which later becomes darkened and sclerotized, a homogeneous inner layer of exocuticle (4), and a narrow layer, of "mesoculticle" (5); each layer is indicated by an arrow. By stage 9 these sequences are completed in giant, tenent, and trichogen cells. Cytolysis is now observable within the tenent cell nuclei and cytoplasm. This is seen at the electron microscope level, in which centers of focal degradation can be clearly seen. The hemocytes, which at the time of first evagination of the legs were gorged with larval tissue fragments, have subsequently been digesting them and passing on the degradation products to the ceveloping epidermal cells. These phagocytic blood cells now, at stage 9, embark on a second phase of phagocytosis. By stage 10 the phcnotic tenent cell nuclei are nearly all engulfed. In fact, the exact age of the foot

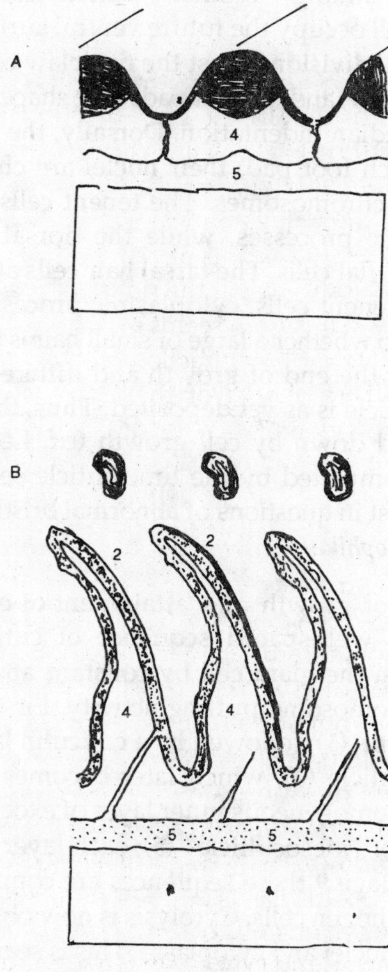

Fig. 2.15. The cuticular layers determined from electron microscope observations in A, dorsal cuticle, B, ventral tenent hairs which occur in very precisely arranged rows. The same layers are laid down of comparable times in dorsal cuticle, ventral tenent hairs and tarsal trichogen, hair, cells. *Sarcophaga*. Layers—1: ecdysial membrane; 2 : cuticulin layer of epicuticle; 3: dense exocuticle, later melanized and sclerotized; 4: homogeneous amber coloured exocuticle, not melanized later; 5: "mesocuticle," 6: extensive layer of endocuticle.

can be determined by three means: (a) by close inspection of the extent of the darkening process and actual shade of specific areas, (b) by close investigation of the relative degree of phagocytosis of tenent cell nuclei, and (c) by close inspection of the puffing pattern of the chromosomes. All are equally exact and reliable. The tracheoles and giant cells remain unattacked at this time. Endocuticle secretion within the giant cell cytoplasm occurs after this at a time when darkening of the exocuticle is occurring over the ridges of the dorsal cuticle. Endocuticle (layer 6) secretion seems to occur by a combination of intracellular secretion and retraction of cytoplasm (and nuclei and blood cells) to the base of the foot, so that the space formerly occupied by giant and tenent cells is now occupied by endocuticle. The position of blood cells in relation to that or giant nuclei and the former position of tenent nuclei. Even at the light microscope level, association of blood cells with the giant cytoplasm can be seen. The association becomes more close, and at emergence the giant nuclei and cytoplasm also become ingested by the blood cells. The blood cells finally come to lie as a layer over the inner border of the foot pad cuticle, at the base of the pad.

There is no innervation of individual cells in the foot, though possibly a basal stimulus is received that can be transmitted from cell to cell. What determines that cell division should cease in the giant cells and cells of the future claw at a time when mitoses are still occurring among the future tenent cells? What determines the regular pattern and uniform deposition of the different cuticular layers in the various epidermal cells? That the layers are similar and are deposited at comparable times in tenent, trichogen, and giant cells. The biochemistry must be similar, yet the spatial distribution of the various layers is different, and this is essentially what gives each product its characteristic "differentiation." What initiates cell death among the tenent hair cells? Is it simultaneous or is there a focal point at which the process starts? Focal degradation begins before any phagocytosis on the part of the blood cells is to be observed. Does this prompt the blood cells to attack the tenent cells and not the giant or the tracheolar cells? Interestingly, the final deposition of endocuticle by the cytoplasm of the tenent cells is carried out in the absence of their nuclei.

What of the chromosome activity while these sequences of growth, cuticle secretion, and finally death are occurring in the giant cells? For the first time, we seem to have a system where one should be able to correlate closely chromosome activity and puffing sequences with the activity of a single cell through growth, various phases of clearly defined activity, and finally cell death. Detailed analysis will be made in the vast amount of information that has accumulated as a result of work carried out on dipteran salivary glands and to a lesser extent on the chromosomes of cells of the Malpighian tubules, gut, or seminal vesicles. The Malpighian tubules persist through pupation but are of small dimension while the salivary glands degenerate shortly after pupation. Since epidermal cells are specifically the "target organs" for the molting hormone ecdysone, repetition of much of the work carried out on salivary glands should produce particularly interesting results with foot pad cells, particularly work such as that of Karlson and his colleagues (e.g. 1965). It would be exciting to see if and how hormone changes can affect puffing activity and in what way they may alter the biochemistry and ultrascopic structure of the cell.

These discussions might seem to have taken us far from considerations of the whole metamorphosing insect, but in fact it is visible changes in the epidermal cells and their products that constitute external evidence of metamorphosis.

What determines the differences between the various cells of the epidermis of the developing insect? They are variants of the same cell type — the epidermal cell. When the layers of cuticle that are secreted are identical in two or more variants, what determines the spatial distribution of these cell products, and what determines the cell shape? If one could answer these questions one would be nearer to answering the problems posed by the phenomena of transdetermination of Hadorn and of the problems of patterns and prepatterns discussed at length by Sondhi (1963) and in a broader context by Waddington (1962).

It will be seen from the foregoing necessarily superficial survey of insect metamorphosis that each system in turn has its own facinating problems, which, to name but a few, include general considerations of body change, diurnal rhythms, problems of cell determination, cell differentiation, cell death, and cell interaction. Many of the problems are common to those found in other metamorphosing animals, as will be seen in succeeding chapters. Nevertheless, in the insects alone one cannot but be impressed by the infinite number of problems that the one phenomenon of metamorphosis poses to the inquiring developmental biologist.

3

Biochemistry of Insect Metamorphosis

Introduction

The metamorphosis of the higher insects embodies the most profound reorganization of a grown animal that is known. It represents advanced polymorphism, for a single genome determines the development of a differentiated free-living organism which, when fully grown, is largely destroyed and reconstituted in a new form adapted for a totally different life. Such a phenomenon, under the control of hormones, should provide prime material for the study of the regulation of genetic expression and its role in morphogenesis.

This challenge has inspired biochemical study of insect metamorphosis for many years. However, since all organ systems of the insect are to a greater or lesser extent modified in structure and function during metamorphosis, some change may be expected in almost any biochemical feature that is examined. Accordingly, the literature of the subject is extensive and diffuse, and reviews of it have tended to give weight to each author's own predilections (Gilbert and Schneiderman, 1961; Karlson and Sekeris, 1964; Agrell, 1964). The contemporary emphasis on molecular genetics and nucleic acid and protein synthesis seems most likely to yield understanding of the fundamental mechanisms in metamorphosis, but investigation of these fields has not yet progressed far with insects. The earlier attention to nutrient reserves and pathways of energy conversion now seems less close to the central question. Nevertheless, these studies have contributed to knowledge of the milieu in which the genes are working and of the nature of the biochemical systems for which they are responsible, and we are not yet in a position to say confidently what is essential and what is not. Therefore, I shall attempt to summarize the present state of understanding in several biochemical areas where enough work has been done to permit some coherent discussion. Other subjects will be arbitrarily omitted; among the more important of these are the development and functions of the silk gland and the ovaries, although much work has been done on both. References to the literature will necessarily be selective, and most examples will be drawn from two insect groups which have received a great share of research attention. The higher Diptera undergo the most complete metamorphosis among insects, replacing most of their organ systems. Some are easy to raise in the laboratory with a short generation time, but they have the disadvantage of small size which has discouraged biochemical work with discrete tissues (except to some extent flight muscle). In the Lepidoptera, there is far less abandonment of larval cellular structure, but this order includes many species with the experimental advantage of large size, and more work has been done with discrete tissues. Biochemically, there is probably unity in the underlying mechanisms of histolysis and of histogenesis in all insect groups. Processes which are not strictly part of metamorphosis may be relevant—e.g., the degeneration and reconstruction of muscles in the larval molting cycle of the bug *Rhodnius* or in the adult diapause of the Colorado potato beetle. Metamorphosis is not wholly confined to the transformation from larva to adult, as there is often a small degree of change in form at each larval molt.

Use and Interconversion of Reserves

During their last larval stage, in preparation for metamorphosis, holometabolous insects accumulate substantial nutrient reserves. These are chiefly deposited as fat and glycogen in the fat body, a cellular tissue which is a center both for storage and for intermediary metabolism and which proliferates greatly during the last larval instar (Kilby, 1963). In the larva of the wax moth, *Galleria mellonella,* this preparation for metamorphosis can be detected early in the last instar, which differs metabolically from the preceding larval stages (Janda et al., 1966). Peak values for glycogen content are often in the neighbourhood of 10 per cent (but sometimes as high as 30 per cent) of total body dry weight, and the lipid reserves are generally greater, values of 20 to 50 per cent of dry weight before metamorphosis being common.

Analytically studies on changes and conversions in the insect pupa have been stimulated by the fact that it is a closed system which exchanges only gases with the environment. A body of the early work is summarized and discussed in an admirable review by Needham (1929), but few conclusions can be drawn from it because of inadequacies in the analyses. In particular, trehalose was not recognized as an ubiquitous sugar in insects until 1957, and the extent to which the chitin and protein of the cuticle are resorbed in the molt and reutilized was generally underestimated or ignored. References to more recent studies can be found in several reviews (Wigglesworth, 1965; Gilbert, 1967; Wyatt, 1967a).

Among the most comprehensive data on a single species are those recently obtained by Birt and colleagues on the sheep blowfly, *Lucilia cuprina*. Several conclusions may be drawn. The principal source of energy is fat, the content of which at adult emergence is about one third of that in the mature larva. The rate of fat oxidation is more rapid at the beginning and the end of metamorphosis than in the middle, in accord with the changing rate of respiration. At the beginning of metamorphosis before the pupal molt, there is a fall in total protein and in free amino acids, which are partly oxidized and partly converted to carbohydrate and fat. The utilization of protein leads to accumulation of much uric acid in the early fly pupa (Russo-Caia, 1960). There is also a fall, before the pupal molt, in chitin as the inner layers of the larval cuticle are digested and resorbed; some of this is apparently deposited as glycogen, which is later reconverted to chitin in the growing adult cuticle. Thus,

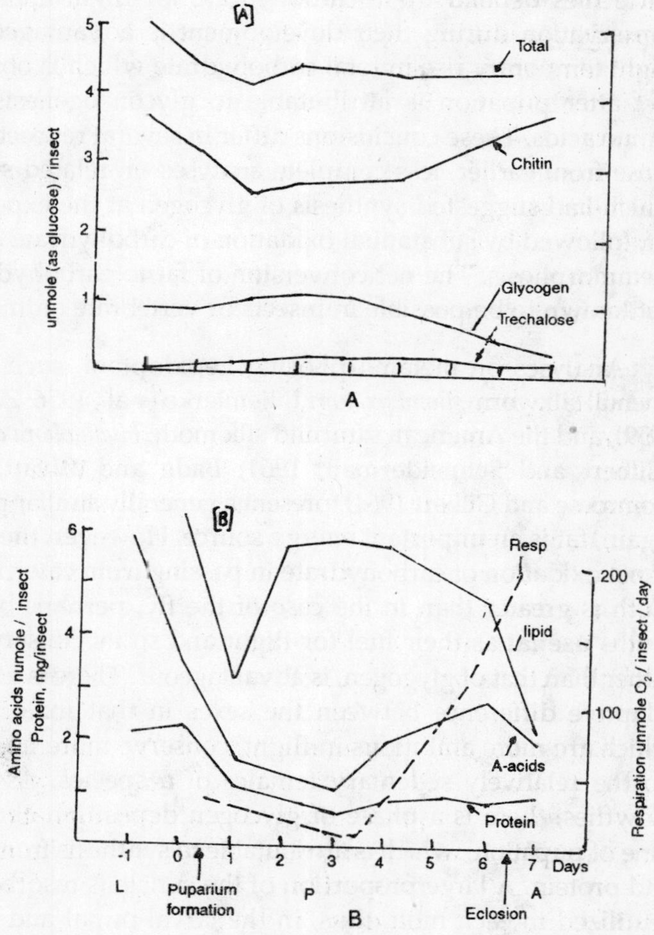

Fig. 3.1. Changes in composition and respiration of the blowfly, Lucilia cuprina, during metamorphosis. A: Carbohydrates (after Crompton and Birt, 1967). B: Respiration, neutral lipid, free amino acids, and soluble protein (after Howells and Birt, 1964, and D'Costa and Birt, 1966). The scale for protein has been corrected according to personal communication from Dr. Howells, Development stages: L, larva; P, pupa and pharate adult; A, adult.

when chitin is included, the total carbohydrate content undergoes only a slight net fall throughout metamorphosis.

Since flies depend upon carbohydrate for flight energy, its conservation during their development is advantageous. A slight temporary rise in total carbohydrate which is observed just after pupation is attributable to glyconeogenesis from amino acids. These conclusions differ in several respects from those from earlier, less complete analyses on related species, which had suggested synthesis of glycogen at the expense of fat, followed by substantial oxidation of carbohydrate during metamorphosis. The net conversion of fat to carbohydrate is not known to be possible in insects or vertebrate animals.

Analyses on metamorphosing Lepidoptera, such as the oriental silkworm, *Bombyx mori* (Niemierko et al., 1956; Zaluska, 1959), and the American saturniid silk moth, *Hyalophora cecropia* (Gilbert and Schneiderman, 1961; Bade and Wyatt, 1962; Domroese and Gilbert 1964), present a generally similar picture. Again, fat is an important energy source. However, the extent of net oxidation of carbohydrate in passing from caterpillar to moth is greater than in the case of the fly, perhaps because moths use fat as their fuel for flight and so its conservation, rather than that of glygogen, is advantageous. There is a further adaptive difference between the sexes in that male moths, which are more ambitious in flight, conserve more lipid than do the relatively sedentary females of a species. As in the blowflies, there is a phase of glycogen deposition about the time of pupation,, which is attributable to synthesis from chitin and protein. A large proportion of the cuticle is resorbed and reutilized in each molt; thus, in the larval-pupal and pupal-adult milts of the Cecropia silk moth, the resorbed fraction is more than 80 per cent of the cuticular chitin.

Evidence concerning the balance of metabolism in insects also often has been sought by measurement of respiratory quotients. Such measurements, especially with pupae, however, are subject to a number of pitfalls (e.g., release of CO_2 in bursts, trapping of CO_2 in nongaseous forms, disturbance of metabolism by accumulated CO_2), and in general more direct methods of study are preferable. The more reliable estimates of respiratory quotients are in reasonable agreement with analytical results.

Along with the quantitative changes in body composition are certain cases of apparently qualitative change (Wyatt, 1967a). In blowflies, e.g., trehalose cannot be detected in the feeding larva, but makes its first appearance just before pupation and then increases to play an important role in the flight metabolism of the adult. In pupae of the Cecropia silkworm, which pass the winter in diapause, most of the glycogen is converted to glycerol, and part of this is reconverted to glycogen at the beginning of adult development in the spring. But this is an adaptation to survival in cold weather rather than a process characteristic of metamorphosis, for glycerol is accumulated by many insects which survive cold winters, and the life-cycle stage in which it appears is always that which passes the winter, be this the embryo, larva, pupa, or adult. In short, the use and interconversions of reserves in the metamorphosis of different species is variously adapted to their respective needs.

Little is known about the regulation of these conversions. A recent review on metabolic control mechanisms in insects cited much evidence for the existence of control but few instances where enough is known to justify designation as a mechanism (Harvey and Haskell, 1966). Fat and carbohydrate metabolism, like other aspects of insect development, must be more or less directly under control by hormones, and cases can be cited in which metabolic change follows change in hormonal state (Wyatt, 1967a). Certain rapid adaptive changes can result from hormonally released activation of preexisting enzymes, such as glycogen phosphorylase. The longer term changes in metamorphosis are presumably associated with changing enzyme levels, but there exists little satisfactory biochemical evidence on this point. One example is the elevation of chitinase, a participant in digestion of the cuticle, at each molt in the silkworm, *Bombyx mori* (Jeuniaux, 1961).

RESPIRATORY METABOLISM

Respiration and Energy Supply During Metamorphosis

One of the earliest and most frequently recorded metabolic observations concerning insect metamorphosis is that the rate of respiration changes in a manner that can be described as a U-shaped curve. In the earlier papers, the significance of this general course of respiration was much discussed. The first suggestion was that it reflected successive histolysis and histogenesis, the rate of respiration changing in accord with the amount of organized tissue present at any time, but it then became clear that the histological picture often did not support this simple view. When the activities of respiratory enzymes, such as cytochrome oxidase and dehydrogenases, were measured, a somewhat better correlation with respiratory rate was found (Agrell, 1949; Sacktor, 1951). No more than a rough agreement between respiratory rate and activities can be expected from such measurements, however, for the course of change during metamorphosis varies for different enzymes and different tissues.

In any case, it is now clear that tissue respiration is generally not limited by the levels of respiratory enzymes. The capacity for respiratory depends on the content of mitochondria and their electron transport system, which is relatively low in tissues such as fat body and high in others such as muscles. But the actual respiration is generally substantially less than this capacity, as a result of respiratory control by the coupled phosphorylating system, which is subject to the levels of ADP, ATP, and inorganic phosphate in the cell (Harvey and Haskell, 1966). Thus, the rate of respiration adjusts itself to supply the demand, in the form of high-energy phosphate, of the work done by the tissue, which includes biosynthesis, active transport, muscular activity, and so on. In the pupa and the pharate adult insect, where there is little muscular activity, one may expect protein synthesis to be a major energy user, and indeed the rates of respiration often show some correlation with those of protein synthesis.

Some evidence on the balance of energy metabolism in insect metamorphosis can be obtained by analysis of metabolite and coenzyme levels. It was suggested as early as 1893 by Bataillon, on the basis of crude carbohydrate analyses, that the metabolism of the silkworm pupa might somehow be anaerobic, and similar ideas have recurred in various forms (Wyatt, 1963). Several compounds are now recognized as products of experimentally induced anaerobic glycolysis in insects: lactate, alanine, pyruvate, and glycerol-1-phosphate (Gilmour, 1965), but recent analyses show none of these accumulating during metamorphosis of the blowfly (Crompton and Birt, 1967). There is, in fact, much more lactate in the actively respiring larva than in the pupa. In addition, analyses of pyridine nucleotide coenzymes show that the ratio NADH/NAD is somewhat lower in midpupal life than in the active larval and adult stages, whereas anaerobic mechanisms would be expected to lead, on the contrary, to accumulation of the reduced form (Birt, 1966). Thus, there is no evidence for anaerobic processes in the blow-fly pupa. Indeed, it would be surprising to find them, for the supply of oxygen is ample to support the depressed rate of metabolism, and aerobic processes give more efficient use of reserves which the pupa needs to conserve for the construction and initial activity of the imago.

The levels of "high-energy" and related phosphate compounds are also of interest. In the blowfly, ATP and the phosphagen, arginine phosphate, show a decrease in the midpupa, and inorganic phosphate shows an almost reciprocal increase. The doubtless corresponds to changes in the body content of tissues which are characteristically rich in high-energy phosphate, notably muscle. The rise in AMP presumably represents degradation of ATP and ADP from such tissues. That the ATP/ADP ratio remains high through metamorphosis, and that phosphagen is never exhausted, show that the ATP-generating processes in the pupa are at all times fully able to keep up with the energy demand.

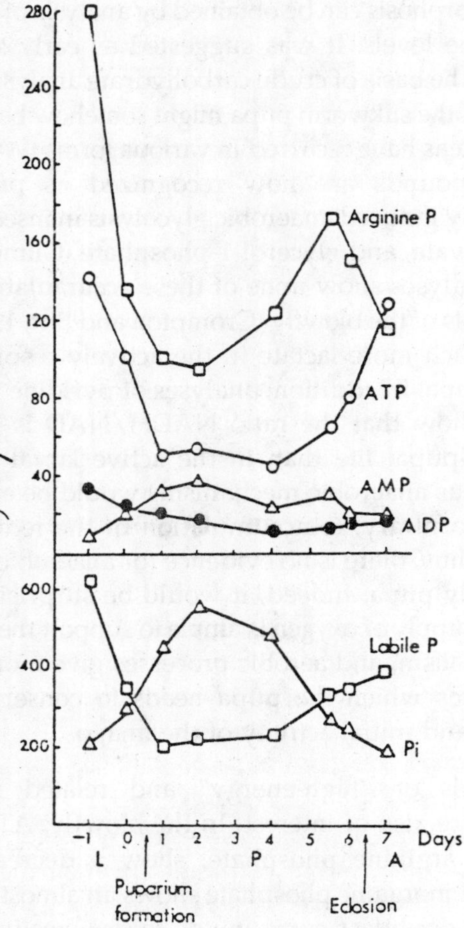

Fig. 3.2. Acid-soluble phosphorus compounds during metamorphosis of the blowfly, Lucilia cuprina. The adenine nucleotides were estimated by enzymic assays. Labile phosphate is that released in 10 min by 1 N H_2SO_4 at 100°.

Respiratory Metabolism and Diapause

When the pupal stage enters the resting state of diapause, the depression of metabolism is extreme and prolonged. This is the case with *Hyalophora cecropia* and some related saturniid moths, which have received much attention directed toward testing the relationships between respiration, respiratory enzymes, tissues growth, and hormone action (reviewed by Shappirio, 1960; Harvey, 1962; Wyatt, 1963; Harvey and Haskell, 1966). Since diapause is terminated and adult development is initiated by the prothoracic gland hormone, ecdysone, this system has been used for analyzing its action. A provocative observation was that respiration of pupae in diapause is highly resistant to the cytochrome oxidase inhibitors carbon monoxide and cyanide, a fact previously noted for the embryonic diapause of the grasshopper, *Melanoplus differentialts*. In spectroscopic and enzymic assays, the cytochrome content of tissues in diapause was found to be much reduced—cytochrome oxidase is present at low levels, but cytochromes *b* and *c* could not be detected at all. Accordingly, it appeared possible that terminal electron transport in diapause might proceed by some pathway other than the usual cytochrome system and that resumption of growth might depend upon the resynthesis of cytochromes which in turn might be closely linked to the action of the hormone. Further experiments showed, however, that when appropriately tested, diapause respiration is indeed sensitive to cyanide and carbon monoxide and since the latter inhibition can be reserved by light, participation of cytochrome oxidase is indicated. Apparently, the tissues contain substantially more cytochrome oxidase than is needed to support the normal rate of electron transport so that the oxidase can be largely inhibited with no effect on net pupal respiration. When the demand upon this enzyme is increased, however, either by lowering the oxygen tension or by enhancing the rate of respiration (e.g., with 2, 4-dinitrophenol), inhibition of the pathway becomes demonstrable.

This conclusions raises the question: How is diapause respiration held to its low rate? Since cytochromes *b* and *c* were not detected spectrtscopically, it seemed at first that a limiting step might exist here. Oxygen uptake by the pupae, however, is strongly stimulated by dinitrophenol, which shows

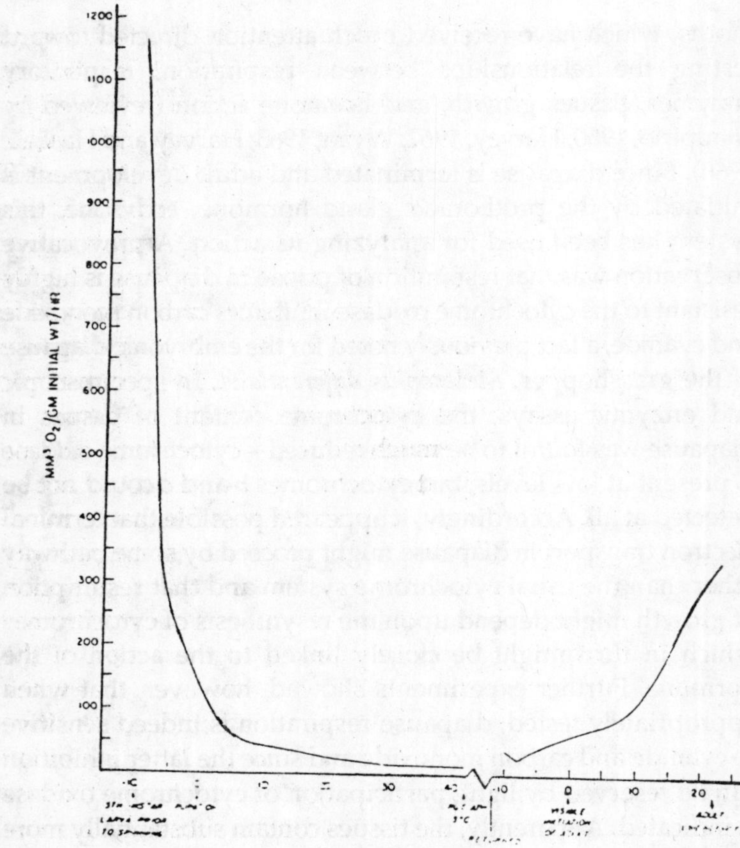

Fig. 3.3. Respiratory rate in the silkmoth, *Hyalophora cecropia*, in the mature larva, the pupal apause, and the developing adult.

that the capacity of the electron transport enzymes exceeds their normal load and indicates respiratory control via the coupled phosphorylation system, as is usual in active tissues. In addition, analyses of adenine nucleotides in diapause and developing tissues show that the ATP/ADP ratio changes little, even though the turnover rate is much higher in development, so that a nice co-ordination between phosphorylation and dephosphorylation is indicated, just as in the nondiapausing blowfly pupa discussed above. The ATP level in diapause tissues is high enough so that deficiency of metabolic energy can scarcely be responsible for restraint of growth.

The suggestion has also arisen that diapause pupae possess partially anaerobic metabolism as a consequence of restricted electron transport. This seemed to be supported by the gradual conversion of glycogen to glycerol in the Cecropia pupa. It now appears, however, that the characteristic glycolytic products glycerol-1-phosphate, lactate, and pyruvate are not elevated in the normal diapause pupa, though they are produced in experimental anaerobiosis. In addition, the NADH/NAD ratio is not elevated in diapause. The production of glycerol, therefore, is probably subject to special regulatory mechanisms as an adaptation to cold, and the diapause silkmoth pupa is no more anaerobic than is the blowfly pupa.

The present indication from these extended studies on the extremely suppressed metabolism of diapause is that the pathways do not differ qualitatively from those of other tissues, but there is a co-ordinated decrease in activities. The breakdown of the cytochrome system is impressive, yet the actual rate of respiration appears to be limited by energy demand which, in the pupa, must largely represent biosynthesis of protein and other tissue constituents. Furthermore, resynthesis of cytochromes, though necessary, is not sufficient to release resumption of development, for injury to diapause pupae is followed by the former but not the latter. Certain problems remain: e.g., cytochromes *b* and *c* have not been detected in tissues other than muscle of the diapausing Cecropia pupa, and the evidence for their participation in electron transport remains indirect. The measurements of respiration all have been made with whole pupae, and it is not known how much is contributed by muscle, epidermis, fat body, and other tissues whose activities and enzyme complements are quite distinct.

Respiration and the Corpus Allatum Hormone

In addition to the effect of ecdysone in increasing respiration when diapause is broken, enhanced respiration is often attendant upon the action of the hormone of the corpora allata. Increased respiration has been noted both during the "juvenile" action of this hormone, when corpora allata are implanted into larval insects, and during its gonadotrophic action in the adult. When corpora allata were implanted into *Galleria* larvae of different ages, however, and development was modified to various extents, the effect on respiration was always correlated with the degree of morphogenetic change. It appears that this hormone, like ecdysone, may influence respiratory metabolism indirectly by inducing growth and structural change in the tissues (Sehnal and Slama, 1966).

Development of Flight Muscle

A picture of the co-ordinated synthesis of enzymes in the construction of a specialized tissue has been gained from studies on the biochemical development of flight muscle. The indirect flight muscles of insects are of the fibrillar type. Between the bundles of large fibrils are packed rows or giant mitochondria ("sarcosomes") which may be 1 to 3 μ in diameter and make up 30 to 40 per cent of the weight of the tissue. These are directly supplied with oxygen by tracheoles. The flight muscle has almost no lactate dehydrogenase and in its normal activity is totally aerobic. The contents of soluble NAD-linked α-glycerophosphate dehydrogenase and of mitochondrial flavoprotein a-glycerophosphate oxidase are usually high and provide a shuttle for the oxidation of extramitochondrial NADH by passage of glycerol-1-phosphate and dihydroxyacetone phosphate in and out of the mitochondria (Sacktor, 1965). Flight muscle is thus constructed for sustained aerobic activity with exceptionally high energy output. Insect leg muscle, such as the jumping muscles of the locust, on the other hand, has a smaller content of mitochondria and possesses lactate dehydrogenase in adaptation for sporadic violent activity followed by rest.

The most comprehensive picture of the structural and boichemical development of flight muscle is that due to the

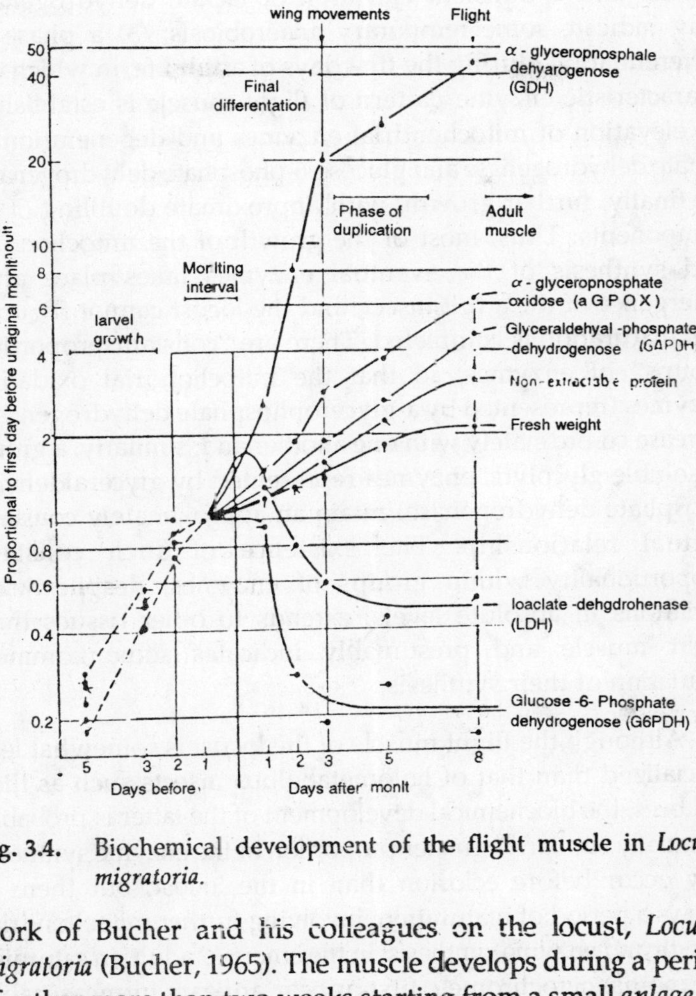

Fig. 3.4. Biochemical development of the flight muscle in *Locusta migratoria.*

work of Bucher and his colleagues on the locust, *Locusta migratoria* (Bucher, 1965). The muscle develops during a period of rather more than two weeks starting from a small *anlage*, or precursor muscle, that is present just after the fourth molt. Even this precursor muscle resembles flight muscle in having only a low content of lactate dehydrogenase. Four phases of growth are described : (1) larval growth, with increase in size (but not number of nuclei) with little differentiation; (2) a phase about the time of the final molt when the larval tracheal supply breaks down and new tracheoles penetrate into the muscle and a transient elevation of lactate dehydrogenase may indicate some temporary anaerobiosis; (3) a phase of differentiation, during the first days of adult life, in which the characteristic enzyme pattern of flight muscle is established by elevation of mitochondrial enzymes and degeneration of lactate dehydrogenase and glucose-6-phosphate dehydrogenase; (4) finally, further growth, with approximate doubling of all components. Thus, most of the growth of the mitochondria and synthesis of the eventual enzymes takes place after emergence of the adult insect, and the locust cannot fly until this maturation is completed. There are "constant-proportion groups" of enzymes, so that the mitochondrial oxidative enzymes (represented by a-glycerophosphate dehydrogenase) increase co-ordinately with one another and, similarly, a group of soluble glycolytic enzymes represented by glyceraldehyde phosphate dehydrogenase) retain an approximately constant mutual relationship. The existence of such constant proportionality within groups of enzymes, despite wide variations in absolute levels, extends to other tissues than flight muscle and presumably indicates some common regulation of their synthesis.

Although the flight muscle of the locust is somewhat less specialized than that of holometabolous insects such as flies and bees, the biochemical development of the latter is probably essentially similar. A greater proportion of the enzyme synthesis may occur before eclosion than in the locust, but there is always a period of maturation, involving further mitochondrial growth and enzyme synthesis, in the emerged adult. In saturniid silk moths, cytochrome *c* first appear appears in measurable amounts in the flight muscle on about the sixteenth day of the 21-day period of adult development, and its synthesis thereafter is very rapid (Chan and Margoliash, 1966). In the blowfly, *Phormia*, the cytochrome content rises about fourfold during the first week of adult life, which reflects an increase in the size of the mitochondria while their number remains constant (Levenbook and Williams, 1956). Growth of the flight muscle mitochondria is also reflected in a 50 per cent increase in phospholipid content of the thorax of *Lucilia*, during the week after emergence (D'Costa and Birt, 1966). In *Lucilia*, the distribution of enzymes and protein in particles of various sizes has been determined, and inasmuch as the specific activities of several respiratory enzymes relative to mitochondrial protein change greatly during development of the fly, it is proposed that the respiratory enzymes and the nonrespiratory, structural protein follow different courses of synthesis (Lennie and Birt, 1967). This is quite likely, since it appears that the nonextractable, structural protein of mitochondria may be synthesized within them, whereas synthesis of the extractable enzymes is a function of extramitochondrial ribosomes.

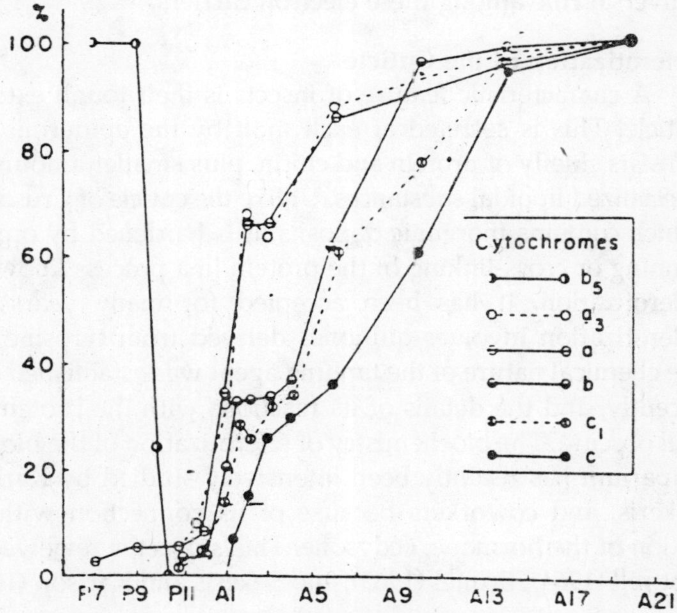

Fig. 3.5. Changes in concentrations of cytochrome components during development of flight muscle in the honeybee. For each cytochrome, the maximum concentration achieved is put equal to 100 per cent.

In the honeybee, the mean volume of a flight-muscle sarcosome increases from 0.015 μ^3 in the 10-day pupa (pharate adult) to 0.041 μ^3 on the day of emergence and to 0.52 μ^3 in the 20-day-old adult. The increase in content of several cytochromes, determined by the sensitive technique of spectroscopy at the temperature of liquid nitrogen. It is evident that their synthesis is by no means synchronous. Cytochrome b_5, which occurs in the microsomal fraction and is believed to participate in biosyntheses, is found in the pupa and declines sharply in adult development. Among the mitochondrial cytochromes, although the general pattern is similar, there are distinct differences in time of first detectability and course of synthesis, so it is clear that constant mutual proportionality is not a universal rule among these electron carriers.

Sclerotization of the Cuticle

A characteristic feature of insects is their tough external cuticle. This is secreted at each molt by the epidermis and consists chiefly of protein and chitin, plus smaller amounts of specialized lipoidal substances. Unlike the cuticle of Crustacea, which contains inorganic deposits, it is hardened by organic tanning or cross-linking of the protein in a process known as sclerotization. It has been accepted for many years that sclerotization involves quinones derived from tyrosine, but the chemical nature of the tanning agent was established only recently, and the details of its reactions with the protein are still obscure. The biochemistry of sclerotization of the blowfly puparium has recently been intensively studied by Karlson, Sekeris, and coworkers because of its connection with the action of the hormone, ecdysone. This subject is reviewed by Cottrell (1964), Brunet (1965), and Sekeris and Karlson (1966).

When the blowfly maggot is full grown and ready to pupate, its soft, white cuticle is inflated into a barrel shape and then becomes hard and dark brown, as the puparium, before the insect molts to a pupa within it. The principal tanning agent has been identified as *N*-acetyldopamine its acetylation has the chemical role of preventing cyclization into an indole,

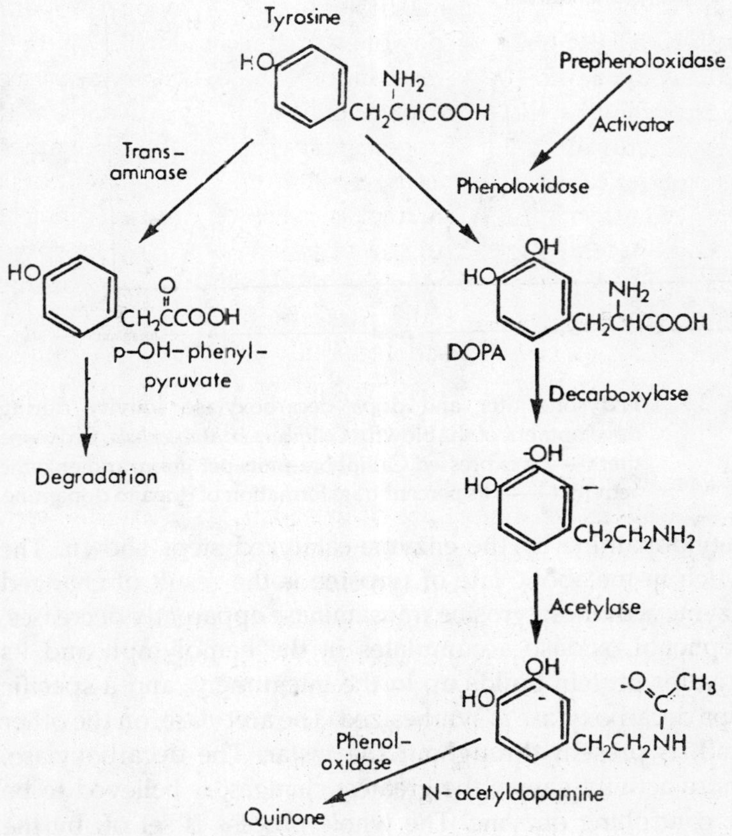

Fig. 3.6. The metabolism of tyrosine in the tanning of the blowfly puparium, according to the work of Karlson, Sekeris, and colleagues.

and when the phenolic ring is further, oxidized to an o-quinone, it is reactive with protein amino groups. In the early third-instar larva, metabolism of tyrosine is largely by a degradative pathway that starts with transamination. During the instar, the level of tyrosine mounts until it is one of the most abundant free amino acids in the hemolymph. Then, at the time of puparium formation, it is largely used in the production of *N*-

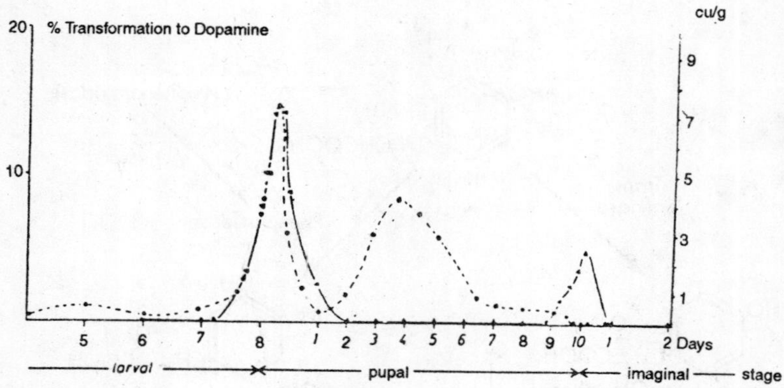

Fig. 3.7. Ecdysone titer and dopa decarboxylase activity during development of the blowfly *Calliphora erythrocephala*. Ecdysone titer (----) is expressed Calliphora-units per gram, and enzyme activity (——) as percent transformation of dopa to dopamine.

acetyldopamine by the enzyme-catalyzed steps shown. The switch in metabolic fate of tyrosine is the result of changed enzyme activities: tyrosine transaminase apparently decreases, prephenol oxidase accumulates in the hemolymph and its activator protein builds up in the integument, and a specific dopa decarboxylase is synthesized. The acetylase, on the other hand, is present throughout the instar. The decarboxylase, whose activity shows the greatest changes, is believed to be the controlling enzyme. The whole process is set off by the increasing level of ecdysone at this stage of the life cycle; it can be prevented by ligation and induced by injection of the hormone, and this forms the basis of the usual assay for ecdysone. Because of the close correlation between production of dopa decarboxylase and ecdysone titer, and other evidence to be discussed below, it has been proposed that the hormone "induces" this enzyme by an effect on nucleic acid metabolism.

The tanning of the dipteran puparium just before the pupal molt is a special case in insect ontogeny, as sclerotization commonly affects newly deposited cuticle just after ecdysis. In several such cases recently examined in Karlson's laboratory (the, blowfly adult, the locust, and the mealworm, *Tenebrio molitor*) the mechanism appeared to be similar, for dopa decarboxylase was active, and tyrosine was converted to *N*-acetyldopamine. close to times of sclerotization. In the blowfly adult, a further source of tanning agent is apparently the hydrolysis of *N*-acetyldopamine -4-0-β-glucoside, in which form excess *N*-acetyldopamine is conserved after puparium formation. After the adult molt, ecdysone titer is low, but there is now recognized another hormone specifically associated with cuticular hardening and designated bursicon (Cottrell, 1964; Fraenkel, Hsiao and Seligman, 1966). It is required for sclerotization of the cuticles of newly emerged flies and newly molted cockroaches and will effect the darkening of discs of colourless integument from the latter in vitro. Little is known about the biochemistry of its action; it has a molecular weight of about 40,000 and possibly may be itself an enzyme. It appears to act at a later stage than the production of *N*-acetyldopamine. Thus, ecdysone and bursicon could have complementary roles in stimulating the production and use, respectively, of the tanning quinones.

Pigment Production

Insects possess many kinds of pigment, including the ubiquitous and refractory melanin, and creation of the characteristic adult colour pattern is often one of the most striking features of metamorphosis. In the period butterflies, the wing patterns are due chiefly to pteridines, and the several pteridines of *Colias eurytheme* wings follows characteristic courses of synthesis, beginning on the third day of pupal life (Watt, 1967). But little is yet known about the regulation of such processes.

One instance in which there is a relation to hormone action involves certain ommochromes. These are phenoxazone compounds derived from tryptophan via kynurenine, which occur widely in the eyes, integument, and internal organs of insects. They are of historic interest because it was experiments with certain eye-colour mutants of *Drosophila* and *Ephestia* blocked in tryptophan metabolism that some 30 years ago first demonstrated the association of specific genes with successive steps in a metabolic sequence. The structure and distribution of ommochromes was worked out in a monumental series of investigations in Butenandt's laboratory. In the hawk moth, *Dicranura vinula,* the green caterpillar develops a characteristic red-brown pattern six days before pupation as a result of the deposition of xanthommatin and rhodommatin. Ligation prevents this change in the abdomen, and it can then be induced by injection of small doses of ecdysone. The enzymic basis of the change has not been established, but the hemolymph levels of tryptophan and 3-hydroxykynurenine rise sharply at the time of pigment synthesis, and it is proposed that ommochrome formation may serve in the removal of tryptophan liberated by proteolysis (Buckmann, Willig and Linzen, 1966).

BIOSYNTHESIS AND DEGRADATION OF MACROMOLECULES

The Proteins of Larval and Adult Insects

A question of fundamental importance is to what extent to protein constitution of an adult insect differs from that of a larva of the same species. No all-out attack on this problem has been described, but there has recently been a good deal of skirmishing with it.

Most attention has been given to insect blood (hemolymph), because this protein solution is easy to obtain and it undergoes large quantitative changes during ontogeny (Wyatt, 1961). The usual pattern with respect to protein is a relatively low level often 1 to 2 per cent), during the earlier larval stages, then a rapid rise just before pupation (often 6 to 8 per cent, but 18 and 20 per cent have been recorded), then there is some decline in the pupa and a further fall during the later part of adult development. Qualitatively, a number of analyses by electrophoresis on filter paper (summarized by Wyatt, 1961) gave too little resolution to be of much significance. More recently, starch gel and acrylamide gel electrophoresis have been applied to the hemolymph of a number of species, and between 10 and 20 bands can generally be recognized. These have no correspondence with the components of mammalian serum and differ characteristically with the species. There is often a yellow or green chromoprotein band and a dense band of low mobility which may represent the major hemolymph component of very high molecular weight (more than 500,000) which has been detected in several insects by other methods. In the transformation from larva to adult, most of the electrophoretic components generally persist, but there are certain qualitative differences. Thus, in the tent caterpillar *Malacosoma americanum,* out of 17 bands detected in the full-grown larva and early pupa, two to five were apparently missing in young larvae and in adult moths, and the adult did not a possess any obvious new components (Loughton and West, 1965). When blood from fourth-instar tomato hornworms was concentrated fourfold before electrophoresis, several otherwise undetected components were revealed, and in this insect the adults apparently lack some components characteristic of larvae and contain several fresh ones (Hudson, 1966). A study on the blowfly, *Phormia regina,* in which polyacrylamide gel was used with equal amounts of protein from each developmental stage, led to the conclusion that "the newly emerged adult fly has an entirely different picture" from the larva and pupa (Chen and Levenbook, 1966a), though inspection of the published diagrams suggests that about the half of the bands in the adult pattern may correspond to larval components.

Bands observed in zone electrophoresis are, of course, rarely pure proteins, nor is their mobility an adequate criterion of identity. The technique of immunodiffusion, in which antigen and antibody diffuse toward one another in an agar gel, can assist with establishing identity, although when applied to whole hemolymph, it has generally revealed fewer components than has electrophoresis. In an immunological study on Cecropia silkworm hemolymph, Telfer and Williams (1953) observed nine distinct antigens and followed six of them quantitatively from the fifth-instar larva to the adult. Each followed its own course of variation, which shows that their levels are individually regulated. One protein found in female pharate adults was passed into the ovaries and their contained eggs, and was present only as traces in the male (Telfer, 1954). But the appearance of uniquely adult antigens, distinct from those of the larva, is not reported. By a technique of immunoelectrophoresis, which should combine high resolution with definitive identification, Terando and Feir (1967) detected 11 antigens in hemolymph of the milkweed bug, *Oncopeltus fasciatus*, and finding no evidence for stage-specific antigens, arrived at the conclusion that nymphs and adults of both sexes showed only quantitative differences; but it should be noted that this concerns a hemimetabolous insect. Laufer (1963) has applied both electrophoretic and immunological methods to the examination of hemolymph and tissue protein of the Cynthia silk moth and several other species and concludes that there exist both stage-specific and tissue-specific antigens. Enzyme activity (esterases and dehydrogenases) was found associated with a number of the protein zones, but as the enzyme assay is very sensitive it may reveal proteins other than those made visible by staining.

Extracts of the cellular tissues would be expected to be more representative than hemolymph of the overall protein makeup of an insect. Butler and Leone (1966) have prepared whole-insect extracts of several developmental stages of the beetle mealworm) *Tenebrio molitor* and examined these by several types of electrophoretic and serological test. By immunoelectrophoresis, 10 precipitin zones were obtained; in several zones, larval and adult protein were distinguished serologically despite similar electrophoretic mobility. These authors conclude that during ontogeny there are gradual and sequential qualitative and quantitative changes in protein complement.

The electrophoretic pattern of basic proteins extracted from an acetone powder with acid also changes during blowfly metamorphosis

From these various studies, it is very difficult to feel confident in any conclusions as to the extent of changeover from a larval set of proteins to an adult set in metamorphosis. For the enzymic requirements for their metabolism, insect cells, like other cells, must contain at least several hundreds of distinct proteins. Yet the methods that have been used have detected at most 20. These are no doubt the most abundant or the most antigenic components but are not necessarily the most significant in ontogeny. In only a few studies has there been any attempt at quantitation, and in view of the limited sensitivity of the detection methods, it is very difficult to distinguish a qualitative from a quantitative change. There is suggestive evidence (with respect to hemolymph proteins) that, as might be expected, some correlation exists between the completeness of metamorphosis in an insect group and the extent of change in its proteins. Thus, a bug shows no qualitative change, a moth changes a small proportion of its complement, and flies exhibit extensive change. In all cases, however, a proportion of the demonstrable protein components are found at all stages of development.

A quite different, and more demanding, approach to the question of identity of larval and adult proteins is to isolate homologous proteins from each stage and compare them chemically. This has been done with tropomyosin, which was obtained crystalline and apparently pure from both larvae and adults of the blowfly, *Phormia regina* (Kominz et al., 1962). The two preparations were very similar but showed slight differences in amino acid and peptide composition and physical properties. These small differences could conceivably result from impurity or secondary modifications of the protein rather than genetic difference, and thus the question of genetic identity is unfortunately unresolved. Another indication of similarity of larval and adult proteins, though not conclusive evidence of identity, comes from low-temperature spectra of mitochondria from larvae and adult thoraces of *Drosophila;* the two show no significant differences, while both differ characteristically from mammalian mitochondria.

Free Amino Acids and Peptides

A peculiar feature of insects that always has to be taken into account in a discussion of their protein metabolism is their high level of free amino acids. This is highest in the blood, and the tissue pool is generally somewhat less concentrated and differently constituted with respect to individual amino acids. In the more advanced insects, the hemolymph amino acid level is commonly in the range 5 to 20 mg/ml—many times that of mammalian plasma. In addition to the amino acids of proteins, some less usual ones have been identified from certain insects, including β-alanine, tyrosine-0-phosphate, methionine sulfoxide, D-alanine, D-serine and others. The literature, of this subject, as well as other aspects of amino acid and protein metabolism in insect development, is reviewed by Chen (1966). The significance of the high amino acid level is not well understood, although it provides a pool for protein synthesis and appears to have an osmotic role. During metamorphosis, while individual amino acids change in concentration, the variation in total free amino acid pool is remarkably small when one considers the extent of tissue reconstruction that is taking place.

Along with free amino acids, insects also exhibit a variable content of peptides. In *Drosophila* larvae, intensive chromatographic analysis led to the conclusion that more than 600 peptides were present, and these incorporated injected C^{14} - glutamate much more rapidly than did the total body protein (Mitchell and Simmons, 1962). In *Phormia,* a group of peptides is transiently abundant during the early third instar; these are mostly composed of only two amino acids one of which, in each that was analyzed, was either lysine or histidine. Their metabolism and significance is unknown.

The Biochemistry of Histolysis

It repeatedly has been observed since the early histological studies on insect metamorphosis that at certain developmental stages the fat body, and to a lesser extent other tissues, become loaded win protein granules or "albuminoid spheres". These have been regarded as a form of protein reserve and appeared to constitute an exception to the generalization from mammalian biochemistry that animal tissues have no specifically storage proteins. Their probable nature is now becoming apparent with the use of the electron microscope and with growing understanding of the nature and functions of lysosomes. In Lepidoptera, such as the skipper *Calpodes ethlius*, protein granules appear in the fat body about one day before pupation, at the time when the larval cuticle is being resorbed. Since injected radioactive amino acid showed little incorporation into these granules, whereas injected plant peroxidase (detectable histochemically) appeared specifically in them, it was inferred that they contain protein sequestered from the hemolymph and not newly synthesized. Similar uptake occurs in other tissues, particularly the epidermis and especially at times in the molting cycle when there is active protein synthesis. The granules are surrounded by membranes and are believed to transform into multivesicular bodies in which digestion takes place, presumably by addition of lysosomal enzymes, as a source of amino acids for new protein synthesis. Cell organelles such as mitochondria due for destruction, are believed to be isolated in a similar way. The uptake of hemolymph proteins into fat body, gut, and to some extent other tissues of Lepidoptera is independently demonstrated by immunodiffusion tests. Which show several blood antigens to be present in them during the prepupal and early pupal stages.

As an example of programmed histolysis, have studies the breakdown of the abdominal intersegmental muscles of saturniid silk moths, whose function is to pump blood into the wings of the newly emerged moth and which lyse when this role is completed. The trigger for lysis is apparently the cessation of nervous motor stimuli to the muscle. Lysosomelike bodies are observed to undergo change at this time. Proteolytić activity of a catheptic type (pH optimum 4.0) builds up is these muscles at first gradually, then rapidly, during the 3 weeks of pharate adult development, and is associated with lysosomelise particles.

It can be activated and partially released from the particles by detergent treatment, but in the emerged moth the activity and the proportion of soluble enzyme are naturally enhanced. Similar catheptic activity is found also in the far body, but not in thoracic muscle (which, of course, is not destined for dissolution). The intersegmental muscles and fat body also contain an acid phosphatase which shows greatest activity in advanced adult development. These enzymes are considered to be directly responsible for the destruction of the intersegmental muscles.

It is remarkable that, despite the dominant place of histolysis in insect metamorphosis, there is almost no other information on the enzymic basis of the process. Not much can be concluded from several reports of hydrolytic activities without evidence as to their source in tissues or cytological fractions.

Rates of Protein Synthesis and Turnover

Linked to the question of changing protein composition of an insect during metamorphosis is that of the rates and extent of protein degradation and synthesis. A number of measurements have been reported of the incoporation of injected amino acid into the proteins of hemolymph or tissues of insects at selected stages. It is found that incorporation rates rise sharply when adult development is initiated in diapausing silk moth pupae, and that fat body qualifies kinetically as a source of hemolymph proteins. Similar results have been obtained with isolated insect tissues incubated in vitro, a technique which avoids complications due to the blood amino acid pool. But the systematic quantitative study of protein metabolism during metamorphosis seems to be confined to the recent work of two laboratories.

Using *Drosophila melanogaster*, Boyd and Mitchell (1966) fed C^{14}-amino acids to larvae, then after appropriate times for incorporation, extracted body fluid (chiefly hemolymph), and separated radioactive protein fractions from it by acrylamide gel electrophoresis. Then, their turnover was measured by reinjecting into larvae or pupae the isolated radioactive protein fractions and, after incubation, subjecting hemolymph samples to the same kind of analysis. The redistribution of activity was exstensive and varied in different fractions: One was stable in the larva, but after pupation had a half-life of 13 hours; another was too stable for the half-life to be measured. Specificity is also indicated by the fact that injected mammalian hemoglobin did not turn over. When enough nonradioactive valine was injected to increase the blood valine pool about thirty-fivefold, the redistribution of activity from valine-labeled protein during 42 hours was cut from 56 to 39 per cent, which suggests that free valine, but not the valine of the hemolymph, may be an intermediate in the interconversion of hemolymph proteins. The authors conclude that hemolymph proteins "may be an important channel through which the massive protein turnover of the metamorphosing insect passes."

Levenbook and colleagues have examined the question of protein synthesis and turnover in *Phormia regina,* The hemolymph protein is shown to be synthesized chiefly in the larva; at pupation there is a rapid fall in blood protein content, and there is little synthesis from injected amino acids or change in level thereafter. Radioactive hemolymph protein was prepared from C^{14}-labeled larvae, and after injection into larvae, pupae, and pharate, adults, its fate was followed. Very little was converted to $C^{14}O_2$, although injected free amino acids were rapidly oxidized, and conversion of hemolymph protein to free amino acids was also slow. It is inferred, in contrast to the conclusions of Boyd and Mitchell, that the hemolymph proteins are relatively stable. When mature larvae were injected with labeled hemolymph protein and kept until they emerged as flies, however, about there quarters of the recovered radioactivity was found in protein in the tissues. Since the blood free amino acids clearly cannot be intermediates, some form of direct transformation of blood proteins into tissue proteins is suggested. The data seem only to rule out *hemolymph* free amino acids as intermediates and do not oppose degradation to free amino acids which remain within the tissues and are rapidly used in resynthesis.

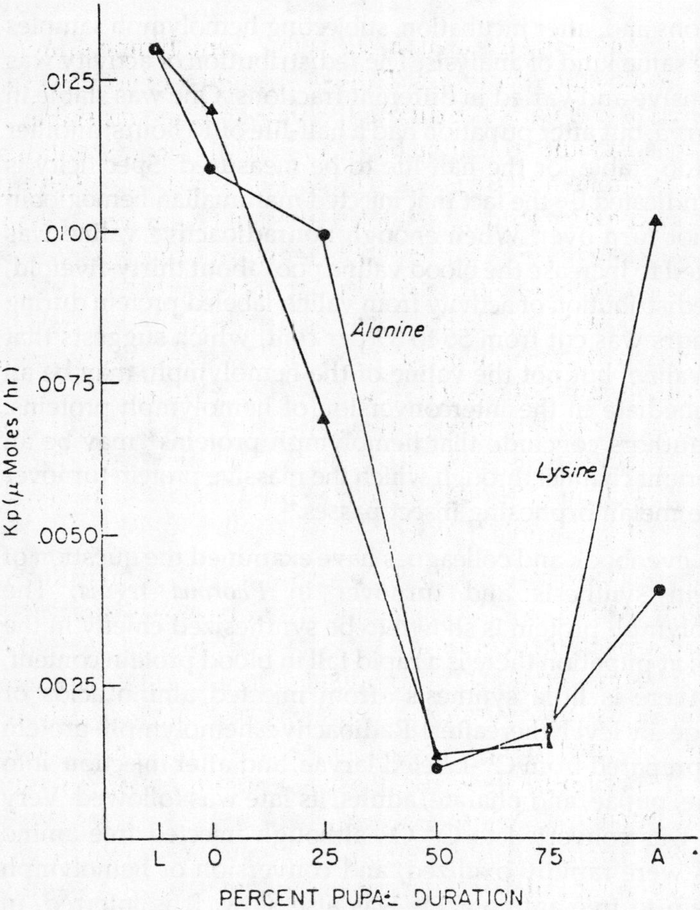

Fig. 3.8. Rates of incorporation of alanine and lysine into protein during metamorphosis of *Phormia regina*, K_p is colculated from the incorporation of radioactive amino acids into protein, their total rate of utilization in the insect, and the pool size.

In a related study, Dinamarca and Levenbook (1966) examined the dynamic state of the free amino acid pool and the rate of protein synthesis from it in *Phormia regina* at different developmental stages. C^{14}-alanine and C^{14}-lysine were injected, and their conversion to C^{14}-O_2, total disappearance from the pool, and incorporation into protein were measured, as well as the pool size for each amino acid in each insect used.

Corrected values for the rate of incorporation into protein could then be calculated and were found to follow a U-shaped curve, closely similar for the two amino acids, except in the adult. The importance of the procedure became clear, for lysine always yielded more of the injected dose into protein than did alanine, but this was a consequence of the more rapid turnover of alanine. With the assumption of a single homogeneous pool of each amino acid in the insect, the further calculation could be made of the total amount of new protein synthesis during metamorphosis, and this came out at the surprisingly small value of 0.16 mg, or about 2 per cent of the total protein of the insect.* As Dinamarca and Levenbook recognize, the assumption is a great oversimplification; nevertheless, they suggest that the total amount of new protein synthesis during metamorphosis is less than has been generally assumed.

The chief conclusions from these studies are, perhaps, to confirm that the hemolymph amino acids (which make up more than half of the total pool) are not in free exchange with those in the tissues and to indicate that the greater part of the protein turnover involves only local intracellular pools. This would be in accord with the cytological evidence that proteins are engulfed from the blood and digested within various tissues that are also synthesizing protein. Calculations based on the total pool of the animal might then fail to give any indication of the true amount of protein synthesis. It seems that still more refined techniques, perhaps involving isolation of specific proteins from individual tissues, are required.

Enzymes Associated with Protein Synthesis

The amino acid activating enzymes, responsible for the first step in protein synthesis, have been assayed and to some extent characterized in different developmental stages of the blowfly, *Lucilia cuprina*, and the fat body of *Sarcophaga bullata*. Activity was found for almost all of the protein amino acids, and where it was not found, this may have been due to inactivation. The total soluble enzyme activity in both series of experiments was minimal in midpupal life and rose to a maximum just before emergence of the adult fly, thus paralleling the changes in rate of protein synthesis in a blowfly pupa. In *Lucilia*, during the early pupal stage, some amino acid activating enzymes, particularly for tyrosine, are bound to a class of cell particle, and it has been suggested that these may have some role in storage of tyrosine for use in cutincular synthesis and tanning.

The activity of glutamate-aspartate transaminase in the housefly declines from a high value at pupation to a minimum just before adult emergence and then rises steeply. A relation to protein synthesis is suggested, but the curve appears more closely correlated with that to be expected of protein degradation. In the Cecropia silk moth, however, this enzyme rises strikingly in thoracic muscle during late adult development, which is certainly an active site of protein synthesis.

Ribonucleic Acid and the Machinery of Protein Synthesis

Increasing attention has been directed recently to the nucleic acids as determinants in the specificity and possible regulators in the rate of protein synthesis, but there is not yet a great deal of information relevant to insect metamorphosis. Measurements of total RNA during metamorphosis of several insect species reveal little change, which is consistent with the histological observation that histolysis and histogenesis are concurrent rather than consecutive, so that the total amount of cellular tissue does not change much. Some changes reported in RNA fractions differing in extractability with phenol from *Calliphora evythrocephala* are very difficult to interpret.

When a single tissue is examined, more significant changes may be seen. Thus, in the fat body of the saturniid silkworm, *Samia cynthia ricini*, during pupation there is a great decrease in cytoplasmic RNA, which is shown by P^{32} incorporation to have been synthesized early in the fifth larval instar. In the Cecropia silk moth, when adult development begins after the pupal diapause there is no discernible net increase in fat body

RNA, but there is an increase in turnover, associated with renewed activity in protein synthesis.

Other recent studies on RNA in insect metamorphosis have concentrated on particular developmental processes related to hormone action, Karlson, Sekeris, and colleagues have examined changes in fully grown *Calliphora* larvae, at the stage when formation and tanning of the puparium follow upon the increasing titer of ecdysone. After injection of ecdysone into larvae taken about two days before pupation, the incorporation of p^{32} into RNA is elevated. With 1.5-hour pulse corporation, the maximal stimulation of 80 per cent over controls was observed at 4 to 7 hours after giving the hormone. A greater effect was observed in RNA from the epidermis, a target tissue for ecdysone, and in RNA prepared with hot phenol from epidermal nuclei there was some stimulation within 1 hour of giving the hormone. The nuclear RNA stimulated amino acid incorporation in microsomal preparations from rat liver or *Calliphora* epidermis. It has been reported from the same laboratory that ecdysone added to isolated epidermal nuclei stimulates their incorporation of C^{14}-uracil into RNA and that RNA prepared from pupating larvae when added to rat liver microsomal preparations acts as a template in the synthesis of dopa decarboxylase, but fuller data are required before one can interpret these observations with confidence. The results obtained with this system have been presented in support of the view that ecdysone and other hormones act by directly causing derepression of DNA in the cell nucleus and release of specific messengers. The evidence, however, does not eliminate the presence of further links in the chain of hormone action.

The synthesis of RNA under the stimulus of ecdysone has also been investigated in the pupal wing epidermis of *Hyalophora cecropia,* during the initiation of adult development after diapause. During natural development, in pupae incubated after a period of chilling, the synthesis of RNA begins well before morphological change, and by the second day of visible adult development the content of RNA in the wing tissue is about fourfold that in diapause. After initiation of development by injection of ecdysone into pupae whose endogenous endocrine sequence is blocked by removal of the brain, stimulation of RNA synthesis is detectable at 8 to 10 hours, intense (about sixfold) at 24 hours, and falls off thereafter. Some effect on protein synthesis begins to appear at about the same time, but the rise is more gradual and more prolonged. The RNA produced under the influence of ecdysone has been characterized in several ways. Its base composition (by pulse labeling with P^{32}) is intermediate between the bulk RNA and the DNA of the species and could be accounted for by synthesis of a mixture of ribosomal RNA and DNA-like RNA. Fractionation of the RNA, after pulse labeling with uridine, showed incorporation in the 4s, 17s, and approximately 30s regions (corresponding to transfer, ribosomal, and ribosomal precursor RNA, respectively) and also in the residue from cold phenol extraction, which would contain nuclear RNA. The extent of stimulation by ecdysome was rather similar in each of these fractions. When assayed for stimulation of amino acid incorporation in an *E. coli* ribosomal system, the RNA produced early in Cecropia wing development was found to be highly active, but there are reasons for believing that such stimulation in a heterologous system may not detectable in the epidermis in diapause, become abundant early in development. It is concluded from this work that new RNA synthesis forms an important early part of the action of ecdysone and contributes to building the stage for increased protein synthesis. But the connection between hormone action and RNA synthesis, and the question of specificity in template RNA production, require further study. The picture in a tissue being awakened from diapause may, of course, differ from that in a tissue which, when already metabolically active, is influenced by ecdysone. But the observations on the Cecropia wing epidermis are consistent with those on the actions of various developmental hormones in vertebrate material.

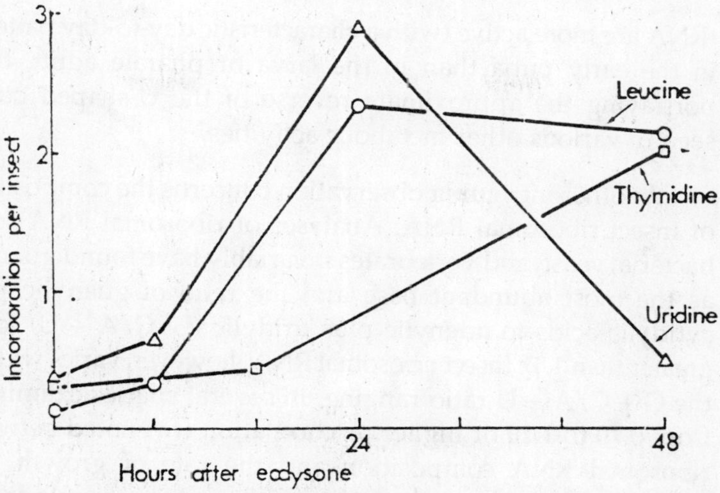

Fig. 3.9. The effects of ecdysone on macromolecular synthesis in wing tissue of Cecropic silk moth pupae. Into pupae from which the brains had been removed, sufficient ecdysone was injected to cause visible initiation of development. Then, of the intervals shown, radioactive leucine was injected as a precursor of protein uridine as a precursor of RNA, and thymidine as a precursor of DNA. After 4 hours for incorporation, the pupal wings were removed and prepared for counting.

The juvenile hormone of the corpora allata, acting in the presence of ecdysone, permits growth and molting in insects but modifies their course, so that the characteristics of immature stages are retained. A plausible basis for this would be a regulatory effect upon RNA synthesis so that larval-specific templates are produced and adult templates repressed, but there is as yet no biochemical evidence for this. When juvenile hormone is injected into *Antheraea polyphemus* pupae at the beginning of adult development, the rate of RNA synthesis is increased, but this may be a result of stimulated output of ecdysone from the prothoracic gland rather than a direct effect on other tissues.

A provocative observation for which no interpretation cab be offered concerns RNA - methylating enzymes. In *Tenebrio molitor,* the enzymes active in adding methyl groups to *E. coli* tRNA are more active (with a characteristic day-to-day pattern) in the early pupa than in the larva or pharate adult, thus portraying the approximate reverse of the U-shaped curve seen in various other metabolic activities.

Another intriguing observation concerns the composition of insect ribosomal RNA. Analyses of ribosomal RNA from bacteria, yeast, and vertebrates invariably have found guanine as the most abundant base and the ratio of guanylic plus cytidylic acids to adenylic plus uridylic (G+C/A + U) to be greater than 1, 0. Insect ribosomal RNA, however, varies widely, the G + C/A + U ratio ranging, in several species examined, from 0.70 to 1.20 or higher. A correlation was noted between ribosomal RNA composition and the rate of growth and development in a number of species, and it was suggested that these may be related because of differences in heat stability of the RNA (Mednikov, 1965). Inspection of the data, however, suggests that there is also a correlation with polygeny. The more primitive insects (Orthoptera) tend to have high G + C, Coleoptera and Lepidoptera are intermediate, and the most specialized group (Diptera) have RNA distinguished by low G + C content. Studies on the properties of ribosomes with such differently constituted RNA would be of interest.

Understanding of the control mechanisms of rate and specificity of protein synthesis in insect metamorphosis will undoubtedly depend on the development and study of cell-free amino acid incoporating systems of insect origin. In such a system for *Tenebrio molitor,* the ratio of incorporation of tyrosine to that of leucine rises greatly during the last two days of adult development, which is believed to reflect the high tyrosine content of the cuticular proteins then being synthesized.

Deoxyribonucleic Acid

When one considers the pivotal position of DNA in determining the characters of organisms, and the wealth of recent critical and ingenious research on this nucleic acid in various living systems, the paucity of information on DNA in insect metamorphosis is remarkable. For very few insect species do we even have reliable estimates of the changing quantity of DNA during development. Nucleic acid determinations with insect material require special attention to technique because of various complications that may arise, e.g., difficulty in securing quantitative extraction, the presence of compounds which interfere in colorimetric reactions, and uric acid which interferes in the ultraviolet range. In *Lucilia cuprina,* the total body DNA falls to about one half just before pupation and then rises rather steadily during the pupal stage and adult development, presumably reflecting breakdown of larval cells, followed by cell multiplication in the imaginal *anlagen.* In the mosquito, on the other hand, there is apparently little change in DNA during metamorphosis, although in larvae there is the unusual situation that up to 30 per cent of the total DNA is found in the soluble supernatant fraction of homogenates.

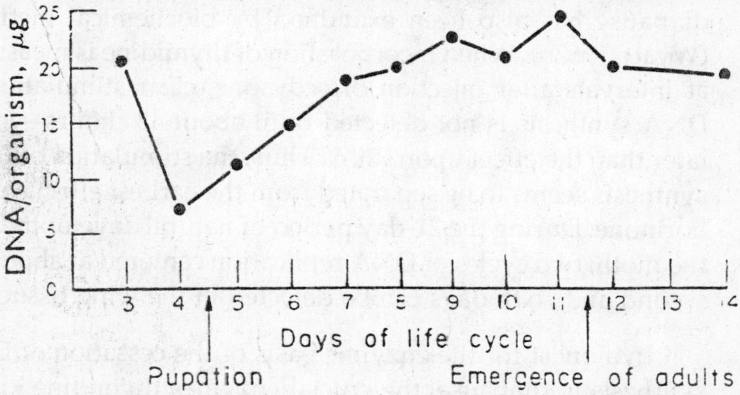

Fig. 3.10. The content of DNA in *Lucilia cuprina* during metamorphosis. From Howells and Birt, 1964. the scale has been corrected according to personal communication from Dr. Howells.

Patterns of DNA synthesis in different tissues are readily surveyed by autoradiography after incorporation of H^3-thymidine, and this technique has been applied to the development of saturiniid silk moths. In the larval molting cycle there is a burst of synthesis in most tissues centered about 3 days before the molt, but in the hemocytes and midgut, synthesis continues at a reduced rate at other times, and in the imaginal wing buds synthesis is apparently unrelated to the molt cycle. Shortly after pupation, in diapausing species such as *Hyalophora cecropia*, DNA synthesis stops in all tissues except the hemocytes and spermatogonia, in which a significant rate of incorporation continues. In a species without diapause (*Samia cynthia ricini*), incorporation in most tissues comes almost to a stop at the time of pupation, but then resumes within a day with the onset of adult development. At the conclusion of a pupal diapause, DNA synthesis appears to be correlated with the stimulation of growth by ecdysone. Adult development of moths is blocked by mitomycin C, an inhibitor of DNA synthesis, and it is clear that DNA replication is an essential part of ecdysone-stimulated development.

The cecropia wing epidermis developing after the pupal diapause has also been examined by biochemical methods (Wyatt, 1967b). When incorporation of thymidine is measured at intervals after injection of ecdysone, clear stimulation of DNA synthesis is not detected until about 48 hours—much later than the effect upon RNA. Thus, the stimulation of DNA synthesis seems to be separated from the earliest effects of the hormone. During the 21-day period of natural development to the moth, two cycles of DNA replication centered at about the second and sixth days can be detected in the wing tissue.

In a quest for the enzymic basis of the cessation of DNA synthesis in diapause, the crucial enzymes thymidine kinase and thymidylate kinase were assayed in silk moth pupal wing epidermis. Both increased manyfold at the initiation of adult development, yet diapause tissue did contain appreciable activity, so that the lack of these enzymes could not be responsible for the total cessation of synthesis.

Inorganic Ions

One further subject which must be mentioned is the remarkable inorganic ion content of many insects. In contrast to the usual composition of body fluids of vertebrates and most other animals, the higher insects, particularly phytophagous species, contain much potassium and little sodium, so that the $Na^+ : K^+$ ratio is far below unity. Further, the level of magnesium is often extraordinarily high (often up to 30 mM and occasionally to 100 mM, though part of this is chemically complexed). This situation is regrded as an adaptation to a vegetarian diet high in potassium and magnesium. Normal ion levels are maintained with the aid of active transport mechanisms, including the transport of K^+ through the gut wall.

In metamorphosis, the inorganic levels may be changed, Thus, in the silkworm, *Bombyx mori,* falls in sodium and magnesium at pupation have been reported, though the data of different authors are not consistent. Since the ionic proportions found in different insect species vary greatly, their having an important role in metabolic regulation is perhaps unlikely. Little is known, however, about the distribution of ions between hemolymph and intracellular fluids in insects, and it is interesting that an effect of ion levels on protein production by blowfly larval fat body in vitro recently has been reported. The matter deserves further attention, in view of the indications that potassium can have roles in the regulation of gene activity in insects and in protein synthesis in various living systems.

4

Metamorphosis in Lepidoptera

There is some advantage in commencing the description of the metamorphoses of insects with the *Lepidoptera,* for nearly everybody knows something about their transformations.

No one has any hesitation in distinguishing an insect of this order, for its shape proclaims it to be a butterfiy or a moth at once.

The very name, scale-wing *(Lepidoptera)*, calls up delicate and airy forms with elegant proportions and beautiful colouring.

The wings, usually very large in proportion to the body, give that peculiar jerking flight to most of the *Lepidoptera* which attracts the attention. There are four wings, each formed by a double layer of colourless membrane, traversed by nervures which are defferently arranged according to the species of the insect; and these delicate expansions are covered with microscopic scales fixed on like the tiles of a roof.

The beautiful colouring of the butterfly's wing depends upon the scales. This nature painting, which seems to deal with glaring and strongly contrasted colours, is so perfect that it never offends the eye. Yet, if we place certain tints in direct opposition, an unpleasing effect is invariably produced. Thus, there are many *Lepidoptera* with a red, yellow, or bright blue band crossing a black ground; but there is a beautiful softening off of the intermediate tints produced by rows of diversely-coloured scales, which, although it escapes us in a general view, does away with all harshness, and conveys an excellent lesson to the painter.

Fig. 4.1. A portion of the wing of *Attacus pavania major.*

At first sight the scales look like dust, and they come off on the finger and thumb when the wings are touched; but under the microscope they present most elegant shapes and singularly elaborate details of structure. Variable according to the genera and species of the *Lepidoptera,* and equally so in different parts of the same wing. The scales may be more or less elongated or even fan-tailed in shape, and rounded at the free edge, or incised, so as to resemble dentations or pointed festoons. There is a short stalk at the base of each scale, which resembles of handle under the magnifying power of the microscope, and it is the portion which is implanted into the alar membrane. The surface of the scales ordinarily presents several longitudinal keel-like prominences, which are parallel and equidistant. Some cross markings, which are very close together, form a very delicate network between the keel-shaped lines, and they appear to have something to do with the connection between the neighbouring scales. The scales are formed by two layers of excessively delicate tissue, and a third layer appears to exist between these. The colours are produced by extremely small dots, beads, or corrugations, which act upon the rays of light, and produce the phenomena of "interference." There may be some special colouring matter in the scales, and most of the effect is brought about by the "interference: and decomposition of light. Dr. Pigott and others have lately shown that the so-called ribs of the scales are corrugations of one or both of the layers of tissue, and that the beads--which are so difficult to resolve into such shapes, except with very good microscopes, handled by first-rate histologists—are dried drops of some secretion which is upon the scale when the whole wing is moist, immediately after metamorphosis.

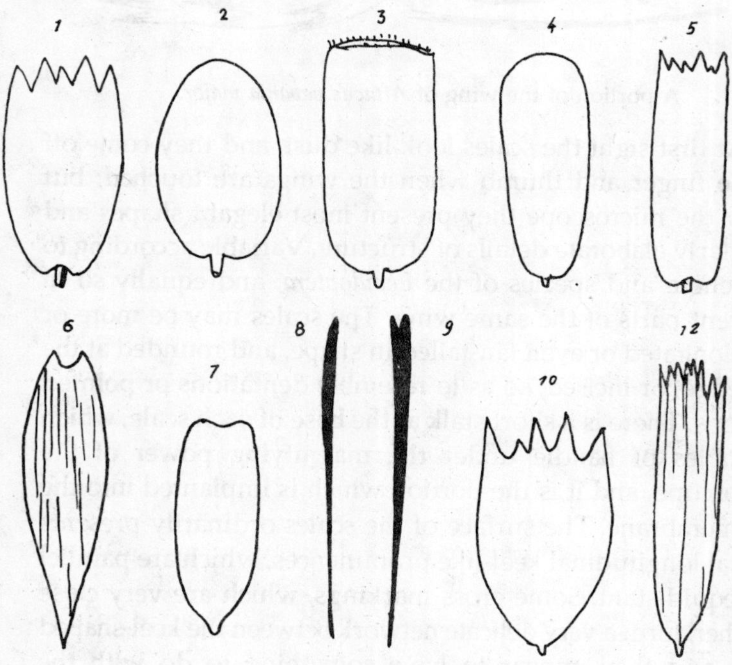

Fig. 4.2. Scale of different Genera of Lepidoptera.

1, 2. Papilio machaon. 3, 4. Morpho menelas, 5. Pamphila aracanthus. 6. Sesia apiformis. 7. Zygaena filipendulae. 8, 9, 10. Sphinx ligustri. 11. Pterophorus pentadctylus.

All the *Lepidoptera,* although the name implies to the contrary, have not their wings perfectly covered with scales. The wings of some are more of less transparent, and this is because they are not covered with scales; nevertheless these elegant structures are to be found on the margins, upon the nervures, and on some parts of the wings. They are never entirely deficient, and must be looked upon in all the *Lepidoptera* as structures superadded during the second metamorphosis.

The legs of the *Lepidoptera* are almost always very slender in proportion to the size and weight of the body. The adult insects, of whose legs we are not treating, do not walk much, and this is particularly the case with regard to those that fly by day—the diurnal or butterfly *Lepidoptera*—which only use their extremities as supports to rest upon. Generally, the three pairs of legs are equally developed, but the first are often atrophied in some kinds of butterflies. These ill-developed legs are smaller than the others, and do not have hooks or claws on their extremities; they are very hairy, and are usually kept close against the chest or thorax.

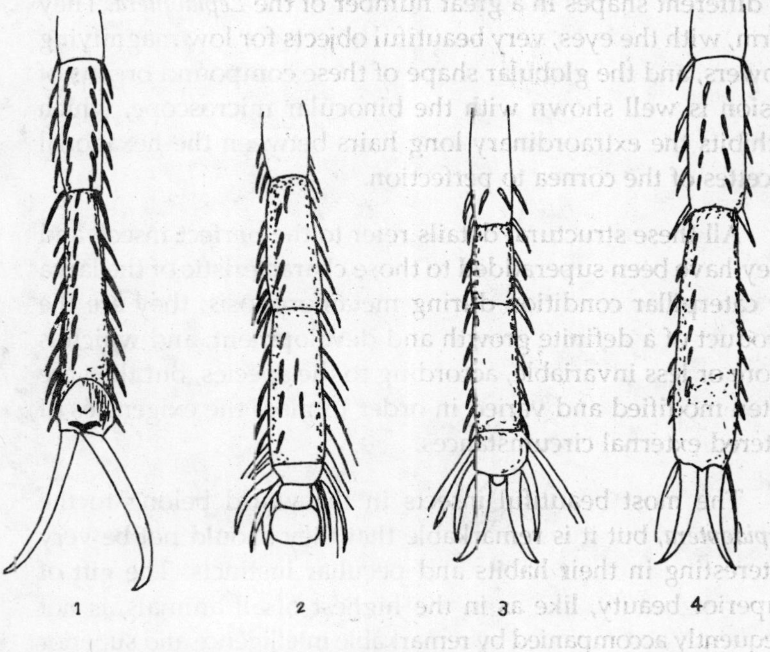

Fig. 4.3. Ends of the feet of Lepidoptera.
1. Papilio machaon. 2. Heliconia. 3. Vanessa Io. 4. Sphinx ligustri.

The legs of the *Lepidoptera* are covered with hairs and scales, and the feet terminate in hooks, which are always of one shape in the moths or nocturnal kinds, and which are greatly modified in the butterflies or diurnal flyers.

Nearly every kind of butterfly has the endings of its feet modified to meet its peculiar habits and the nature of the foliage it crawls over.

We have noticed the wonderful transformation which occurs of the mouth of the caterpillar into the long trunk-like suction tube of the perfect insect, and it is therefore only necessary to state that the lower palpi, which are attached to the lower lip or labrum, are large, and that they probably act as organs of touch.

The antennae are formed by a long row of joints, and are of different shapes in a great number of the *Lepidoptera*. They form, with the eyes, very beautiful objects for low magnifying powers, and the globular shape of these compound organs of vision is well shown with the binocular microscope, which exhibits the extraordinary long hairs between the hexagonal facettes of the cornea to perfection.

All these structural details refer to the perfect insect, and they have been superadded to those characteristic of the larva or caterpillar condition during metamorphosis; they are the product of a definite growth and development, and which is more or less invariable, according to the species, but they are often modified and varied in order to meet the exigencies of altered external circumstances.

The most beautiful insects in the world belong to the *Lepidoptera*, but it is remarkable that they should not be very interesting in their habits and peculiar instincts. The gift of superior beauty, like as in the highest of all animals, is not frequently accompanied by remarkable intelligence and superior sense; and the most gaudy butterfly is a fool in comparison to a dingy-coloured bee. The butterflies have some curious instincts when in the caterpillar state, but they only live to increase and multiply their race, and to embellish Nature. Their existence is usually very brief; it is one of the prettiest of honeymoons, and often love subdues and destroys every other passion. The gormandising caterpillar is never troubled by the ardent flame which consumes even the thought of sipping the nectar of the flowers that rival in beauty the wings of the perfect representation of elegance and love. The early insect lives and eats, and the perfect form loves and dies.

The female butterfly's instincts are principally restricted to selecting proper positions for her eggs. She never sees her hatched progeny, and can have no ideas about the respective merits of cabbages, corrots, or oak leaves as their future nourishment, for, at the most, she dips her long sucker into a few flowers to enjoy their sweetness. Nevertheless she deposits her eggs in a safe place, and as near as possible to the ordinary nourishment of the future caterpillar.

It is impossible to explain the varieties of the colouring of some groups of the *Lepidoptera*, or the remarkable resemblances which exist between the tints of others. Sometimes all the species of the genus have the same shades of colour and markings. The most brilliantly tinted kinds belong to the hottest and moistest climates; for instance, South America, the Moluccas, and certain parts of India.

Several groups of species belonging to the great genus *Papilio* have a very uniform system of decoration. Species of this genus, which are found in the Moluccas, in the neighbourhood of the Straits of Sunda, in India, and in Southern China, have black velvety wings, dusted with a metallic blue or green tint, and ornamented with spots or bands of the same shade. Other species of the genus belonging to South America, to the Antilles, and to Mexico, are distinguished on account of a flaring red spot, or a band with opalescent tints, which ornament the black hinder wings. Multitudes of others have their wings veriegated with yellow and black colours.

The *Pieridi,* whose type is the great Cabbage Butterfly, have usually white wings, and the species of the genus *Colias* have them tinted yellow; moreover, most of those of the genus *Morphos,* of South America, have their wings brilliantly coloured with a metallic blue.

Thus there is a sort of uniformity in the ornamentation and colouring of more or less widely spread natural groups of the *Lepidoptera.* But if we seek the reason why, and endeavour to discover the end that Nature had in view in giving certain shades of colour to certain species, we shall not obtain very satisfactory answers. It is known that the brilliant colouring depends upon the presence fo the scales, so far as the *Lepidoptera* are concerned, and the microscope has proved that no inherent and tangible colouring matters exist in those tiny dust-like particles. The delicate ridges and net-work of the scales act upon the colourless rays of light that fall upon them, and decompose them into a variety of tints that emulate the rainbow.

Many animals, such as birds and insects, live beneath the branches and amongst the sombre foliage of trees, and their colours are correspondingly quiet in tone; they take on a livery which enables them to escape observation, and they thus owe their safety to their mistress Nature. Hundreds of caterpillars escape destruction and extirpation in this manner; but the butterflies of the *Lepidoptera* do not hide, and they rather glory in the brightness of the full light of day.

The inquiry is rather complicated by the fact, elaborated by Wallace, the *Lepidoptera* which belong to different genera have often a similarity of colouring; but a hint from Mr. Bates, that admirable naturalist, who has given us such vivid descriptions of tropical nature in the district watered by the Amazons, throws no little light upon this difficult subject. He noticed immense flocks of *Lepidoptera,* belonging to the genus *Heliconia,* and that there were several other kinds mingling with them which belonged to different genera. Now these strangers in the crowd borrowed the Heliconian decorations, and would have passed muster to superficial observers as having a legitimate right to them. But the eye of the entomologist discovered that some belonged to the *Pieridi,* which we have already noticed as usually having white wings. The insect was clearly one of the *Pieridi,* but its clothing was that of *Heliconia.* It was a dove in peacock's attire, all glowing with red, yellow, and dark colours. Some of the other butterflies of the crowd had the same habits and peculiarities of flight of the *Lepidoptera,* whose garments they imitated. These were not accidental mimicries, for the descendants of these masqueraders do the same thing year after year. Why does Nature then mimic the decorations and colours? Mr. Bates suggests that in clothing a *Pieris* with the garb of a *Heliconia,* Nature has provided it with a means of escape from insectivorous creatures, that do not care for *Heliconia*—natural selection being at the bottom of the matter.

The colours of both sexes are often the same in the *Lepidoptera,* but in some instances the ornamentation and tinting of the males and females differ, as also do their growth and development. Many females have no wings, and others are so imperfectly developed that they can hardly be distinguished from larvae.

It is evident, then, that there is design in the decorations of the *Lepidoptera;* that they are for the good of the insect; that they very under certain conditions of life, the other structures remaining the same; that they have some reference to sex; that they are inherited, variations and all; and that, besides all this, they are proofs of the high art in Nature, and the beauty of God's thoughts.

There is a curious structure which is found on the wings of those *Lepidoptera* that fly very strongly and rapidly, and not in the jerking, irregular manner of butterflies. A sort of hook arises from one of the nervures of the hind wing, and fits into a ring on the large nervure of the front one. Both wings act then very strongly together.

Fig. 4.4. Moth of Sphinx ligustri (Privet Moth), Showing on one side the "catch" of the lower wing fitted into a flat ring in the upper one; and on the other side, the catch out of the ring, and the wings separated.

As a general rule, which is subject to a few exceptions, the diurnal *Lepidoptera*— the butterflies—do not have this structure, and they are therefore classified as *Achalinoptera*—wings without hooks. The rest of the *Lepidoptera* are *Chalinopterous,* and have the strengthening structures.

THE LEPIDOPTERA WITHOUT HOOKS TO THEIR WINGS

Achalinoptera

These *Lepidoptera* are celebrated for the elegance of their shape and brilliant colour. The wings of the butterflies, or diurnal flyers, are usually very large, and very varied in their outlines. Some have them gracefully rounded off, and others have them prolonged backwards symmetrically. When these insects are at rest, their wings are set up, so that only their lower surfaces can be seen; they can hardly walk, and their legs are principally employed to keep the body steady, and to support it. The antennae of all the species are clubbed, and are useful to distinguish the different kinds.

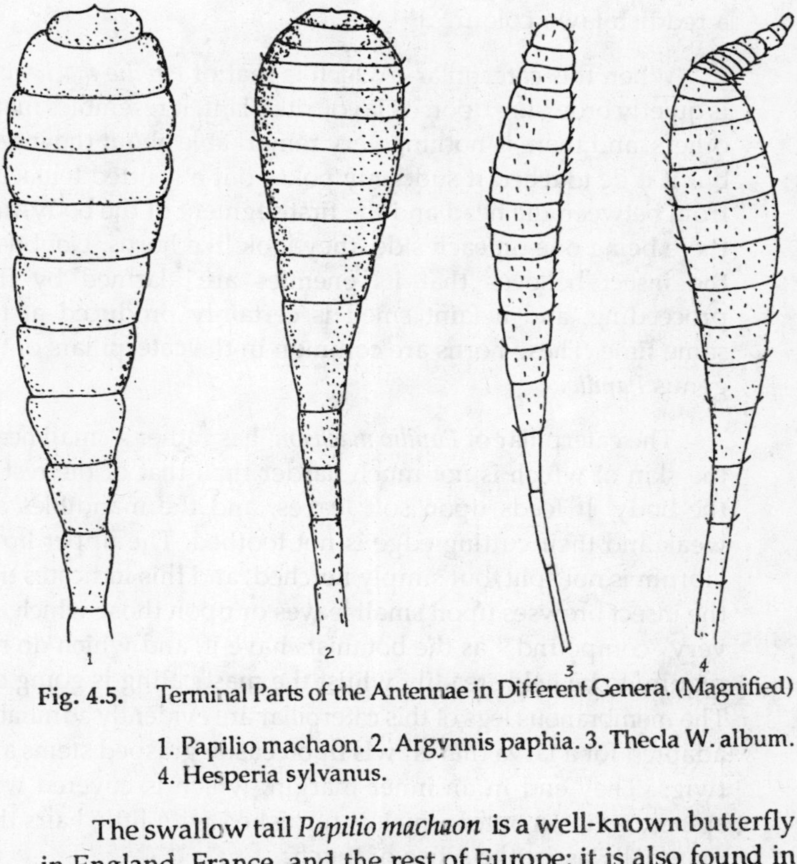

Fig. 4.5. Terminal Parts of the Antennae in Different Genera. (Magnified)

1. Papilio machaon. 2. Argynnis paphia. 3. Thecla W. album. 4. Hesperia sylvanus.

The swallow tail *Papilio machaon* is a well-known butterfly in England, France, and the rest of Europe; it is also found in Asia as far as the Himalayas, and even in South Africa. Every one has seen it flying in the fields and resting on the milk parsley and other umbelliferous plants, and has admired its great size, its yellow wings, marked and splashed with black, and its hinder ones so beautifully ornamented with a row of eye-shaped spots of light blue. The butterfly appears for the first time every year in the month of May and a second time in the month of July.

A very pretty caterpillar, about an inch and a half in length, may be found on the wild fennel and carrot plants during June and September. It is of a very vivid green colour, and is ornamented with black velvety rings and large dabs of a reddish fawn colour.

When this caterpillar—which is that of *Papilio machaon*—is quietly browsing upon its favourite plant, it resembles many others, and there is nothing very remarkable about the insect; but if it be touched it suddenly pokes out a spotted tentacule from between the head and the first segment of the body, and there being one on each side, they look like horns. Doubtless the insect believes that its enemies are alarmed by this proceeding, and a faint smell is certainly produced at the same time. These horns are common in the caterpillars of the genus *Papilio*.

The caterpillar of *Papilio machaon* has rather a small head, the skin of which is not much harder than that of the rest of the body. It feeds upon soft leaves, and the mandibles are weak and their cutting edge is not toothed. The upper lip of labrum is not split, but simply notched, and this indicates that the insect browses upon small leaves or upon those which are very "compound," as the botanists have it, and which do not require to be held steadily whilst the masticating is going on. The membranous legs of this caterpillar are evidently admirably adapted for a larva that crawls upon easily grasped stems and twigs. They end in an inner margin, which is covered with spines, an in an outer which is furnished with little hairs that materially assist the sense of touch.

There are two caterpillars of *Papilio machaon* represented in the engraving, and their peculiar horn-shaped prominences may be noticed behind their heads. An angular chrysalis is represented hung by its tail end and slung also by a girdle of silk to a stem. The caterpillar does not spin a cocoon, but slings itself, and the chrysalis stage is completed in May.

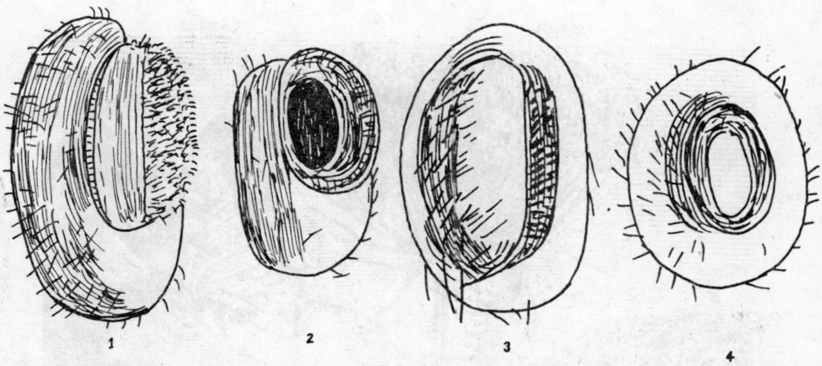

Fig. 4.6. Membranous Feet of Caterpillars.

1. Papilio machaon. 2. Vanessa urticae. 3. Charaxes jasius. 4. Thecla W. album.

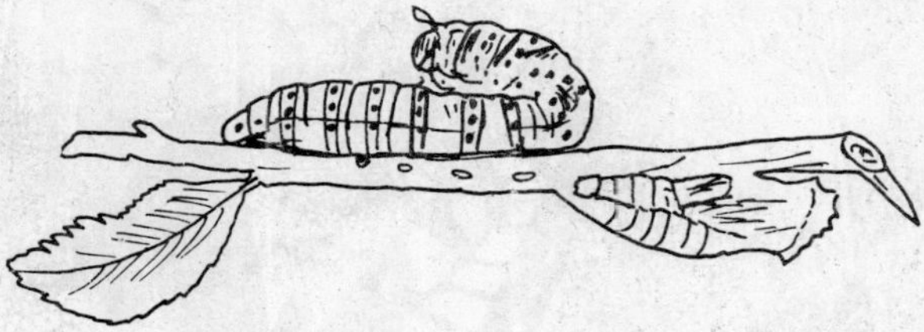

Fig. 4.7. Caterpillar and chrysalis of *Papilio machaon*.

The special adaptation of the external organs of the caterpillar to definite methods of existence is very noticeable. Thus in Carolina, Virgina, and Louisiana, there are swallow-tails *(Papilio asterias)* whose caterpillars live on umbelliferous plants, like those of *Papilio machaon,* and their oral (mouth) structures, and their feet, are like those of the common European insect. In the same districts of *Papilio troilus* lives on laurels in its caterpillar state, and the larva has its labrum or upper lip more split than the others, so as to hold the leaf it gnaws—moreover, the spines of the feet are stronger, so as to enable the creature to walk over the stems and leaves of the plants, which are more difficult to grasp than those of the *Umbelliferae.*

Fig. 4.8. The Metamorphoses of *Papilio machaon*.

These species may be compared with a swallow-tail from Bengal *(Papilio Hector)*, which has its black wings spotted with white and red. Its caterpillar has a very hard head, and its labrum is deeply notched; the mandibles are powerful, and are armed with saw-like edges, and the legs are much more spined than those of the other kinds. The Bengalese caterpillar evidently crawls over vegetable tissues which are not easily climbed, and lives upon large and hard leaves, and it is said that herbaceous *Aristolochiaceae* are its favourite food. This caterpillar's peculiar structures relate distinctly to its habits. It has, however, two rows of tubercles on the back, whilst the body of our species is smooth.

Several very closely-allied genera, which belong to the same little group of *Lepidoptera* as the genus *Papilio,* are very remarkable, for more reasons than one. The *Ornithoptera* are readily distinguished by their great size, their strongly-formed head, their long anterior wings, and their toothed hind ones which have not a vestige of a swallow-tail; they are magnificent butterflies, which inhabit the Moluccas, the Philippines, and the islands of Sumatra, Java and Borneo. Travellers in Amboina are always struck with the beauty of the great *Ornithoptera Priamus,* whose wings have a diameter of form six to eight inches. The black velvety colour and the brilliant and silky green tints of the wings of the male are most harmoniously contrasted, and he is a glorious fellow; but the female, although larger than its beautiful companion, is very quietly clothed in deep brown with white spots.

In most mountainous countries there are species of the genus *Parnassius,* which have a short body, short antennae, and wings as crisp as parchment, and almost without any scales on their surfaces. The *Parnassius Apollo* is common during the summer on the Alps, the Jura mountains, the Puy-de-Dome, the Pyrenees, the Caucasus, and the Sierra Ńevada; and it has rather a striking appearance, on account of its great white wings being semi-transparent, spotted and dotted with black, and ornamented with splashes of vermilion, surrounded by black tints, and marked with white eye-shaped circles. Its caterpillar is velvety black, and has tubercles which are more or less blue, with orange-coloured points. It lives upon the Saxifrages and the Crassulaceous plants of the hills. When it undergoes the transformation into the chrysalis state, it forms a silken net around the body, and attaches the end of the abdomen to its resting-place, and forms the usual girdle by which it is safely anchored.

Most of the pupae or chrysalides of the *Lepidoptera* are naked, but from the above example it is evident that the protecting nature of a silken coverlet bears some reference to the exposure of the quiet insect to cold and rain, which are common enough in Alpine districts, and which would tend to destroy were there no natural clothing. Nevertheless, the protection it gives must be slight, for the insect has to sling itself, after being invested by it.

The *Pieridi*, which are vulgarly termed Cabbage Butterflies, on account of the favourite food of the caterpills of one genus, are found in Europe, Asia, Africa and Australia. The larvae are remarkable on account of the small size of the head and the end of the body, and their hairy skin. The pupae are very angular, and have a pointed projection in front. The butterflies always have wings, without tails, and the legs are long and hairy.

The type of the genus is the *Pieris brassicae,* the great Cabbage Butterfly, which is to be seen in every garden and field in Europe. The butterfly is, moreover, well-known in Asia Minor, in Egypt, in Northern Africa, in Siberia, in Nepaul, and even in Japan. The caterpillar makes sad havoc amongst the cabbages, and it is not a pleasant-looking nor a nice-smelling vegetarian. Its greenish body, marked with three long yellow lines broken into here and there, by little black hairy tubercles, excites indignation in the breast of every gardner. Unfortunately, they are not solitary in their habits, for they generally abound in swarms, and destroy the tenderest leaves of the hardest-hearted cabbages with infinite relish, and with an utter disregard of consequences. Everybody has met with some of these caterpillars wandering about the roads and paths, and crawling up walls and palings, in order to find a cornice, or a safe projection under which they may tun into chrysalides, and rest in safety until the spring-time of the next year. Myriads of them are killed by remorseless market gardners, and devoured by birds; but were it not for a little ichneumon fly *(Microgaster glomerator)* the existence of cabbages would indeed be imperilled.

The caterpillars change their skin several times during their growth, and finally hang themselves up by the tail, and sling a silken girdle round their body before undergoing the metamorphosis into the chrysalis.

Some small *Pieridi,* whose delicate wings are ordinarily variegated with green colours, form the genus *Anthocharis.* This name indicates the grace and exquisite beauty of the little butterflies that rival the flowers they roam over in symmetry. The orangetip *Anthocharis cardamines,* so common in the spring of the year in wood-side lanes, quite deserves its title, and it is the male whose fore wings are marked with an orange spot. In the centre of France, and in Southern Europe, there is another *Anthocharis,* whose males have perfectly yellow wings. But so far as colouring is concerned, there are some remarkable facts is reference to some species of the genus which have white tints on the upper surface of the wings in both sexes. The *Anthocharis belia* of the centre of France sometimes wanders close to Paris, and is found in Southern Europe and in North Africa. There are two generations of it every year. The first butterflies escape from the chrysalis state in the months of March and April, having been in that condition all the winter. Their hinder wings are of a delicate yellow green below or on their under surfaces, and are ornamented with spots of the purest nacreous white. These butterflies produce a new generation, whose adults appear at the end of June or to the beginning of August. These are larger than the first, and their wings are marked with a dull white.

This interesting variation in the colouring of the same species at different times of the year is seen also in one of the *Vanessae (Vanessa prorsa)*, whose successive generations are so different that separate names have been given them. The butterflies of this species, which are common in France, are very small; some of them may be seen towards the end of April or the beginning of May, when Nature has awakened from her long sleep, and the leaves are beautiful from their freshness. They play around the nettles under the shade of the damp underwood. Their wings look new and unworn, for their texture resembles velvet, and they are of a bright fawn colour, being also ornamented with wandering black lines, which give the common name to the insect of the "carte geographique." If the same spots are re-visited during the month of June, none of the pretty little butterflies can be seen, but the nettle leaves are covered here and there with little black caterpillars, which have white dots, and fine branching spines. These are the offspring of the pretty April butterflies. When the caterpillars have attained their full growth they separate, and each, after fastening itself, becomes a grey chrysalis, having the angular form, like those of all *Vanessae*. It is now the month of June, and the temperature of the air is high, so that the development and the metamorphoses are rapid. A fortnight passes, and the little "cartes geographiques" burst from the chrysalis state. Now this second generation differs materially, so far as colouring is concerned, from the first which produced it. The wings of the July butterflies are black, and are scratched with whitish lines.

But this is not all. The *Vanessae* with the black wings lay their eggs, and little caterpillars are hatched, in August and September. These devour the favourite nettles, and resemble those which did the same thing in June. Like the others, these autumnal caterpillars hang themselves up, and are metamorphosed into pupae. If the autumn is fortunately a prolonged summer, some butterflies will escape from the chrysalis condition in October. Strange to say, their wings

Fig. 4.9. The Metamorphoses of *Venessa Io.*

have neither the black colours of the July butterflies, nor the fawn tint of the April brood, but an intermediate ornamentation. This late metamorphosis is rare, but it can be produced artificially by keeping the chrysalides in a warm place. Most of the chrysalides, and sometimes all of them, live in this state through the winter, and are metamorphosed into the true "cartes geographiques" during the spring, and the butterflies are then tinted with fawn colour.

An *Anthocharis* with a larger body than those just mentioned has been noticed in Orenburg, and is not very remarkable in the adult state, but Dr. Rambur having met with it in Andalusia, in the environs of Malaga and Granada, studied its metamorphosis. The caterpillar lives in the fields on coniferous plants. It makes a sort of delicate shroud of silk for itself before becoming a pupa, the shape of which is not at all angular, like those of the other *Papilionidi.*

The most favourite, perhaps, of all the butterflies is the Peacock—*Vanessa Io.* If it came from the tropics alone its value would be incalculable, but although it is so common its beautiful ornamentation and splendid colouring always make it prized. This *Vanessa* appears in the spring-time and also in the summer, and a third generation may ornament the autumn. Nettles are the food of the caterpillar, as they were with the small *Vanessa* just described. In their early days the caterpillars hatched from one set of eggs keep together and do not separate until the time for metamorphosis is at hand. They crowd upon the same leaf or twig, and when this is eaten they move *en masse* to others. Their colour is velvety black, pointed with white, and each segment of their bodies, except the first, carries six branching spines of stiff hairs. The membranous feet have a circle of very fine spines upon them, and they are constructed so as to climb the rather tender twigs and leaves of the nettle. Before turning into the chrysalis the caterpillars fix themselves usually to the leaves of their favourite plant or to some other in its immediate neighbourhood. Fifteen days after this the butterflies escape. The different generations do not exhibit any alterations in their colouring, so that those noticed in the case of the other *Vanessae* must have some other cause than the influence of heat and light. Some of the latest metamorphosed peacocks hybernate and fly early in the spring. The pendant chrysalis and the butterfly of *Vanessa Io* are shown in the plate with several caterpillars. The angular shape of the chrysalis is very evident.

The favourite White Admiral butterfly frequents woods and forest glades in preference to the open fields and gardens, and flies in June, and sometimes as late as September. Its spiny caterpillar is well-known as a honeysuckle eater, and its green colour blends with that of the plant. The caterpillar of the *Papilionidi* already mentioned have their feet especially suited for climbing their favourite plants, but the white admiral's larva is badly off in this respect. Its feet are not sufficiently furnished with spines and hairs to enable it to climb with perfect safety, for the twigs of the honeysuckle are smooth, hard, and often long; the spinning glands are, therefore, brought into requisition, and a fine silken thread placed here and there on the plant, and collected more decidedly on the leaves, protects the caterpillar from falling violently to the ground. The chrysalis is very angular in its aspect, and is of a dark greenish brown, but it is ornamented on the upper part of the abdomen with brilliant silvery spots and a large one of pale green.

The great *Limenitis populi* is rather common in the large forests in France, and its caterpillar choses very inaccessible places, such as the very top of poplar trees, so as to feed safely and well. Many a tired butterfly hunter has given up the chase after this splendid insect, whose flight is very strong and high, and were it not for a peculiar habit of this exquisite and delicate-looking creature specimens of it would be very rare in collections. Many are the evils of dirty habits and common testes, and curious are the longings of the beautiful for the

Fig. 4.10. The Metamorphoses of the Marbled White *(Arge galathea)*.

vulgar. The great Poplar Butterfly loves the moisture of horse dung, and leaves the tree tops to dip its trunk in the nauseous mess: then is the time for the butterfly net.

A large greenish caterpillar which moves very slowly, and looks remarkably stupid and dull, lives during the month of May upon the tops of aspen, willow, and poplar trees, whose constant motion must be very dangerous to it. The feet of the larva are not well furnished with the means of clinging on, but the creature makes use of its spinners and completely covers the leaves upon which it feeds, so that the wind may blow as much as it likes, providing the leaves do not come off. This *Nymphalis ilia* is a large butterfly with dark wings and yellowish white spots upon them, the whole having beautiful violet tints in certain lights. It flies near long rows of poplars, especially if there is any water near, and loves the shade of the meadows close to great woods.

The fine Purple Emperor, *Apatura iris,* whose caterpillars live on the oak trees, is one of the gems of all collections. Its magnificence is greater than its gentility, for Mr. Stainton accuses this monarch of the *Lepidoptera* of having a penchant for homemade gooseberry wine, and for leaving his realms in the air in order to seek the "shades" of a gooseberry bush. M. Pierret says it condescends to still lower tastes, and the only satisfaction is that they lead it to destruction. All these species of the old genus *Nymphalis* prefer sucking very nasty things or the sweeter sap and gum of plants and trees to enjoying the nectar of flowers.

The *Satyridi* are world-wide butterflies. In Europe they are to be found on open plains, arid downs, the road side, forests, and even on the mountains amidst the everlasting snow. From spring to autumn the tawny, grey, and brown butterflies are constantly flying here and there, but for short distances, wherever there is the rankest and commonest vegetation. Their caterpillars live on the different kinds of grasses, and although they must be very common they are very rarely seen except by those who can look for them scientifically. They do not feed upon the slender grass leaves at all during the day, but hide themselves up at the bottom of the plant, with whose colours their tints harmonise. At night-time, however, they move for short distances and eat. This is an interesting habit, and contrasts with that of the majority of their fellow caterpillars. The larvae of the *Satyridi* are all shaped alike. The body is rather hairy, narrowed at each end, and the last segment is furnished with two hooks. The pupae are short and well set, and do not exhibit any of the strange-shaped projections which are common in the chrysalides of the *Nymphalidi.*

A pretty butterfly *(Arge galathea)*, the Marbled White, is rather common in the middle of summer in the whole of central and northern Europe, and it is our commonest representative of the *Satyridi.* The caterpillar lives upon some of the simplest grasses, and frequents the Timothy grass in England. The butterfly has the very delicate antennae hardly swollen at their ends. The black and white ornamentation of the perfect insect is very distinctive, and the accompanying plate shows the fusiform caterpillars and a chrysalis, attended by marbled white, upon one of the Compositae and by two others in different positions.

The genus *Erebia* has its species in abundance in mountainous districts. They are small butterflies, whose black wings are ornamented with eye-shaped spots of fawn or red colour; and these negroes, as they are termed, are found upon the Alps, the Pyrenees, the Caucasus, the Himalayas, and even upon the Rocky Mountains. Some closely allied forms, belonging to the genus *Chionobas,* are of a tawny or pale greyish yellow colour, and are found in the remotest north of Europe, America, Iceland, Siberia, and Kamschatka, and one kind lives in the Alps.

There are some *Satyridi* in South America, the forms and colour of wings of which are very different to those of the

Fig. 4.11. Caterpillars and Butterflies of *Thecla W. Album*.

European kinds. Thus, the *Haetrae* of Brazil and Guiana have festoon-shaped wings, which are partly transparent. The scales on the wings, scantily sown, as it were, are almost wanting in many spots, where they are often replaced by pretty little hairs. The membrane of the wing is, however, very iridescent, and produces very pretty effects of light and shade.

The caterpillars of the *Thecla W. album,* one of the *Lycanidae,* and so called on account of a curious white mark near the end of the hinder wings like the letter W, feeds on the elm and hawthorn, and owes its safety to its colour being so very exactly like that of the foliage upon which it lives. The general shape of the larva and the short legs impede its movements, and the insect is rarely found off a leaf.

After having devoured one leaf the caterpillar crawls on to the next, and this is about the extent of its travels. The caterpillar usually lies very closely upon the under surface of leaves, and its tints so resemble• those of the elm that it is not readily perceived. The crysalis is found attached to the leaves, which have partly contributed to the maintenance of the larva, and the butterfly escapes about a fortnight after the first transformation. One of the species of *Thecla*—the Green Hair-streak, or *Thecla rubi*—has an ornamentation, so far as colour is concerned, which is, perhaps, exceptional in the *Lepidoptera.* Its wings are brown above, but of a uniform light blue beneath. A Californian species has its wings of the same colour, and affords an example of the close relationship between the butterflies of North America and the temperate tracts of Europe.

The *Hesperidae* are butterflies that have many points of resemblance with moths, and therefore are unlike those we have noticed hitherto. They have a thick body, a large and strong head, and the antennae swollen and spindle-shaped at their ends, or hooked. Their legs are well-developed; the hind ones are particularly strong, and the others are spined. The wings are small in proportion to the size of the bodies, in these *Lepidopera,* and they are generally grey, brownish, or fawn in colour. The butterflies have a curious and hesitating flight, and are very properly called Skippers. The caterpillars differ from those of the other *Lepidoptera* already noticed as much as the butterflies do; they have a long and thin shape, with a large rounded head, and they are remarkable for their brisk movements.

Fig. 4.12. The Metamorphoses of *Pamphila aracynthus.*

In order to hide safely, the caterpillars roll up the side of a leaf, and fasten it with some of their silk. Before the first metamorphosis they choose a safe hiding-place; then they attach themselves by their extremity like the other diurnal forms. Before transformation proceeds, however, they swathe themselves with a set of silken lines, which cross and re-cross the body, and finally the thin, long, and angular chrysalis rests in a slight cocoon. The European species are all small, and the name of *Pamphila* is given to those of a genus which delight in underwood and thick forests. Their bodies are thinner than those of the other *Hesperidae,* and the ends of their antennae are egg-shaped.

The engraving represents the long and slender caterpillar and the elongate chrysalis, this being swathed with crossing fibres of silk. The butterfly is brownish black above, and the male is almost without spots, but the female is slightly marked with yellow. The posterior wings are ornamented beneath in both sexes with white spots, encircled with brown, and which resemble so many mirrors. This insect inhabits large woody tracts during June and July. The caterpillar, as shown in the engraving, is found on grasses in the spring; it is of a green colour, like the plant it feeds upon, and as it generally keeps near the stalk, is not readily discovered. The transformation into the pupa form is preceded by the formation of the silken swathing already noticed.

The leaf-rolling propensity is especially noticed in a species of *Syrichtus,* one of the *Hesperidae,* which curls up the mallow leaves upon which it feeds.

The caterpillars of most of the *Papilionidae* do not swathe their bodies, but place a silken girdle over themselves very cleverly, just before the first metamorphosis. The cabbage butterfly caterpillar, when it has shed its skin several times and has attained its full growth, retires to a quiet place—on palings, or under the eaves, and fixes its tail-end to the surface upon which it is to hang for many months. It applies its mouth to the the wood-work, for instance, and draws out some silk. After doing this several times, and thus producing a small silken tangle, the insect attaches the end of its body to the silk, and hangs head downwards. The caterpillar has thus fixed itself, and the end of the body grasps the silk so perfectly, that the whole hangs safely. But to prevent the wind moving the future chrysalis, and rubbing it against the wood, another operation is commenced. As it hangs head downwards, the catepillar throws back its head, and twists it sideways, so as to touch the wood on one side of the body. The touch pulls out a silken thread, and then the insect carries the head to the other side, and fixes the silk. This is repeated several times, until a sort of sling is placed over the body. Then the first metamorphosis takes place, and the chrysalis finds itself slung by the tail, and girded to the wood-work by a silken sling.

The first step to cocoon making, which is so usual in the next division of the *Lepidoptera*, is shown in the habit that *Pamphila aracynthus* has of swathing itself with a network of silk before undergoing the first metamorphosis, and the leaf rolling propensities of many caterpillars and moths are foreshadowed by a species of *Syrichtus*.

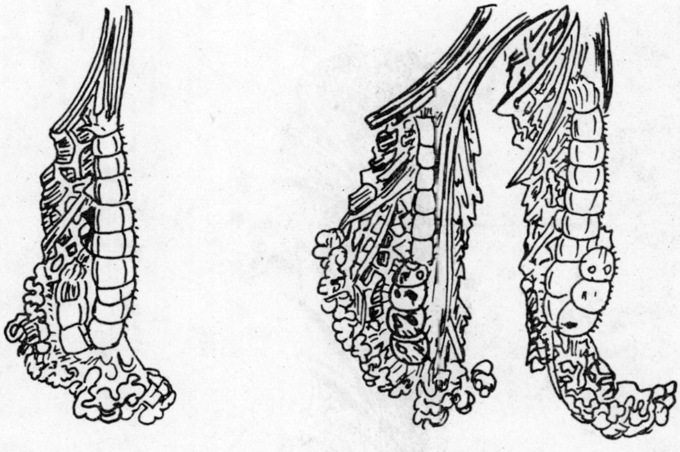

Fig. 4.13. Caterpillars of the Cabbage Butterfly Slinging Themselves Before the First Metamorphosis.

The angular shape of the butterfly chrysalides may have something to do with their comparatively unprotected and uncovered state, for this sharpness of outline which causes them to differ in appearance from most living things is not so perfectly seen in the case of the *Anthocharis*, from Spain, which spins a slight cocoon.

It has already been stated that the repeated sheddings of

the skin and mucous (epithelial) tissues of the caterpillars during their growth really form a most important part of the phenomena of evolution, for some very necessary alterations in the internal structures progress especially during the last stage of quiet retirement which accompanies the final change of skin before the commencement of the chrysalis state.

Although several broods of caterpillars are born in a year in such species as *Anthocharis belia* and *Vanessa prorsa,* yet the successive larvae present no distinctions in their external ornamentation, but the perfect insects which are metamorphosed from these similar larvae do differ very materially. Yet the butterflies of *Vanessa Io,* a species closely allied to *Vanessa prorsa,* are always the same, notwithstanding there are successive broods of them in the year.

Fig. 4.14. Chrysalides Slung and Suspended.

This variation in the colouring and ornamentation of the butterflies of successive broods is most interesting, especially when it is considered in relation to the opposite state of things so common in some species of moths, whose generations of caterpillars differ, the perfect insects remaining unlatered. The Change in the butterfly must be determined in the chrysalis state, and during the progress of the second metamorphosis. It has been proved that the duration of the chrysalis condition can be lengthened by placing the pupa in cold and dark places, or in ice, and shortened by increasing the usual amount of warmth. But no variation in the ornamentation of the perfect insect has been produced by this artificial lengthening or shortening of chrysalis life. So that most probably the disposition of these butterflies to vary has some deep significance, which refers to the duration of the species in time—moribund genera and species varying much, and new forms striving to adapt themselves to peculiar conditions.

The hybernation of some of the large butterflies through the winter in temperate climates is very interesting.

5

Metamorphoses in Hymenoptera

However much we may become interest in the habits, structures, and metamorhoses of other insects, there is always a sense of satisfaction in studying the *Hymenoptera*. The infinitely numerous insects which are classified under this important order are not celebrated for great elegance of shape, magnificent colours, and elaborate ornamentation, and they are usually small in size, and modest in their decoration. Some, it is true, are brilliant enough, but this is very exceptional, and they never attain a large size.

When the *Hymenoptera* are examined no one expects to meet with unusual beauties; but the interest they convey is far higher in its origin than that of simple decoration. The butterfly is admired on account of the richness of its colours, its elegant shape, and the magnificent wings; and a beetle is attractive on account of the dazzling brilliancy of its elytra, which may resemble the most lustrous metals or the finest jewels. But the butterfly and the beetle, which revel in all the decorative wealth of nature, lead useless and apparently purposeless lives, and their external beauty is their sole attraction.

The hymenopterous insect pillages the flowers; it has quiet tints and sombre colours; and its form and shape are simple in the extreme. This insect, which has not very elaborate ornamentation to make it attractive, soon commends itself to the close observer, whose delight in the tiny thing increases to a sense of great admiration, as its agile movements, indefatigable perseverance, and independent spirit gradually unveil themselves. What a wonderful contrast there is between the gaudy butterfly or the magnificent *Buprestis* beetle, and the bee! In the one case there is an existence of decorated idleness, and in the other a modest exterior enhances the charms of a laborious life, in which all the resources of the highest instinct are combined to a common end. Contrasts of this kind are not uncommon in Nature, and are not restricted to families, but are to be noticed in species, and even amongst individuals. There is a great amount of uniformity in the shapes, the physiology, and in the phases of the development of the *Lepidoptera*, and also in their methods of life and instincts. There are no grand differences, although there is an infinitude of peculiarities.

But the distinctions between the principal types or kinds of the *Hymenoptera* are most striking. These insects, when perfect, usually feed upon vegetable matters, but during their larval condition some live upon plants, and others remain within excrescences which they produce upon the leaves, roots, and stems of trees. Many larvae are absolutely carnivorous, and live within the bodies of all kinds of insects; whilst a vast number are provided for by their parents, and receive, as legacies, either a living thing to be fed upon, or a delicacy composed of honey and the pollen of flowers. There is every possible variety of food, and every imaginable method of life.

The hymenopterous larvae differ amongst themselves in their conformation and in their development. Those of an entire family are hatched sufficiently well developed to take care of themselves; they live in full daylight, and can walk, so that they may be compared to caterpillars. But others pass a hidden existence, beyond the help of the outside world, because they must live in the place where the eggs out of which they have come were deposited, and, indeed, they never care to wander. There are some kinds which are so ill-developed that they have no means of locomotion, and cannot even take their food themselves, so that they require mothers, or a staff of nurses gifted with industry and sufficient intelligence to find a safe place and plenty of proper nourishment for their charges.

Many *Hymenoptera* — unlike almost all the other insects—never abandon then their eggs, and the love of then offspring has been implanted in them by Nature. This obligation could not exist without bringing with it the necessity for, and the desire, of, labour; without exciting maternal love, and many instincts verging upon reason. The smallest amount of intelligence requires a very perfect organisation for its production, and so does a high order of instinct. Mechanical appliances, implements, tools, and industry are requisite to produce good work, and they are only given by Nature to those creatures which are intelligent enough to use them.

The *Hymenoptera* are the most highly endowed of all insects, yet they often live for a time in apparently most miserable conditions; and parallel instances are not uncommon in the vertebrate animal kingdom—for the most gifted birds can only at first exist, thanks to their parents; but many less favoured kinds, like the gallinaceous birds, for instance, run about and take care of themselves soon after they escape from the egg.

In spite of the great differences which exist between the principal types of the order now under consideration—differences in the shape and external character of the larvae and the adults; differences in the internal organisation; differences in the progress of development, in the kind of life, in the diet, and in the condition of existence of the species—all the insects of this great zoological division have common characters, which are readily distinguishable.

The *Hymenoptera* have four membranous wings (whence their name—υμηυ, membrane, πτεπου, wing), which are free from scales, and they are marked with more or less numerous nervures, which usually do not form a reticulation. The wings, which are rather small in relation to the bulk of the body, are usually perfectly transparent. Sometimes they have a violet hue; but more frequently they are dusky in colour. The wings characterise the *Hymenoptera;* and the ancients called them—as the moderns do—"flies with four wings."

These insects have strongly formed heads, and very large eyes upon the sides of them. Usually there are three ocelli on the forehead. The antennae differ in their general structure, even amongst the species of a genus, and, as might be anticipated, in families also. Their mouths are formed for nibbling or crushing, and also for suction. All the structures are free, and are more or less well-developed. The mandibles are always strong, but differ much in form, according to the habits of the species. The jaws and the lower lip, which are usually rather short, elongate in some kinds which pump up honey, and they then form a kind of trunk, which is doubled up against the thorax when it is not wanted. The thorax is swollen out, as it were, and its integument is solid and dense, like a cuirass; and the muscles which produce the movements of the wings, and which are attached to it within, are very large and strong. The legs have moderate powers; and, although these insects are better able to walk than the *Lepidoptera,* still they cannot be called, in comparison with others, either walking or jumping insects, their best means of locomotion are their wings; and the insects fully understand this, for even when they have to walk a short distance the wings assist the legs, and help the not very heavy body to move along. The only *Hymenoptera* that can be called true walkers are those that have no wings, like the neuter ants, and their agility is very wonderful. The legs are often provided with accessory structures, and are, moreover, modified so as to enable certain species to live particular lives.

Most of these insects have a very movable abdomen attached to the thorax by a very narrow waist, which is notoriously small, in the wasp tribe for instance. The females have a long and slender tube at the end of the abdomen, which is of great importance to them, either as an ovipositor or as a sting. The weapon of offence is invariably formed by the same kind of structures in all insects that possess it, and Lacaze Duthiers has discovered it in a rudimentary condition even amongst the *Lepidoptera*. These structures are modified in various manners, but the same end is always kept in view, and the resulting organ is of the greatest important.

The Hymenoptera undergo complete metamorphoses. Some larvae are usually soft and white, and, as they have no legs, they look like worms; but some exceptional kinds have small fore and hind legs like caterpillars. Almost all the larvae, when they have attained their full growth, shut themselves up in a cocoon, which they make up with the silken stuff they are able to produce, and then are transformed into pupae.

The principal divisions of the *Hymenoptera* are readily distinguished by characteristic species, so that their classification is comparatively simple, and some remarkable habits which are prevalent amongst some peculiarly constructed kinds appear to have definite relations with their metamorphosis.

THE HYMENOPTERA WITH OVIPOSITORS

Tenthredinidae and Siricidae

The celebrated entomologist Latreille distinguished the *Hymenoptera* which had the abdomen attached to the rest of the body by its entire width from those which had a slender, waist-like junction. This distinction, although not very important, is very handy. The species with thick waists are collected into two families, one very numerous in species and the other much less so. The principal, or first family, is that of the saw-flies, or *Tenthredinidae*, and the other the *Siricidae*.

The saw-flies are common everywhere, and are to be seen and watched whilst they fly, work, and crawl about in nearly every garden and wood. Once seen, the peculiar habits and odd manners of these active flies are never forgotten. Their body is short and compact, their mandibles are long, flat, and cutting, the jaws are short, feeble, and furnished with six-jointed palpi, and the antennae are either thread-shaped or are thick at the end or branched. The *Tenthredinidae* are particularly distinguishable on account of the movable, dentate, saw-like ovipositor which is attached to the abdomen of the females.

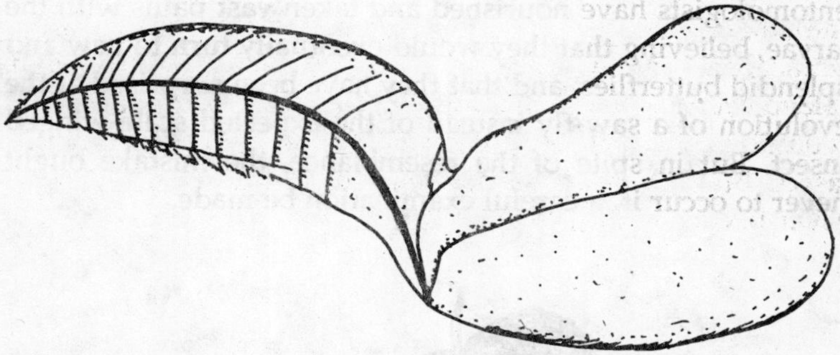

Fig. 5.1. The Ovipositor of the *Hylotoma rosea.*

This is a wonderful little instrument. It is in the form of a double saw, the length of which varies according to the species, and there are from fifteen to twenty or more regularly placed teeth added to the cutting edge. But the little saw of the *Tenthredo* is not always so simple in its construction as that of a carpenter, for the teeth or dentations are covered with others, so that the teeth of the saw are themselves toothed. The perfection of this minute weapon suggests that manufacturers might take a lesson from it and invent a cutting saw with double serration which might be of infinite use in the arts.

The double saw corresponds to the stylets of the sting of the wasp and bee. The two valves are kept fixed by being inserted or fixed into a scale which adheres to the sides of the abdomen; and the process that surrounds the stylets of the sting may be noticed in its rudimentary condition as a thin membrane in the saw-flies, but it is of no importance to them.

The *Tenthredinidae* have active dispositions, and are ornamented with various colours, and their bodies are smooth and shining. They do not consume much nourishment in the adult stage, and they then appear to prefer fruit, which they snip in pieces with the aid of their cutting mandibles.

The larvae resemble caterpillars in their shape and habits, yet it is hardly possible for any one to mistake them for the miniature form of the *Lepidoptera*. It is true that amateur entomologists have nourished and taken vast pains with the larvae, believing that they would eventually turn to new and splendid butterflies, and that they have been surprised at the evolution of a saw-fly instead of the expected scale-winged insect. But in spite of the resemblance, the mistake ought never to occur if a careful examination be made.

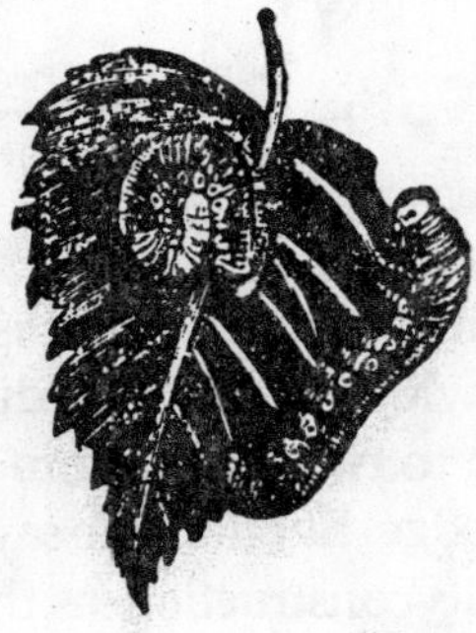

Fig. 5.2. Larva of a saw fly.

Caterpillars have never more than five pairs of membranous or hind legs, but the larvae of the saw-flies have generally seven or eight, and never less than six pairs. The feet have, however, circles of hooks, like the caterpillars. The head is always rounder and more globular, and the eyes are more wide apart and distinct, than in the larvae of the butterflies. The saw-fly larvae have, moreover, peculiar movements, and especially that of curling themselves up, and they do this for hours during the great heat of the day; if they are alarmed, they cock up the end of their bodies in a threatening manner, and the attitude doubtless frightens some of their disturbers.

The larvae eat the leaves of most plants, and they collect together in colonies, and often do a great deal of mischief.

When they have attained their full size, the larvae from cocoons with silk, and glue the threads together so as to make stout and hard cases; and some of them are to be seen stuck between the twigs and leaves or upon the small branches, whilst others are placed in the ground. These ground cocoons, however, contain pieces of stone and dust. Every caterpillar, as soon as its cocoon is made, begins to transform itself into a chrysalis; but the saw-fly larvae remain inert and huddled up for several months after having made their snug and safe hiding-places. The metamorphosis does not take place at once, but occurs sometime or other during the winter or early spring; and when it is complete, the fly escapes, usually by cutting the end of the cocoon with its jaws, and, should it be a female, it soon begins to use the ovipositor. It flies to and fro, seeking a proper place to deposit its eggs in, and takes a great deal of pains to select a spot where the nourishment of the future larvae will be at hand; and it is not simply satisfied, like the butterfly, with laying its eggs on or near the plant upon which the young are to feed. The saw-*fly* places its eggs in the very midst of the food, and protects them with extraordinary care. It makes a series of slits in the stalk of a leaf or in the twig of a shrub with its tiny saw, and deposits an egg at the very bottom of the cut, covering the opening with a secretion that hardens and keeps out the wet and the air. Some saw-flies only attack the petioles of leaves; others like sawing the hard twigs of trees; and each species has its particular method of egg-laying and wood-sawing, and each chooses a particular plant. Some large *Tenthredinidae* only carry a very delicate and small saw, for the eggs are to be laid in the very tissues of the leaves, which must not be torn or pierced through; and others, that are small in size, have great saws which are strong and double-toothed, for they have to cut into hard tissues; so that with a little trouble the naturalist can tell what sort of a saw a tenthredinate insect must have, by observing the habits of the fly; and he can satisfy himself, on the other hand, when he sees the saw, concerning the peculiar habits of its possessor.

All the *Tenthredinidae* are closely allied by their structural peculiarities; and some external characters have caused them to be grouped in two great tribes, which contain many genera.

The tribe, the species of which have short and stout bodies, contains the greatest number of saw-flies, and it is subdivided into four groups—the *Cimbicides,* the *Hylotomites,* the *Tenthredites* and the *Lydites.*

The *Cimbicides* are the largest of the saw-flies. They have thick bodies, and their antennae are swollen at the end so as to resemble a club, and have never more than eight joints. Their flight is heavy, and they make a great buzzing. Their larvae have a peculiar tough-looking skin, and have no less than nine pairs of membranous legs. The common German species, the yellow Cimbex *(Cimbex lutea)* is a large form, and may be considered to be the representative insect of the genus.

The difference between the males and females of this saw-fly is so great that many naturalists considered them to belong to separate species. The male is rather long in shape, and has large legs, its colour is brown, and the insect is marked with a bright yellow spot on the first segment of the abdomen. The female, on the country, is plump and rounded in figure, and has a yellow colour, the fore parts of the body being blackish. The adult insect flies in the month of May. The females, although they are large insects, have a very small ovipositor saw; but they do not require any other, for, instead of using it upon dense vegetable fibre and hard twigs, they simply incise the stalks of leaves.

The larvae are of a pretty green colour, and have a black ray down the back; they do not indulge in the threatening posture of lifting up their tails, but curl themselves up just like a sleeping cat.

There are some small saw-flies which are well-known, and which belong to the *Hylotomites;* their antennae are slightly swollen, and have from three to seven joints. The genus *Hylotoma* is the most important, and its species are very common in Europe. They usually have the greater part of the body coloured yellow, and the rest black; and the common rose saw-fly is a well-known example, and one that is especially hated by rose fanciers. It is to be found in every garden, and its habits can be watched with ease.

When all is bright and full of sunshine early in the summer morning, and when the rosebuds are longing to open and to display their magnificent colours, then is the time to watch the restless saw-flies that move here and there with great velocity, and rest every now and then on the rose trees. These pretty flies are about one-third of an inch long, and when they fly their wings measure nearly half an inch across. Their bodies are smooth, gleaming, appear highly polished, and have an orange tint, whilst their heads, antennae, and much of the body, are black, as are the ends of the legs and the pieces of the tarsus or foot. These *Hylotomae* belong to the species *Rosea,* but at first nothing unusual is observed respecting their habits. After a while the flies become more numerous, and as the day passes on some begin to think of laying their eggs. The females fly to and fro, and soon several may be recognised, for they are crowded and heavy with eggs. The slow-flying saw-fly may be noticed buzzing about the most succulent and healthiest of the roses, and it is so anxious to lay that it cares little for the spectator. It crawls and runs over the leaves and suckers, and looks here and there, and is evidently intently occupied in some important duty. Finally, the saw-fly finds a delicate succulent 'twig or stem, which it likes, and immediately the insect clasps it with its legs, bends the back, and places its head low. Then the saw suddenly appears under the end of the body, and is dug deeply into the green stem like a knife. Immediately the skin of the twig is pierced, the laminae which form the double saw are stretched slightly apart, and the gentle saw-like motion cuts deeper and deeper. Each saw cuts so as to separate from its fellow, and to widen the slit, whilst the rough parts of the outside of the saws scrape and enlarge the wound also. In a few seconds or minutes the slit has become large enough for the purpose of the saw-fly, and when this time has arrived the insect rests perfectly still for a short period, and the saws separate sideways and enlarge the slit; at the same time an egg passes down between them slowly but surely. The egg is laid at the bottom of the slit in safety, but more has to be done to ensure the future comfort of the larva and its freedom from danger. If the egg were simply left at the bottom of the wound the vegetable fibrous cells would soon close over and effectually imprison it for ever, and the young larva would be crushed before it was born, or starved. Every care is taken, however, to prevent this misfortune. The egg is hardly laid before a muscus secretion, provided by special glandular structures, escapes from the insect, and enters the wound and covers its sides; it acts as an irritant to the plant, and soon makes itself felt by the vegetable tissues, which it alters and indurates in a remarkable manner. No sooner is it deposited than the fibres become wider apart, and soon blacken, whilst the deeper-seated tissues become thickened and impaired, so far as their growing powers are concerned. The egg thus remains free at the bottom of the slit, and does not suffer from the growth of the surrounding tissues, thanks to the glandular secretion.

Egg after egg is deposited in the same manner, and with the same precautions, and we may often distinguish the twigs where all this has gone on, and which contain a long series of eggs in separate slits. Sometimes only three or four wounds may be counted, but ten or fifteen or more punctures are often made. The operation of laying is by no means easy, and it requires much care; so the female only deposits a few eggs every day and curiously enough it never attacks a second twig on the same tree. Day after day a new plant is chosen, and there is evidently a plan in this. Why should the insect avoid the tempting stems of the same bushes, and seek those of other rose tree? Simply because were it to lay too many eggs on one tree there would not be leaves enough for the progeny. What a wonderful instinct is this, that causes the creature to think for the future, and to reason more wisely than many men! The *Hylotoma* takes care of all its little ones in the most clever manner possible, and it leaves them enough to live on without starving each other—an example which might be followed by most of the creatures that consider themselves infinitely higher and more able than the tiny insects.

About eight or ten days elapse after the eggs have been fixed into the twigs of the rose, and then the young larvae are hatched, and they immediately crawl on to the leaves, and begin to eat them.

Their growth is rapid, and they change their skins time after time, but without the general shape of the body or its colours undergoing a decided alteration. The body of the larva is of a more or less dark yellow colour, and the sides are green. It is marked very generally with numerous black and shining tubercles, which are covered with hairs. The head is yellow, and the black eyes are surrounded with spots of the same colour.

These larvae resemble true caterpillars exceedingly, especially when they move from place to place, but their characteristic attitude is very different from those of the *Lepidoptera*. They are very prone to adopt the curious and striking posture which has been noticed already. Holding on to the plant by their fore legs, they stick up the end of the body, and if they are feeding with the body in the usual horizontal position, and are disturbed suddenly, this threatening attitude is forthwith adopted. Under different circumstances, and when they are at rest, they simply twist the last segments of the body underneath, but do not roll themselve up in a spiral like many other larvae of saw-flies.

By the end of the month of June the larvae of this interesting saw-fly have reached their full growth, and most of them walk off the leaves and twigs that have been so serviceable to them, and descend to the ground and hide themselves in the soil at a slight depth. Others do not leave the stem of their favourite tree, and not a few stop upon walls or upon the trunks of trees. Wherever they may hide or stop they form an oval-shaped cocoon, which is composed of silk joined together by a very glutinous secretion, but it never contains grains of sand or of earth.

The cocoons of the rose saw-fly are of an earthy yellow colour, and their singular construction was noticed in the last century by Reaumur. On the outside there is an elastic tissue which is able to resist very considerable shocks and knocks, and the microscope shows that it is made up of fibres arranged like the network of a racket mace. This is the first envelope, and the cocoon beneath it is made up of a soft, close, and flexible substance which is not adherent to it.

When the second metamorphosis is complete, and the *Hylotoma* has changed its pupa skin within the cocoon, it gnaws the soft layers with its mandibles, and cuts the stout outside threads with them, and escapes.

The second generation of the rose saw-flies appears towards the end of July or the beginning of August. They soon lay, and the larvae may be seen upon the roses during the whole of the autumn. As soon as the temperature of the season begins to decline, and before the fine days are all gone, the larvae descend to the earth and establish themselves in their cocoons in the soil, and none of them remain on trees or walls. They have the instinct to place themselves out of the danger of frost and cold, for the winter time is full of peril to them. After being snugly housed in their cocoons, where they have to remain until the next spring, they become dull and stupid, and hybernate as larvae. The long winter passes and the early spring finds the larvae in much the same state, but a few fine days stimulate their vital processes, and the metamorphosis into the pupa condition proceeds.

The saw-flies are a great pest to horticulturists, and yet a slight knowledge of the method of the metamorphosis of the larva would enable those generally very self-opinionated persons to save many a rose, and prevent much loss to themselves. The habit of crawling down the rose stem into the ground on the approach of winter, and of hybernating in the cocoon at the base of the little tree, should suggest the property of raking and thoroughly disturbing the superficial mould. By these means the cocoons can be easily seen and destroyed. There is another plan, which is not so ready however, and it is to notice where the eggs are laid, and to paint the twig over with some viscid liquid that will fill the cracks, and seal up the unfortunate larvae quite hermetically.

The rose saw-flies do not live in common, and each one makes its own cocoon. In the Brazils, however, there is *Hylotoma formosa,* which, according to Sichel, makes a nest in which many participate.

The *Tenthredites* are the most numerous of the say-fly family, and they may be recognised by their antennae not being swollen and not having less than from nine to fourteen distinct joints. Their habits and metamorphoses resemble those of the *Hylotoma,* and it is therefore, only requisite to select a few examples where there is a shade of difference, and where the structures of the insect are evidently modified for a purpose.

The *Athaliae* are small insects, and their colour resembles that of the *Hylotomae* generally speaking, but their antennae are sufficiently distinct as regards their construction to prevent the genera being confounded. They are finely pectinated in the male *Athaliae,* but they are not composed of more than from nine to ten joints, and this is the case in both sexes.

Althalia centifoliae, so well known to English entomologists, from Newport's magnificent descriptions and studies, is one of the commonest species. The larvae are of a greenish black colour and have eight pairs of membranous legs. They occasionally are to be found in great numbers on culinary vegetables, which they eat and waste remorselessly. United in numerous groups upon each plant, they devour the leaves and even the tenderest twigs. A closely allied species (*Athalia spinarum)* attacks cruciferous plants, and does a great deal of damage during certain years. The larvae of these small saw-flies undergo their metamorphoses in the earth, and as they have not much silk at their command they agglutinate earthy particles together, and make a rude sort of cell or cocoon.

The females lay their eggs by inserting them with their saw-ovipositor between the skin or cuticle of the turnip leaf and its cellular or parenchymatous structure. The egg is therefore only separated from the atmosphere by a most delicate membrane, so that the heat and moisture soon affect it and hatch the little *Athalia centifoliae* very rapidly. It appears necessary that this egg should be exposed to much heat in order to hasten the birth in time for the turnip to be enjoyed by the larva. Newport states that the cocoon of this insect is lined with silk.

The larvae of the genus *Selandria* are of a somewhat peculiar shape. Their legs are extremely short, and their bodies are lubricated with a viscid secretion which makes them look like small slugs. Reaumur, in fact, called them slug larvae. The larvae of *Selandria Aethiops* often become very injurious to cherry, pear, and other fruit trees. They remain perfectly quiet during the heat of the day, and seem to draw their heads within the first segment of their bodies, and they are them covered over with the slimy moisture so as not to look like insects at all. By night and during the early morning they are lively enough, and they then devour the foliage.

The *Lydites* form a small group of saw-flies which are characterised by having a great number of joints to their antennae. They are the saw-flies of the north, and they attack fir-trees and pines.

Fig. 5.3. *Lophyrus pini.*

The *Lophyri,* which belong to this group, have a thick body, and the males have doubly pectinated and feathery antennae. The species are very difficult to separate, and entomologists have had to take into consideration the disposition of the nervures of the wings, in order to decide upon particular forms with certainty.

The commonest species is *Lophyrus pini.* The male is quite black; the female is almost entirely yellow, and is spotted with black. The adult insect flies during the summer months, and the females saw the tissues of the twigs and fine stems of fir-trees, and insert their eggs. The females are so industrious, and are so prolific, that vast numbers of eggs are laid upon a tree, and the parent seems to know that there will be enough and to spare for its larvae. The young larvae live in groups, and collect in great numbers; they eat first of all the young shoots and buds, and finally do not despise the hardest foliage. Many of the trees they attack are soon terribly despoiled of their leaves, and have their existence endangered, and every now and then large forests of firs and pines are nearly destroyed by the ravages of this insect, which certainly chooses a strong tasting diet.

The larvae of the *Lophyri* do not dig into the earth in order to undergo their metamorphosis, like so many of the saw-flies; but they construct their cocoons upon the branches of the tree they have despoiled. The cocoons are grouped together on the stouter twigs of the trees, and are more or less oval in shape.

Fig. 5.4. The Metamorphoses of *Lophyrus pini.*

They are often found empty, and one end opens like a trap-door. This is because the saw-fly, when ready to escape, cuts a nearly circular slit in one end, pushes up the flap, and flies off.

The position of the oval shaped cocoons is shown in the engraving, and there are some perfect insects represented. The females are engaged in depositing eggs, whilst a male is flying. The larvae have the characteristic curl of the tail.

Some larvae of the genus *Lyda* live in groups, and like the *Lepidoptera* under similar circumstances, protect themselves by forming a net-work of silk for the whole colony. The pear-trees are now and then much damaged by *Lyda piri,* whose larvae collect in great numbers upon the leaves, and spin a web in common to protect themselves.

The *Cephines* are the second tribe of the saw-flies, and they have slender bodies. They live in the interior of plants when in the larva form, and thus imitate the peculiar habits of many caterpiliars, and, as might be expected, they present similar peculiarities as regards colour and the development of the legs. The *Cephines* in fact exhibit so great an arrest of the growth of the membranous legs that they surpass nearly all caterpillars in this respect, and, indeed, they are almost like worms. The species of the genus *Cephus* have a compressed abdomen, and the ovipositor of the female is hardly visible, whilst the antennae are swollen at their ends, and composed of twenty-one joints. The commonest kind is *Cephus pygmaeus*—a little insect of a third of an inch in length; it is black, and has a yellow border to its third, fourth, and seventh segments. It is very injurious to cereal crops. When the wheat crop is about to ripen, if some white ears are seen elevated above the others which are heavy, green, and somewhat bent, there is a tolerable certainty of not finding the grains but of discovering one or more larvae of *Cephus pygmaeus*. These small white larvae may be exposed by breaking the white ears, and they live in the powdery dust which they have produced by gnawing the corn, and by their dejections. These larvae appear to have a strong impression that the wheat ear is all very well for a temporary home, but that it is dangerous for a permanent one, for before the harvest the insect crawls down the stem and buries itself in the ground near the roots. There it makes a cocoon, and hybernates through the winter, out of the way of sickle and scythe.

Fig. 5.5. *Sirex gigas.*

The *Siricidae* are a more numerous family than the last, and they may be known by their long bodies and short thick mandibles; the antennae being thread-like. The principal genus is *Sirex.* The females of it have a long, straight saw, which is toothed, for it has to pierce something harder than leaves and rose twigs. The species are more common in Germany, Northern Europe, and North America, than in southern districts; and they frequent the forests of firs and pines. The great *Sirex— Sires gigas—* is a splendid insect : The female is black and yellow, and attains the length of an inch without reckoning the ovipositor, which is half an inch long. This long instrument is used to perforate the bark of fir-trees, and the larva finds itself in the midst of its food.

The genus contains species that have been called by the name of *Urocera;* and they have, of late years, excited no small astonishment in the minds of many military men, for some of these insects took a fancy to eat lead and gnaw bullets. Marshal Vaillant, a Russian colonel, M. Motschulsky, and the director of the museum at Vienna, M. Koilar, have asserted and proved that the leaden bullets made for the French army during the Crimean war were riddled by *Sirex juvencus.* These insects certainly got out of the boxes, in the wood of which—green when they were made—the larvae were included; and they were found in the middle of the bullets, gnawing away, and perforating the lead with their strong mandibles.

Oryssus coronatus is one of the *Siricidae,* and the female has a thin saw, which is folded underneath the abdomen; but, unfortunately, it metamorphoses are unknown.

There is a remarkable saw-fly *(Tenthredo Lucorum)* which lays its eggs in the hawthorn trees soon after they have pushed forth their tenderest leaves. It is a large fly, and is of a dusky-brown colour, and about the size of a wasp. The female has a small and very thin laying apparatus, and she chooses the upper surface of a leaf, and pokes the edge of the ovipositor just under the cuticle; she then moves the saws gently, and expands them, moving them sideways. As the saws are drawn out of this most delicate wound, an eggs is left in their place, and gummed in by a viscid secretion. The female lays several eggs, and they are pale and small. After a while, the leaf grows, and is well nourished with the juices of the plant; and the eggs grow also, being close to the respiratory organs of the leaf, and in the midst of the nutritious fluids. The larva can be seen in the egg, curled up, and turning over and over, after the manner of all embryos. It is hatched after a while, and it crawls upon and devours the leaf. The thorns suffer a great deal from it, and were it not for a fellow hymenopterous insect that lays its eggs in the larva, they would have much more damage done to them. When the larva has attained its full growth, it makes a very woody cocoon, and sticks it to the twigs of the tree upon the leaves of which the insect lives. The cocoon is brown and very hard, and the larva hybernates in it, and is transformed in the spring. The fly gnaws off the top of the cocoon from within, and escapes; its jaws, like those of most others, being made solely for this purpose. The tom-tits get a great many of the larvae by nibbling the cocoons during hard winters.

THE HYMENOPTERA WHICH PRODUCE GALLS

Galls Flies (Cynipsidae)

The *Hymenoptera* that produce galls, or excrescences upon plants, form a group by themselves. They are small insects, very distinguishable, and they stimulate a great number of different kinds of plants to develop thŏse unusual growths of their tissues which are called gall-nuts, galls, oak-apples, and C. Everybody has seen these curious and odd-looking excrescences; but very few have noticed the beings that have grown within them—and which, when they have attained their perfect form, escape and fly. A winged insect not much more than a tenth or a twelfth of an inch long is not readily observed, especially when amongst the host of little flies that continually move around us.

These small *Hymenoptera* belong to one family—that of the *Cynipsidae*—which contains the genus *Cynips*, and others very much like it. The *Cynipsidae* have an oblong and very convex body, the abdomen of which is attached to the thorax by a very thin pedicle. The saw ovipositor is very curiously made; it is very long and slender, and is twisted up in a spiral form within the abdomen of the female when it is not required. But when the insect has made up its mind to lay an egg, having chosen the proper spot, the muscles of the abdomen suddenly unfold this spiral, and the ovipositor is straightened, and thrust instantaneously into the plant.

The *Cynipsidae* are found everywhere in spring and summer. When about to lay, the females take great trouble to seek a good place on a proper plant : they make a puncture with their long ovipositors in a twig or leaf and deposit either one or a considerable number of eggs. When the puncture is made the insect spreads a secretion over the wound which is of a very irritating character, and it is believed to excite the nutrition of the vegetable tissue to develop irregular masses of its cellular structures. There are multitudes of plants attacked, and generally the shape of the gall differs in each, and is produced

Fig. 5.6. Galls of Cynips *termunalis.*

by a different insect. Hartig, Westwood, Giraud, and others, have described some very unusual forms. The oak trees afford a refuge for, and nourish many kinds of gall-flies, which do not spare the trunks, the leaves, the twigs, or the roots. Oak-apples, or the rounded masses attached to oaks in the summer, some small and others large, readily attract the attention; they are galls that have been produced by a *Cynips.* They might be taken to be the fruit of the oak, and they are to the eye miniature apples— oak apples. The largest are always at the base of the leaves, and within them there is a cavity tenanted by one larva; it remains there in a dull and stupid condition during the winter, and is transformed into a pupa in the spring. The adult insect is developed within the cellule, and is obliged to cut its way out into the world with its mandibles. This *Cynips* is of a bright brown colour, and is called *Cynips quercus baccarum.*

The small oak-apples that usually are found in plenty on the lower surface of the leaves are formed by another species— *Cynips guercus folii.*

The branches of the oak in the spring-time are often covered with nodules, some of which are very large and irregular in shape. Their surface is smooth, their colour light green, passing into red in some spots, and there are usually several on a twig close to each other. They are constructed on a different plan to those of the gall-leaf insects, for they contain twelve or fifteen cellules, or more, each of which contains a larva. These kinds of *Cynips* fly in the middle of the summer, and there are probably two generations in each year. They are called *Cynips terminalis,* on account of the peculiar ornamentation of the female. The sexes differ much in their shape and colour; the male has large transparent wings, and its body is of a uniform bright fawn colour, whilst the female has no wings whatever, and is brownish, the end of the abdomen being of a shiny black tint.

The large engraving shows the galls of the insect. They are cellular, and are aggregated in masses upon the twigs.

A very common species of *Cynips*—*Cynips aptera*—produces great galls upon the roots of the oak, and it is a most remarkable thing that neither the males nor the females should have wings.

The gall nuts of commerce, which are used as dyes, and to make ink and tinctures, and whence gallic acid is derived, are produced by the punctures of *Cynips gallae tinctoriae.* These insects affect an Eastern species of oak, *Quercus infectoria,* and their galls are remarkably hard, round and tuberculated. They only contain one larva. The cellule is moderately capacious, but the walls are very thick; nevertheless, the little *Cynips* perforates them without much difficulty, in order to come to the light of day.

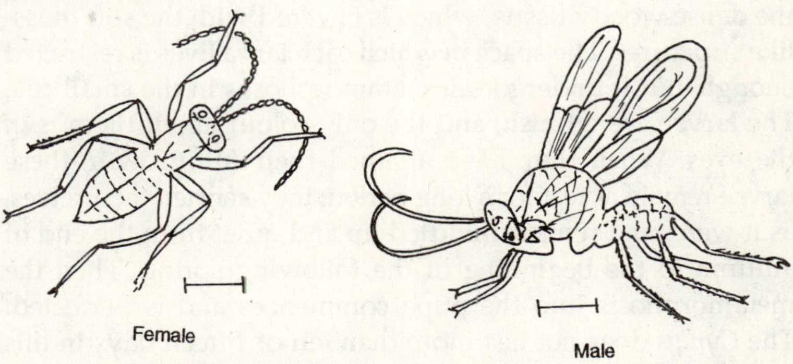

Fig. 5.7. *Cynips ternubakus.*

The rose fungus is one of the most curious galls; it is produced by a *Cynips*, and is found upon the hedge roses and sweet-briars. The adult insect which produces these excrescences is about the fifth of an inch long. Its transparent wings are slightly colouded, and it is of a glossy black colour. The females lay their eggs towards the end of May or the beginning of June, and the galls soon form, but they grow at first very slowly, and subsequently quickly, so that they attain their full size on the approach of cold weather. Sometimes the galls are found, or flat, at others they are irregular in shape, and resemble medlars, and they are about the same size as those fruits. They are moss-like in texture, and appear to be composed of branching filaments, which are placed closely together. Hence their name of "hairy galls." There is a sort of stem or peduncle to the galls, so that when they become variously tinted with green and red colours towards the end of summer they look like pretty fruit. During the winter time these colours disappear, and the galls become uniformly brown.

The mossy surface of the rose fungi is so soft and compressible that it appears at first sight to form the whole of the gall, but this is not the case, for there is a woody texture beneath which is very hard indeed. On cutting the gall across it will be observed that the larvae inside are admirably protected against injuries from without, for their cells are in the midst of the dense woody tissue, which is covered with the soft moss-like structures. The space in which each larva lives is restricted enough, but it undergoes it metamorphoses in the small cell. The larvae are whitish, and the only colour about them is in the eyes. When they have attained their full growth, these larvae remain at rest for a long period; they shorten themselves, as it were, and remain huddled up and quiet from the end of autumn to the beginning of the following spring. Then the metamorphosis into the pupa commences and is perfected. The *Cynips* does not last more than ten or fifteen days in this form, and then the second metamorphsis takes place. But if the spring is not genial, and if the temperature is low, the perfect insect does not get out of its cell, for it remains within waiting for the fine weather. When the warmth of the season is sufficient for the well-being of the *Cynips* it gnaws its way out of the gall and escapes.

There is a *cynips* that undergoes its metamorphosis in the fruit of fir-trees in Southern Europe, Africa, and the East, and the perfect insect eats its way out of the fruit when the ovules within are being fertilised with pollen; consequently, this insect brings out with it a quantity of the pollen in the form of dust.

Now this has been taken advantage of by practical men, and figs that are backward and small are stimulated to grow artifically by fixing a fruit containing the *Cynips* upon them.

The *Cynips* comes out of the fig covered with pollen, and immediately pushes its way into the ill-developed fruit, the ovules of which it fertilises; and then the growth of the fruit proceeds rapidly.

The so-called, Dead Sea fruits are galls which have been produced by a *Cynips*.

THE PARASITIC HYMENOPTERA

Ichneumonida, Ckalcididae, Proctotrupidae

The parasitic *Hymenoptera* from a little world of themselves, and they have the same habits, instincts, and metamorphoses, although their numbers are enormous. They form the insect array which Nature employs to check the superabundant multiplication of the phytophagous or vegetable-feeding creatures of the class. Every parasite attacks a particular species or some closely allied kinds, in obedience to some mysterious law; and many an unfortunate caterpillar is subject to the violence of several of these dangerous enemies. Nearly every insect has more than one parasite, and this fact enables us to estimate the multitudes of *Hymenoptera* that are in the world.

These *Hymenoptera* belong to three very distinct types; but each of them has female which are furnished with a straight, thin, and sharp ovipositor. These parasites are very pretty and elegantly formed insects when in the adult form, and are gifted with great agility and restlessness. But in their early condition they cannot move, having no locomotive organs, and their structures are so soft that they are destroyed with the greatest ease. The larvae look like worms or maggots, and do not attain a great perfection of development during their growth. The contrast between the lazy and sluggish larva of the active and restless adults of the parasitic *Hymenoptera* is almost as great as that of the larva and the imago of the industrious *Hymenoptera*, such as the bees.

All the parasites seek out a caterpillar, a larva, or an insect which suits their purpose, in order to lay an egg within its body. A larva which is born from this egg is nourished by the blood and fat of the victim, whose vital organs it does not touch or injure in any way, for were it to die the parasite would come to an end also. It is only when the larva is nearly full grown, and is about to undergo its metamorphosis into a pupa, that it appears to know that the life of the victim is not likely to be of much further use. It then devours the internal organs of the unfortunate insect, and undergoes its transformation. The skin of the victim protects some of the pupae of its destroyers after all the inside has been eaten.

Nearly all—if not quite all—insects are subject to the attacks of parasitic *Hymenoptera*. Fine, smooth, and brightly-coloured caterpillars often have a black spot upon their skin, and this is the healed wound of the ovipositor of one of the parasites.

Sooner or later the creature is sure to die, and it never reaches the stage of growth when it can lay eggs or reproduce its kind, for before this time the growing larvae within destroy it, as it were, by a slow consumption. Some affected caterpillars die soon, others nearly reach their full growth, and a few undergo their transformation into the chrysalis state before death. It is, therefore, not an uncommon thing for a butterfly collector, who hopes to see a fine moth disengage itself from its pupal covering, to be disappointed by the appearance of several little parasitic *Hymenoptera* that have been living within the chrysalis he has been keeping.

Although we know a great deal about the economy of the parasitic *Hymenoptera*, still there are some points of the greatest possible interest in it, and which are really very provocative of wonder. When a parasitic insect of this family discovers a caterpillar feeding on a leaf in broad daylight, there is nothing very wonderful about it, because we could do as much ourselves; but when it is evident that the female parasitic *Hymenoptera* finds out a larva which is situated inside a fruit, and within a branch or trunk of a tree, and perfectly out of sight, we may well-wonder how this is done. When one sense fails, another is supposed to supply its place, and the sense of smell of the insect is thought to do what the eyes evidently cannot perform. It is probable that the sense of hearing may assist in the discovery, but it is by no means proved. There is another instinct which is very remarkable, and, indeed, as curious as that just mentioned. A large parasite deposits only one egg under the skin of a caterpillar or other insect, for its larva, as it increases in size, requires all the juices of the victim. If two or three eggs had been laid, and larvae had been born from them, they would have been starved after a while, for the body of the infected insect would have been consumed twice or three times too quickly. But if a moderate sized parasite attacks a tolerably large caterpillar it introduces two or three eggs, and if a very small *Ichneumon* meets with a fine larva it may lay as many as fifty of sixty eggs in it. There is always enough and to spare in these instances, and it would appear that the insects considered the future. But there is a great sameness in the habits of insects, and these parasites choose their victims upon the same plan generation after generation. If the large parasite came across a new caterpillar three times as big as any it ever saw before, it either would pass it by, or would still only lay one egg. On the other hand, a very small parasite would always lay the same number, whatever the size of the victim might be.

All insects are not equally exposed to the attacks of parasites; for instance, those that know how to form shelters are less so than those which live in broad daylight on leaves, and hairy caterpillars are less liable to suffer than the smooth kinds. It would appear that the long tufts of hairs and the branching spines which cover the bodies of so many larvae are the structures which protect them, like an armour, from the sudden and eventually fatal stab of the *Ichneumons.* The peculiar movements given to the hairs by the contraction of the segments of the body evidently prevent the parasites from giving the stab with certainty.

Larvae are especially subject to the attacks of the parasitic *Hymenoptera,* and there is no difficulty in understanding why they should be chosen as the homes for the future young of such creatures as the *Ichneumons.* The life of adult insects is usually brief, and is cut short by many accidents, so that were the larvae of the parasites developed within them, their existence would be constantly in danger, and very few of them would come to perfection. It is very wonderful, however, that those insects which do live longer than others in the adult form, should be subject to the attacks of parasites, whilst the others, whose existence is invariably short, should never be pierced by the *Hymenoptera.* Many *Coleoptera,* which are well covered with a hard integument, and which, like men in armour, care little for the attacks of their fellows, succumb to the active lancer-like parasitic flies. The weevils, for instance, are long-lived insects, and are encased in armour, but their active enemy finds out the joints, and sticks in its ovipositor between the articulations of the body, and lays the egg, which grows like a canker within.

The parasitic *Hymenoptera* are of all dimensions; many are of considerable size, but the name of the minute insects is indeed legion. So small are some of their larvae that several can be accommodated within the body of an *Aphis* or plant louse, and even in an egg of an insect. The eggs of many *Lepidoptera* are destroyed by the larvae of the parasitic *Hymenoptera* that have become developed within them from the eggs introduced by the ovipositor.

Almost all these parasites introduce their eggs under the skin of their victims, so that their larvae are never seen, but there are some kinds which only lay their eggs on the outside of caterpillars, and other insects. When the larvae of these last-mentioned parasites are hatched they pinch up the skin with their mandibles, and force the front part of their heads within the victim's body, and remain in that position. They then suck the juices of their prey, and the greater part of their body is exposed to the air.

Since nearly all the parasitic *Hymenoptera* lead the same sort of life, we might expect not to find any great structural differences amongst them. There is, however, a considerable variety in the length of the ovipositor of the females. The females which deposit their eggs within the bodies of caterpillars or larvae that live in the open air have very small ovipositors; those which attack larvae which are underground, or are protected by some shelter, have longer instruments; and those that have to reach insects which live within the trunks of trees, have very large and greatly developed egg depositors; so that the *Ichneumon's* habits and prey may be guessed by the length of the ovipositor.

Nature employs these parasites constantly when insect life is too luxuriant; they stop the multiplication of a crowd of creatures that would do an infinitude of injury, and they limit the numbers and the geographical distribution of many species. When the insects which are most injurious to agriculturists are studied, the important office of the parasitic *Hymenoptera* becomes most evident, and they evidently render us most important services, for they check the ravages of the larvae which destroy cereals, vines, and vegetables of all kinds. The balance of life between the vegetables, the cterpillars, and the parasites is most extraordinary. A few caterpillars come from the egg of one butterfly, perhaps in a newly cultivated district, and they are not noticed. The next year hundreds of caterpillers are found devouring everything, and then one or two stray *Ichneumons* come upon the scene. They are so clever that they never miss their aim, and they lay thousands of eggs in as many caterpillars. The year following there are fewer caterpillars, and a great host of *Ichneumons,* that riddle every insect that comes in their way. Spring comes round again, and there are no caterpillars, vegetation flourishes, and then comes the turn of the parasites. Their office is completed, and they pass away. It is thus that armies of caterpillars suddenly appear and disappear.

There is some difficulty in applying the word parasite to all insects that live in intimate relation with others. The parasite just noticed lives at the expense of the vital juices of its victim or host. It does not devour the body, and then remain glutted. But many insects, like human parasites, only eat the provisions which others have laid up in store for the rainy day, or for their offspring, and these, of course, should be called the true parasites. Others live with their strange companions, and look out for scarps, and what happens to come in their way, these are comrades, and are very common in many families of the *Articulata.*

The *Ichneumonidae* have, as a rule, long thin bodies, slender legs, very large wings, which are much veined, and long, slender, vibrating, thread-shaped antennae. Agility is written in every structure of the *Ichneumons;* the jaws have long palps; and the slender body, large wings, and the the legs, which can be used for running, indicate unusual locomotive powers. The lively appearance of the insects is heightened by their projecting eyes and the long and restless antennae. They are not to be caught readily, but rush off with wonderful rapidity out of the way of danger, and if a female is seized she endeavours to stick her ovipositor into the fingers of her captor.

There are two very distinct tribes of *Ichneumonidae;* the *Ichneumons* proper and the *Braconidae.* The insects of the first tribe have four joints to the palps of their lower lip, and those of the last have only three.

The species contained in the first tribe are usually tolerably large, and the abdomen is rounded at its sides; their ovipositor is very small, and does not project from the hind part of the body when the insect is at rest. They introduce their eggs into the bodies of caterpillars, and other larvae which live in full daylight. The genus *Ichneumon* forms part of this tribe, and its species have the abdomen thick, and attached to the thorax by a waist. These insects are elongate in form, and are usually decorated with yellow or red bands and spots, which are rendered striking by their black ground. The *Pimplitae* are those *Ichneumons* which have the abdomen slightly narrowed at its commencement, and very projecting ovipositors; and the females of some kinds, especially of the *Ephialtae,* have them longer than the rest of the body.

The black *Ephialtes,* a common parasitic *Hymenoptera* of the centre and north of Europe *(Ephialtes manifestator),* may be seen during the summer time in the avenues of woods, and resting on flowers, or running upon the trees. The whole of its body is of a shining black colour; its transparent wings are slightly clouded, its legs are long and slender, and are of a bright red colour, the hind ones being brownish. The male is thin, and rather small, but the female is very robust, and comparatively large. The larvae of the *Buprestis* beetles, which live inside the roots of trees, are afflicted by the females of this fine parasite, and there are few things more interesting than the sight of the egg laying. In very fine weather, when the sun in shining in great power, the insect appears full of animation. Its antennae are perpetually vibrating and moving in different directions; its wings are trembling, and it is constantly running over the trunk of a tree with great rapidity. Every now and then the insect stops, runs to the right hand and to the left, and goes over the ground like a hound that has lost the scent. Then it seeks another locality, and the same vivacious proceeding is repeated a hundred times. Suddenly the insect appears to be satisfied that it has made a discovery it stops, and commences to dig its long and curved ovipositor into the bark; there is a larva of *Chalcophora Mariana* in the wood, and the *Ephialtes* has become aware of it. The insect lowers its head and grasps the bark with its legs; it then lifts up its abdomen, bringing the ovipositor beneath, and plunges the point within a narrow fissure, so that it reaches the larva, which is grubbing away, in apparent safety, at the rotten wood. When the stroke is given, the valves of the ovipositor separate, and curve, so that the thrust, when it overcomes the resistance of the wood, penetrates far inside. In the engraving, which is, of course, a diagram,

ICHNEUMONS.

Fig. 5.8. *Ephialttes manifestator*. The male is flying on the left, and the female is introducing an egg into the body of a larva. Another *Ephialtes, Rhyssa persuasoria,* a female, is on the branch to the lef hand.

this operation is shown, the wood being removed over the larva. The curve of the ovipositor beneath the abdomen is apparent. All this is very surprising, for the active insect manages to discover the deeply secreted larva, which it cannot see, and to find out a convenient fissure in the wood down which to thurst its ovipositor with certain effect.

The *Rhyssa persuasoria* has the same habits as the *Ephialtes,* and is a much prettier insect, being marked with little spots, or lines of a yellow colour.

The *Ophionidae* are *Ichneumons,* which can be distinguished at once by the peculiar shape of the abdomen, for it is bent, and compressed, literally like a sickle. They have a small ovipositor, and deposit their eggs either within caterpillars that feed on leaves in broad daylight and are unsheltered, or upon their skins. The eggs are somewhat remarkable, and have been carefully examined. They are oblong, and have a long and twisted peduncle, and this is fixed on to the skin of the victim. The young larva, on being hatched, breaks its egg shell on the side remote from the peduncle, allows its body still to remain within the pedunculated shell, and thus attacks the caterpillar in safety. Sometimes the female misses the caterpillar, and the egg sticks on her own body, so that when the larva is born it at once attacks its own parent. Most of these kinds are large insects, and *Ophion luteus* is very common in Europe.

The *Braconidae* are a very large tribe, and are the commonest of the parasitic *Ichneumonidae.* They multiply at a great rate, and most of them are very small indeed. Many of them frequent flowers during fine weather, and are enemies of the beetles, and the species of the genus *Alysia* are parasitic upon the great family of the flies *Muscidae.*

The Microgasters belong to a group which differs from that of the *Braconites,* and are very small insects, being about the twelfth of an inch long. They have no less than eighteen joints in their antennae, and they deposit a quantity of eggs inside many species of caterpillars.

After having lived in the juices of the caterpillars, the larvae of the *Microgasters* having attained their full growth, begin to devour the viscera also, and finally they eat through the skin. They then spin silky cocoons, and undergo their metamorphosis. One of the species, *Microgaster glomeratus*, which is black in colour, and which has fawn coloured legs, can be recognised easily, and its metamorphosis examined. The caterpillars of the cabbage butterfly, *Pieris brassicae*, are frequently killed by this little parasite, and occasionally nearly every one of these destructive vegetable feeders falls a prey. They usually live until about the time of the first metamorphosis, and then they begin to look out for a safe place to hang themselves up, and to undergo the transformation into the chrysalis state. But it is just then that they die, for then the *Microgasters* which have had enough of the caterpillar, pierce through its skin, and form their cocoons on and around it. They complete their metamorphosis in a few days, and the perfect parasites fly. The cocoons of these *Microgasters* are usually united in packets, are oval shaped, and being composed of very fine yellow silk, look like those of silkworms in miniature. The smallest of the ichneumons are the species of the genus *Hybrizon*, and they wage war against the *Aplides*, one of them laying its egg in the *Aplis*, which is so common on the delicate stems of roses.

The *Proctotupidae* are distinguished from the *Ichncumonidae* by their oblong body, the simple condition of the nervures of the wing, the moderately long antennae, and the long pendant palps. They are such little insects that they must be examined with a good lens or a microscope, if they are to be studied. One tribe of them attacks the maggots of flies, especially of the wheat fly and another lays its eggs in the larvae of the *Tipulae*—the daddy longlegs tribe. Some of the parasites of the maggots spin cocoons beneath the skins of their victims. The *Teleas ovulorum* is a tiny insect of about a fiftieth of an inch long, and it lays its eggs in the eggs of moths, and when its larvae are developed they enjoy their curious and small habituation.

The species of a most important family of the *Hymenoptera* have small palpi, the antennae consist to twelve or thirteen joints, and the nervure of the wing is simple and bifurcate; they are numerous, and the members of them are very common. These *Chalcididae* are small creatures, and attack all sorts of insects, and do not even spare the other parasites.

The species of the genus *Chalcis* are common, and the individuals are larger than those of the other kinds. They may be known by the peculiar shape of the antennae, by the enlargement in the thighs of the hind legs, and by the prominent ovipositor. *Chalcis minuta*, which is very common in some localities, lays its eggs in the bodies of different kinds of caterpillars, and it is especially fond of the *Pyralis*, which is so troublesome to the vine growers. Nine other species attack this pest also.

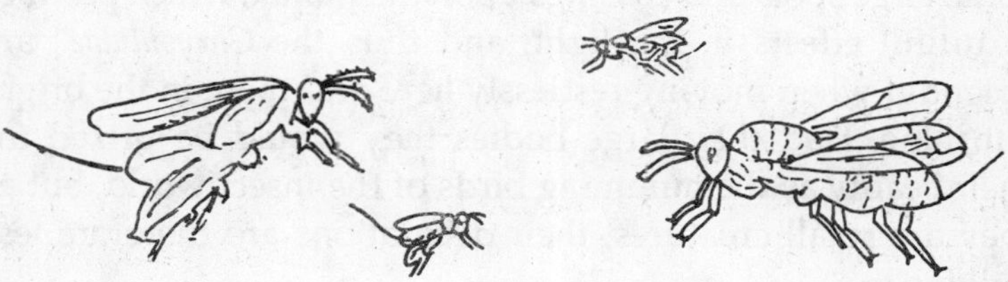

Fig. 5.9. The Adult Female of *Diplolepis bedeguaris* — The Adult Females of *Chalcis minuta*.

The *Diplolepidae* have the body magnificently coloured with golden green tints, and their ovipositor is very long. The females usually seek out the larvae of the species of *Cynips* which are enclosed in galls. The larvae of the rose fungus upon the sweet briar are frequently attacked by the insect represented above, and a genus of beetles *(Anobium)* has its members killed and preyed upon by a *Perilampis*.

A small host of *Pteromalidae* may be seen to issue from the body of a chrysalis of a *Vanessa;* and the tiny *Eulophae*, which are perfect insect gems, glowing with golden green and bronze tints attack caterpillars, maggots, and the eggs of butterflies and moths indiscriminately.

HYMENOPTERA WITH DENSE INTEGUMENTS

(Chrysidida)

This family of *Hymenoptera* cannot be very readily associated with the others. The *Chrysididae* are commonly called ruby-tail flies or gilded wasps, and they have a close resemblance on account of their peculiar shape to tiny wasp-like flies, but they never attain a third of the size of the common *Vespa*. The brilliant colours of the *chrysididae* strike the observer at once, and their fiery tints look like a blaze of jewels. Some species have the body tinted with a beautiful golden green, others shine in ultramarine, and many have this splendid blue relieved by a scintillating ruby red colour on the abdomen. There are some large spots and chasings upon the thorax, which produce beautiful effects in the light, and thus the *Chrysididae* are splendid when moving restlessly here and there in the bright sunshine. Had they large bodies they would be prized for their beauty as the humming birds of the insect world, but as they are small creatures, their decorations are therefore less striking.

The integuments of the *Chrysididae* are very thick and hard; and the insects have the power of rolling themselves up like a ball, by bringing the abdomen against the underneath part of the thorax. This is quite a peculiarity amongst these *Hymenoptera;* moreover, they differ from others by having a cylindrical body, the abdomen being concave below, and very convex above, and attached to the thorax by a very short waist. The rings of the abdomen appear to shut up more or less within each other, like a telescope slide. The females carry a very sharp sting, which gives great pain. These insects have filiform attennae, long mandibles; the jaws have palpi, formed of five joints; and the lower lip projects, and is membranous, so that they can suck. The wings are moderately veined, and the legs are slender. These insects exhibit a wonderful amount of activity under a broiling sun, and then they are never still, but are constantly flying here and there, and resting a while upon the flowers. In some spots of central Europe—where these insects are much more common than in the north—they collect in numbers, and afford a magnificent spectacle, and they resemble fiery pearls rushing to and fro. They have no industrial habits, and do not lay eggs within other insects. Their larvae, however, only live upon living prey, and the instinct of the female leads her to lay her eggs where the future offspring will be protected against danger. She has no hesitation about laying them in the nests of the other *Hymenoptera*, and especially where there is some provision being laid up for the larvae. She does not care for the solitary bee or a *Bembex*, for she has a terrible sting and a hard cuirass, which is not to be pierced by ordinary insect weapons.

The female introduces herself into the nest of the bee, for instance, and rolls herself up if attacked, and then, being victorious, lays her eggs on the food which has been stored up. The little parasites are hatched, and live upon this food with the larvae of the original tenant, and when the store has been devoured they attack the larvae, and suck their juices. They make a cocoon before undergoing their metamorphosis, but very little is known before their transformations.

The *Chrysis ignita* is very common in England, and may be seen in constant motion, running upon walls and palings; it lays its eggs in the nests of hornets, sand wasps, and solitary wasps.

Some beautiful insects of the genus *Hedychrum* have habits like those of the *Chrysis* just mentioned. Lepeletier de Saint-Fargeau describes the attack of a sand bee upon one of them which had invaded its dwelling in order to lay eggs. The *Osmia*, a sand bee, discovered the intruder in the act, and immediately proceeded to turn her out by laying hold of her with its mandibles. But the *Hedychrum* rolled herself up like a

Fig. 5.10. Chrysidid/Æ
Chrysis ignita, Stilbum calens, Parnopes carnea, whose cocoon is to be seen enclosed in the larva of a *Bombyx.*

ball, and was invulnerable. The bee carried her out, gave her a good shaking, bit her wings off, and left her. She had her way for all that, and crawled back again into the nest, and laid her egg.

The *Stilbi* have a very convex abdomen, and are inhabitants of warm climates, but a species is found in Central France. The species of *Parnopes* are green insects, with flesh-coloured abdomens. *Parnopes carnei* lays its eggs in the nests of the *Bembex*.

THE ANT FAMILY (FORMICIDAE)

Ants are so common in nearly every part of the globe that they are very well-known insects; nevertheless, hey have all sorts of characters given them. They are considered destructive and encroaching, as well as disagreeable things, and they have, as a rule, a very doubtful reputation amongst the public in nearly every country; but naturalists and those who inquire a little more carefully into the mysteries of the insect world, are always ready to admire their wonderful instincts. In fact, the ants posses those marvellous gifts in as great a degree as any other beings. The ancients were struck with their admirable underground works, and from the earliest times imagination has assisted those who look superficially into the natural history of these little creatures, at the expense, however, of truth. The curious idea — which appears to have commenced in very remote times, and to have been carried down by tradition, and which was assisted by the results of careless observations — concerning the habits of the ants in collecting and storing up provisions, as it were, under the influence of a wise foresight, is evidently incorrect. But although this foresight is not developed in the ants, they still have a few resemblances by which they can be compared with men, so far as social habits are concerned. The same is true for wasps and bees, and many other insects, but when the matter is carefully looked into, the analogy between the insect and human society is much less than might have been supposed. It is evident that all the individuals of a tribe of ants act in concert to carry on a common work, which is invariably effected by means of all, or at least of the greatest possible number of the insects. This is certainly wonderful, but there is nothing like what is observed and which appears to be inevitable in human society, namely, a hierarchy composed of individuals which command, and of others who obey. To all appearance, the most perfect equality exists between all the individuals of insect societies; no one commands and does not work, for they all work, and each insect appears to understand what is required of it, and affords assistance to others when it is necessary — being influenced by the instinct of doing a certain definite duty; at least this appears to be consistent with correct observation. Many authors have imagined that the industrious insects that live in communities obey certain chiefs, and have, as it were, a sort of government. The opinion appears to have originated in a very simple manner—because such a state of things exists amongst men, and appears to be necessary; but all the best established facts relating to insects which live in communities, such as ants, wasps, and bees, prove that every one has an equal share in the government of the commonwealth, and that the so-called kings and queens, which are the objects of particular attentions, have no authority, and are in no way concerned with the occupations and labours of the rest of the tribe. The ants, like almost all the *Hymenoptera*, are born in a very rudimentary condition, for their development within the egg has not progressed much. The larvae are even incapable of taking the food which is within their reach, for their mouth-pieces are very small and feeble, so that it is necessary that the nourishment should be placed inside their mouths. Thus they may be said to be constant objects of care and anxiety to others of the community. The nursing, which is attended to with wonderful perseverance, is carried out by the workers, who are first-rate nurses, and at the same time admirable architects and builders of spacious and comfortable nurseries.

The united species of ants constitute the family of the

Formicidae, and are very readily distinguished from other *Hymenoptera.* They have a triangularly-shaped head, bent antennae, the first joints of which, always very long, are stalk-like; they have a large upper lip, strong mandibles, and the jaws and lower lip are short; the legs are long and slender, and the abdomen, which is more or less oval, is attached to the thorax by a short and narrow waist. The ants without wings, which are most commonly observed by everybody, and which run about roads and paths, and which have a very narrow thorax, are the workers, or neuters; they crawl over plants and up walls, and are found everywhere. But the males and females are only to be noticed at certain times of the year, and can be distinguished at once from the workers by their possessing wings, and a very broad thorax. There are a great number of species of ants, and they may be divided into three groups, which are characterised by some rather important structural differences. These groups are the *formicites, Ponerites,* and the *Myrmicites.* The segment at the base of the abdomen forms a single knot or swelling in the first and second groups, but the females and the workers of the *Ponerites* have stings, and those of the *Formicites* have not. The *Myrmicites* have a sting, but their abdominal segment is formed into two knot-like divisions.

The common genus *Formica,* the true ants, belongs to the *Formicites,* and its individuals may be distinguished by their triangular and very dentate mandibles, which are their tools. As every species of ant has peculiar habits, it is best to take one as the type. The ant to which we intend to devote our attention is by no means uncommon, for although it may not be found in woods and copses, yet it is sure to be met with upon nearly every road or path, and very few people that live in Central Europe can say they have never seen a long train or column of the red ant. This ant, *Formica rufa,* or the Red Ant, is principally known through its workers, which are nearly red, and are very interesting to watch. The workers have a black spot on the head, and sometimes on the thorax, and the abdomen is blackish above, except at the waist. The female, which has the same general tint as the worker, is shining as if it had been polished, and it has the upper parts of the head, the thorax, and the abdomen of a beautiful black colour, and the wings are rather dusky at their bases. The male, which is almost as large as the female, is quite black and hairy.

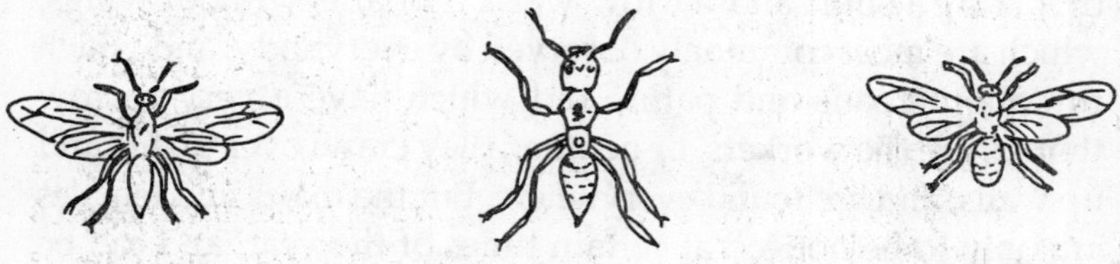

Fig. 5.11. Ashy Ant.

Now, let us examine the dwelling-places of these red ants. They may be found in woods where the underwood is not very dense, and they constitute slightly elevated and rounded little patches of earth at the foot of oaks for instance, but sometimes they are of considerable width and length. The surface of the hills is made up of a mass of little pieces of wood, bits of straw, flints, grains of wheat or of oats, and of earth, and it does not require much trouble to become aware that they are all arranged with a purpose, although they appear to be laid about confusedly. If a portion of the roof be broken down a crowd of agitated ants comes to the breach, and begins to repair the mischief at once, working with the most intelligent and persevering determination. A glimpse may be had of the interior of the building whilst the roof is being re-constructed. Sometimes the nest descends to a considerable depth, and it will be noticed that the ants have had to clear out and excavate portions one after the other, as the swarm increased in numbers. It appears to be made up of a tangle of pieces of wood all about the same size, and which appear to be piled up anyhow; but by examining the confused mass with some attention it will be noticed that the morsels of wood are so arranged that they form chambers, galleries, and more or less irregular tracks or runs. Moreover, they are so arranged as to permit a free passage in every part of the nest. In fact, the pieces of wood are disposed after a definite plan, and with much care. Those in the lowest part of the ant's nests are stuck into the ground, and are so piled up one upon the other as to prevent the effects of a sudden and violent shock; above all this the tiny beams are laid in stages, one over the other, and in some parts of the nest the intervals between them are filled up with such substances as earth, grains, and dried leaves. It is the presence of these grains—which are used by ants as materials with which to construct their nests — that has given rise to the idea that these insects collect provisions for the rainy day. But ants do not eat wheat or grains of any kind; their nourishment consists of more or less fluid or soft substances, and these supposed wise and economical creatures do not store up, but live from day to day from hand to mouth, and, like most insects, become stupid and hybernate when winter arrives.

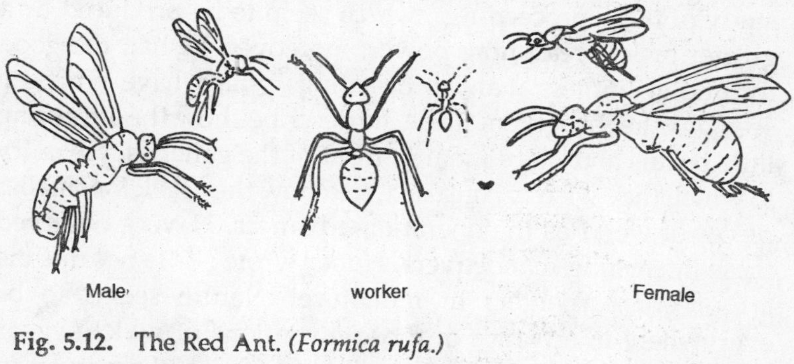

Fig. 5.12. The Red Ant. *(Formica rufa.)*

If a nest of red ants is examined at different times during the day, a constant succession of alterations will be seen to be going on. Very early in the morning everything seems quiet about the nest, and there are no openings visible, but there are some small cracks which might allow the ants to get out by undergoing a tight squeeze. Then a few of the insects begin to show themselves, and run over the dome of the nest, and they appear to be just awake; gradually they increase in numbers, and then some of them are noticed carrying and bringing little bits of stick, whilst others are clearing out and sweeping the passages to the nest. Should it happen to be fine weather, many large openings, which communicate with the principal galleries, are soon made outside the nest, and then the whole population becomes actively engaged, and each individual appears to have something to do. When the evening comes, the insects shut up the nest, and close the passages, and do everything to enable themselves to pass a quiet night after their labours, by rendering their house secure against violence and invasion. Should heavy rain come on during the daytime, the ants, fearing that water might get inside their nest, hasten to close its openings, and every member of the swarm appears to be impressed with the necessity of doing its best, and all sorts of materials are brought at once to shut the water out, and to place the erection in safety.

Pierre Huber was the first naturalist who made a careful study of these proceedings, which seem to be produced by a greater or less reasoning power; moreover, he had the good fortune to witness many proceedings which have not been seen by other observers. Thus, he describes how the ants bring little beams and place them in front of the galleries, the entry to which they wish to construct, and how they seek for smaller pieces of wood as the operation advances. Having observed these interesting manoeuvers, Huber wrote :—"Is not this the art of our carpenters in miniature? Nature seems to be everywhere in advance of those inventions of which we, as men, are so proud." Huber is correct, for if careful observers of Nature had existed amongst the early races of men, many important branches of knowledge and mechanical ideas, which civilised nations have taken centuries to discover and to complete, would have been found out very soon.

The engraving of the nest of the red ant was drawn from Nature in the forest of d'Aunay, near Paris. The rounded form of the nest is very evident, and there is a great rent in it, which occupies the centre of the mass. The dark holes into which working ants are carrying large oval-shaped cocoons, which

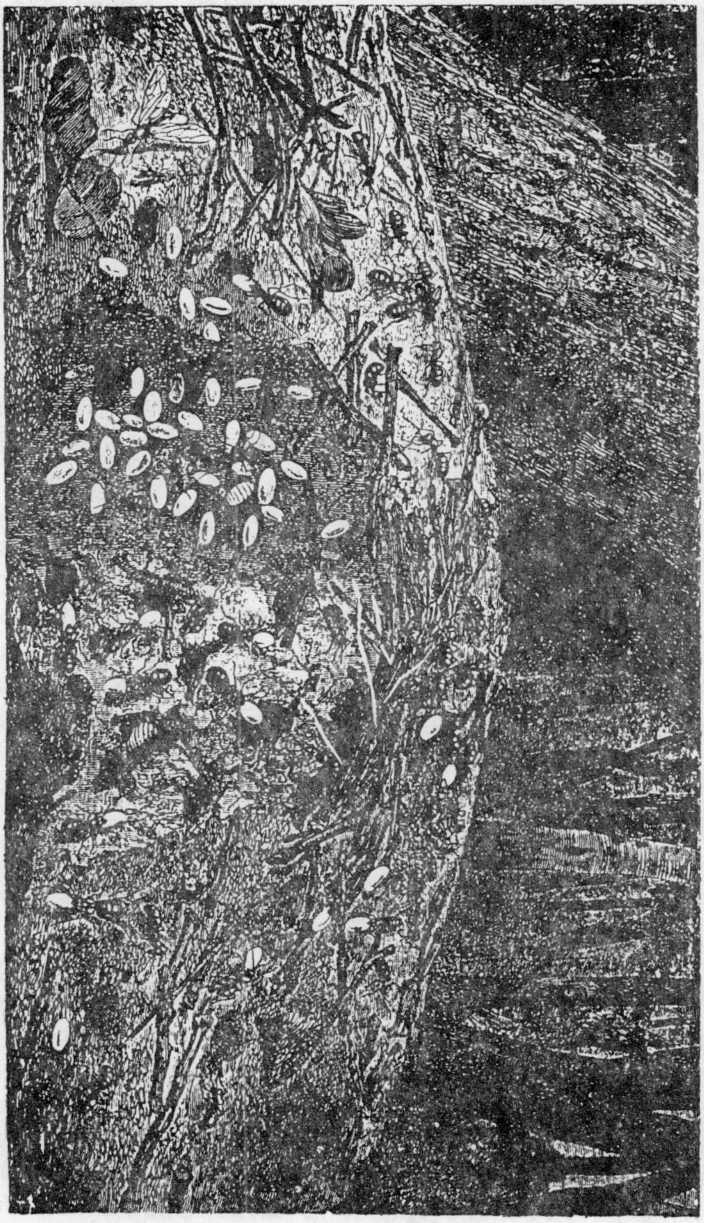

Fig. 5.13. The Nest of *Formica rufa*.

are commonly called eggs, are the commencement of galleries, over which small pieces of wood may be noticed fixed in the earth. On the left hand side there are some male and female ants, with wings, and some workers may be seen close to the tree, pulling at a fly which has come within their reach. The whole nest is in a great state of activity in consequence of a portion of it having lately been broken into.

It is very interesting to observe how the ants commence to build their nests, and the beginning of the work may be seen when the overplus of inhabitants of a nest of obliged to leave it in order that a new colony may be founded at the foot of some other tree. First of all the ants have to do mining work, and they dig into the soil with their mandibles, and after prolonged labour they manage to make a cavity. Then the little bits of wood and the other building materials are sought for, and brought in and stuck into the earth, being crossed one over the other in a most ingenious manner, so that the chambers. The greater or less sized cells, and the galleries, are constructed and strengthened in the lowest part of the nest, and then the upper stories are built. If a large ant's nest is opened early in the spring-time, no winged insects will be found, but a multitude of workers, and a few females, which have lost their wings; they are readily distinguishable, however, by their thick thorax and colour. All the female ants are born with wings, and very shortly afterwards they leave home, and fly off with a number of males, but they soon settle down to be careful and steady housekeepers; for whether they stray back to their old nest, or whether they come near to otehrs, they begin to tear off their wings, and should any workers be in the neighbourhood they assist them in completing this extraordinary operation. Nature does not care to perpetuate useless structures, and they generally drop off, or become smaller, but in this instance the ant appear to know that the wings are about to be useless to them, for the females will live a perfectly sedientary life for the future, consequently they act instead of Nature, and destroy their useless organs of flight. The wings of the female ants have nervures which are more or less divided close to their origin, so that they can be cut through very easily, and without the insect suffering any pain. The females begin to lay during the first fine days of the year, and their white and very tiny eggs increase sensibly in size until the larvae escape from them. The workers take wonderful care of the eggs; they place them in special chambers, and appear every now and then to lick them, and they carry them alternately to the upper and lower stories of the nest, and manage so that they are never exposed to too great a heat, and that they are kept in a uniform temperature. When the larvae are hatched, more care than ever is required on the part of the laborious insects, for then the ants, which we have noticed as able and industrious architects, have to take their turn in the nursing; and certainly more attentive, vigilant, and devoted servants could not be found.

Fig. 5.14. The Larva, Nymph, and Cocoon of the Red Ant.

A larva magnified and of the natural size, seen sideways and from above. The nymph seen from below, actual size and magnified. The cocoon of the nymph enlarged, and of the natural size.

The little vermiform larvae cannot move, but they have the instinct to lift up their heads, and to open their mouths, so as to receive their subsistence from the jaws of the nurses, and they are thus fed like little birds lately hatched. It is not for themselves alone that the ants seek honey, sugary liquids, and the juice of fruits, with such avidity, but it is quite as much for the larvae , to whom these nice things are carried. The particular delights of ants are known to everybody; every one complains in the country of the constant visits of these insects and they show themselves wherever fruit can be obtained, and where there are syrups and sugar. They may be seen upon flowers, the nectaries of which they visit for the honey, and they are almost always upon the stems of those plants which are crowded with *Aphides.* Not that they do any harm to them, on the contrary, they rub them gently with their atennae, and then the *Aphides,* apparently pleased, cause a little drop of sugary liquid to appear at the end of the small cylindrical tubes, which are situated at the extremity of their bodies. This is what the ants require, and they soon sip it up. The *Aphides* thus act as cows to the ants, and it is very remarkable that as yet no one has been able to recognise the use that this secretion may be to the Aphides *themselves.* As soon as an ant discovers a spot where there is some plunder to be had, bands of them will constantly come and visit the place. The ants return to the nest gorged with food, and give up a portion of it, which they have retained in their gullet-pouch, to those which have remained at home, and especially to the larvae. Hundreds of observations have proved that these industrious insects communicate with the understand each other. Frequently they may be seen to stop and to touch each other with their antennae, and many naturalists have thought that these appendages are the special organs of a particular insect language. They may be noticed singly or in numbers succouring wounded brethren, and leading them to the nest, and when the inhabitants of two nests encroach upon each other, fight in military array, and with great skills.

It is very interesting to watch the workers in their almost incessant nursing of the larvae. They clean them by rubbing them with their palpi, they carry them in the early morning up into the higher stories of the nest, so that they may be warmer, and they bring them back again into the depths of the cells when the sun becomes to hot for these frail little creatures. It is very wonderful how the worker ants manage to carry these very soft larvae in their very sharp mandibles, but accidents never happen, and such a things is unknown as a wound produced by unusual pressure, or by tumbling up against the walls of the long corridors and chambers of the nest. When the larvae become full grown they spin a silky cocoon in order that they may undergo their metamorphosis in it. The cocoons are oval in shape, and indeed so much so that they are usually called ant's eggs; and it is certainly curious that a great number of intelligent and highly educated people believe that ants lay eggs much larger than themselves. In fact, it does not strike them as being at all wonderful. Does not nearly every one talks about the ant's eggs, which are collected to feed young pheasants? but these so-called eggs are really cocoons manufactured by larvae, and which contains them or nymphs. The nymphs are perfectly enclosed in their silky cocoons; they are white, and resemble the adult form swathed up. The workers carry them about as they do the larvae, and with the same intentions.

When the metamorphosis is completed the perfect ant is absolutely too weak to tear open the silk of its cocoon, and would inevitably perish in this natural prison were it not for the vigilance of the workers, which, like excellent nurses, never let their charges out of their sight. Perceiving that a change has taken place inside the cocoon, these admirable attendants open it with their mandibles, and set free their new companion. But even when thus safely born the lately metamorphosed insects are not in a condition either to take care of themselves, or to take a part in the labours of the community, so their nurses do not leave them, but give them nourishment, and then lead them all over the nest and thus appear to introduce them to their new life. Fortunately, not much education is required, and the young are soon able to participate in the labours of the old. Whenever large numbers of workers are born or are metamorphosed, some males and females come to the light also. The workers love the old house, but the males and females appear to have but one wish, and that is to fly away from home as soon as possible. Some do not fly very far, and when they slight, as has already been noticed, the workers assist the females, and carry them to their nest; most of them, however, fly, or are carried by the wind to great distances, and when they settle they form new colonies. The males appear to die soon after they see their pretty companions remorselessly sniping off their wings, and settling down to a quiet humdrum life.

It was formerly supposed that the females which alighted at a great distance from their old nests returned again, but Huber, having great doubts upon this subject, found that some of them after having left the males, fell on to the ground in out-of-the-way places, whence they could not possibly return to the original nest. It was evident that either they must do something for themselves or else die in obscurity. Huber observed a solitary female go down into a small underground hole, take off her own wings, and become, as it were, a worker; then she constructed a small nest, laid a few eggs, and brought up the larvae by acting as mother and nurse at the same time. The larvae were a generation of workers, and when they grew to adult age, and began to execute all their usual works, from that moment the mother ant became lazy, and did not more. In order to make himself doubly sure upon this point this excellent naturalist imprisoned a single female ant in a little cage, and observed its proceedings. It is quite evident that the greater part of the intelligence and instinct of the ants is devoted to the care of their young before and during metamorphosis, and it is always interesting to notice the agitation of the workers when a hole is made in a nest, and a number of larvae and nymphs are exposed to daylight, or when they fall from the top to the bottom of the breach. The workers immediately rush to succour these tender little things, and the first care is to carry them off to the deep passages out of sight. This having been done, the reparation of the nest is thought about, and the ants, working energetically, will often repair the mischief in the course of an hour. Although most of the species of ants resemble each other in their habits, instincts, and intelligence, each one chooses a particular position for its nest, and establishes it upon a special plan, which has greater or less reference to some peculiar habit. Huber calls those kinds which only use earth as the material for their nests, Mason Ants; there are two kinds of them which are common over a great part of Europe, the Brown Ant, *Formica* fusca, and the Mining Ant, *Formica cunicularia*. The neuters of workers of the first species are of a blackish brown colour, and their bodies are covered with fine black hairs, the antennae and legs being of a reddish tint; the females are of a brilliant black colour, and the males have fawn coloured legs, but are otherwise dark. The workers of the second species are of an iron-red colour, and the females have the same tint, but the males are black. The brown ants, or those of the first species, hollow out the soil and make in the interior of a great cavity cells, galleries, and more or less spacious avenues, all in stories, one over the other, and their nest looks like the dome of a vault outside. The mining ant constructs its dwelling very much in the same manner, but the roof never appears above the level of the soil. These insects work the earth with their mandibles, and make pillars, columns, and partitions out of it, and do not neglect to fill up cracks, or to level irregularities, and in fact they act like very able builders. The mason ants cannot work during very dry weather, and they wait patiently, with their buildings half finished, for rain, but should some parts of them which were built up as the weather became very dry show any signs of cracking, the intelligent insects pull them down, and wait for better times. Huber gave some mason ants a good watering from the rose of a waterpot when they were in this somewhat desperate condition, and they set to work immediately, and built up walls and cells, and completed a story in a few hours.

Other ants live in old trunks of trees, and cut and work the wood in a wonderful manner. Of these, *Formica fuliginosa*, which is of a brilliant black colour, with pale reddish tarsi, is the most common example, and its works almost defy description. It builds many stories, which are almost always horizontal, forms platforms and beams of about the thickness of paper, and makes upright partitions of the same delicacy, and these include and partition off rooms and cells, the number of which may be really said to be immense. In some places they make little columns instead of the partitions, which give a miniature palatial aspect to the excavation. Other ants settle down inside the beams of houses, and interfere with the safety of large buildings in consequence of their very intelligent and destructively excavating habits. All these different species, which have such diverse constructive instincts, appear to take care of their larvae in very much the same way; but there is one kind, *Formica sanguinea*, which is of an iron-red colour, that settles down in the nests of other species, especially in those of the brown and mining ants, and singularly enough it attacks these kinds in order to carry away their nymphs, so as to make slaves of them. This is an extraordinary fact, but there are other examples of the same kind, which we shall notice presently. We have already mentioned the curious use that some ants make of the *Aphides*. The red ants visit the plants which are covered by them but they do not hesitate to look after cochineal insects, scale insects, and several *Hemiptera*; moreover, in those parts of the world where there are no *Aphides* or scale insects these ants seek the *Cicadellae*. But under certain circumstances the ants do not care about passing good food in order to enjoy the tiny drops of aphis liquor, so in order to save time they bring the authors of these little luxuries into the immediate neighbourhood of their nests, and sometimes carry them inside. This is peculiarly the case amongst the ants which lead very sedentary lives, and they select those *Aphides* which live upon grass and roots. Such species of the genus *Formica* are always very attentive to the captive *Aphides*, which they collect together in considerable flocks. Huber says that an ant's nest is all the richer according to the number of its captives; these are the cattle, the cows, and the goats of the ant tribe. Besides these objects of care, many little insects live with the ants, and receive very good treatment from them. Some of them will be noticed in the next chapter, so it must suffice at present to state that the young of the beetles of the genus

Clavigera have tubular hairs, which the ants are very fond of licking, and they thus appear to have some special secretion which pleases their hosts. It is a very remarkable fact that the larvae of these beetles cannot take their own nourishment, but they receive it from the mouths of the ants, and these extraordinary little beings thus have the instinct to nourish and to keep alive the beetle larvae, which afford them some pleasant luxury.

There are some ants with arched mandibles, which are narrow and sharp, and which resemble warlike weapons rather than implements of work. They belong to the genus *Polyergus*. The common species of this genus, *Polyergus rufescens*, is a small insect about a quarter of an inch long, and is a warrior, and does not care for building or any such trouble. It inhabits underground nests, where the brown and mining ants are found, but it does not make its own places of refuge, as it has no structures with which it can work, so it captures the workers of other colonies and employs them as slaves. Huber describes a curious scene in the life history of these ants, which he saw in the neighbourhood of Geneva in 1804. He observed a great mass of ants which were red or russet in colour, and of tolerably large size, traversing a road. They marched in a body, and with great rapidity, and they took up a space of from eight to ten feet in length by three or four inches in breadth, and they passed over the road in a very few minutes, penetrated through a thick hedge, and moved on to a grass field, over which they marched in regular array, in spite of numerous obstacles which came in their way. They soon came near a nest of blackish coloured ants, which was situated about twenty paces from the hedge, the dome of it being higher than the grass. Several ants were round about the door of their nest, and immediately that they observed the approaching army they rushed upon the head of the column; the alarm was given within, and a crowd came out from the underground works. The red ants quickened their pace, and soon pushed the black ones into their nest, they then clambered up the dome of the nest, many forcing themselves into the largest avenues, whilst other worked hard with their mandibles, so as to open a breach in the walls. This was soon done, and the rest of the invading army penetrated through the breach. Three or four minutes afterwards the red ants came out again in great haste, each one holding between its mandibles a larva or a nymph. They took the same route homewards, but not in the same military array. This is a clear case of slave making; and after a time the red ants live in peace with and enjoy the results of the labours of their black workers. Under these circumstances he red ant's nest is said to consist of amazons, or legionaries and workers. It is a wonderful instinct that causes the red *Polyergi*--which cannot work the soil and construct underground edifices, and which cannot nourish their own larvae—to select workers from the nests of other species to do all this. They never seek adult workers, for they would never stop in a strange home, so they look out for nymphs and larvae, which never attempt to depart from the nest when they have completed their metamorphosis; but set to work to build cells as well as to feed the larvae of their captors. A very curious ant is found in Mexico, and was described nearly thirty years since by a Belgian naturalist, M. Wesmael. The abdomen of this ant *(Myrmecocystus Mexicanus)* can be distended to a prodigious extent with a sort of honey. They are common near the town of Dolores, and are known there under the name of *Busileras,* and they live in underground nests, which are not distinguishable from without. These insects do not present any unusual distension of the body in early life, but after a while it is observed in a certain number of individuals. When the accumulation of syrup is very great in the intestines, the body becomes so distended that it resembles a transparent bubble. When the ants are thus blown out they cannot walk, but remain fixed or suspended to the floors of the galleries of their nests. The women and children of the country dig them up and enjoy their honey, and it is by no means unusual for these insects to be served at table, the head and the thorax being removed, so that the sweet portions resemble little isolated bladders on a plate. One would almost fancy that this syrupy secretion must collect for some purpose, and probably these very corpulent individuals are the nurses of the establishment.

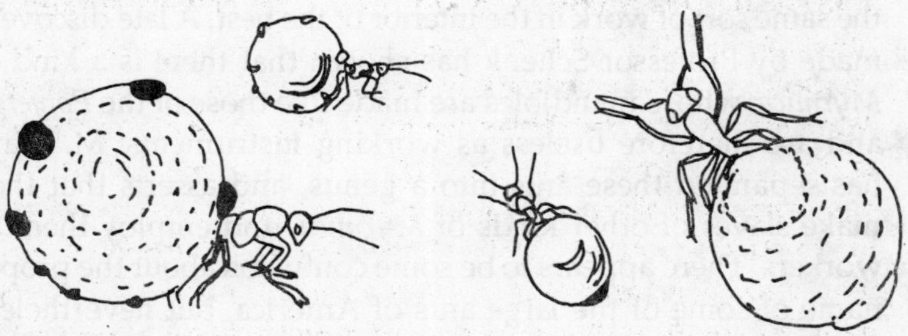

Fig. 5.15. *Myrmecocystus Mexicanus.*
The ants are represented as seen from above and sideways, natural size and magnified.

Many of the ants belonging to tropical countries are classified under the group of the *Ponerites,* and one of them *Typhlopona Oraniensis,* which is found in Algiers, forms small colonies; it is of a pale red colour, and lives under stones. The males and females are unknown, but the workers all appear to be blind, they never come to the light, and they labour constantly in their dark galleries. How they obtain their food, and of what it may consist, are matters for future observation.

The *Myrmicites* are found in Europe and in America, and the ants of this division are usually very small in temperate climates, but in the inter-tropical regions of America many of the species attain a very considerable bulk, and are also very abundant. There are two forms of workers in the nests of the different kinds of *Myrmicites;* in one the insects have a small head, and in the other a very large one. This difference in the size of the head has been frequently recognised amongst the workers of the nests in South America, and it appears to be the case also with a European ant. The ants with large heads are called Soldier Ants, in consequence of the expressed opinion of travellers, who assert that they are particularly devoted to the defence of the nest. The supposition that these robust and well-armed individuals might be males has not yet been justified by anatomical investigations. M. Lespes has noticed in the European species that both kinds of these neuter workers do the same sort of work in the interior of the nest. A late discovery made by Professor Schenk has shown that there is a kind of *Myrmica* whose mandibles are made like those of the *Polyergi,* and are therefore useless as working instruments. M. Mayr has separated these ants into a genus, and asserts that they make slaves of other kinds of *Myrmica,* and employ them as workers. There appears to be some confusion about the proper name of some of the large ants of America, but nevertheless these insects are known by popular terms; and one, which is called the Travelling Ant, and which lives in Brazil, does some good, and at the same time a great deal of mischief. One kind, *Acadoma cephalotes,* is known by the great size of its workers, which have an enormous head and spine behind, and a tuberculated thorax, the whole insect being of a blackish-brown colour; the female, however, although very large and of a brown colour, has an ordinary sized head. These ants unite in immense numbers, and travel long distances, and suddenly appear in houses, and destroy beetles, rats, and all the disagreeable things that collect in tropical habitations. They make their nests with leaves, and Mr. Lund says he was once passing near to a tree, the foliage of which was perfect, when he was surprised by the leaves falling off it like so much rain, although there was no wind; he went to look at the cause of this strange phenomenon, and found an ant working away with all its energies upon each leaf stalk. The stalk was soon cut through, and the leaf fell to the ground. There was a most curious scene going on, however, at the foot of the tree, for the ground was covered with ants, which were cutting the leaves in little bits as soon as they fell down. In less then an hour the tree was stripped of its foliage, the leaves were cut up, and their pieces carried off.

The Driver Ants of West Africa and stated by Dr. Baird to have no permanent dwellings, as other ants have, but to range about in vast armies in quest of prey. They make their excursions during the night or in very cloudy weather, the hot rays of the sun being fatal to them. If obliged to march during the day they construct underground arches of clay cemented by a fluid secreted from the mouth, and under these arches the pupae and the prey are carried by the workers. If on their march they are intercepted by a stream of water it is said that they throw across it a bridge of their own bodies, over which the main column marches with freedom and safety. It is to be hoped that if this is true the obliging bridge makers can swim. The community is divided into three classes, soldiers to attack and disable their prey, assistants to divide the prey into portable portions, and the labourers or workers. The bite of these insects is very severe, and so intense does it become by accumulation, that large animals, when confined, have been overpowered and destroyed by them. They have been known to kill the *Python Natalensis*, the largest serpent in that part of the world; and they are decidedly aggressive in their habits. Their entrance into dwellings, is known by a simultaneous movement of all the rats, mice, lizards, and cockroaches with which they may be infested. Even man has to get out of their way, and the driver ant chases him from his home. They are most useful in the economy of Nature; they consume much dead animal and vegetable matter which might otherwise taint the atmosphere; they tend to keep down the rapid increase of noxious insects and vermin; and they compel the inhabitants to observe habits of comparative cleanliness, as a filthy town or house is sure to be visited by them.

FOSSORIAL HYMENOPTERA

The *Hymenoptera* thus termed belong to several well characterised families. Some resemble the ants both in appearance and in their habits; others have a peculiar look about them, which particularies them, and the rest resemble wasps. But although these fossorial or excavating *Hymenoptera* differ so far as their external characters are concerned, they have many mutual resemblances in their habits and methods of life. They are generally very industrious insects, are gifted with great constructive powers, and are animated with a wonderful mounts of instinct and fore-sight in the preparation for the subsistence of their young. But it must be understood that this last gift only pertains to the females, which are always armed with a sting, for the males do not trouble themselves about their young in any way. These *Hymenoptera* excavate the earth and walls, and sometimes penetrate branches of trees and stems of bushes, so as to make safe places and shelters for their larvae. They are furnished with cutting and toothed mandibles, and their legs are covered with spines, which can act like rakes, and they use these working implements with great dexterity. When they are adults their nourishment consists of fluids, such as the honey from flowers and the sap or gun of trees, but when they are larvae they are carnivorous, and can only exist upon living prey. Consequently the mothers are obliged to chase other insects, and to catch them for their larvae, and it is therefore a curious and suggestive sight to witness a non-carnivorous insect hunting prey for its carnivorous young, which it is never destined to see. There are only two kinds of individuals amongst these *Hymenoptera*, namely, males and females and there are no special workers or neuters, as there are amongst wasps, bees and ants. The females are always solitary, and work away at the building of their nests, and supply the young with food without any help whatever. Every female chooses an appropriate place, and each species has its particular fancies with regard to the future position of its young; the locality may be chosen in the earth, a cliff, a wall, or in the stump of a bush. They often avail themselves of cavities formed by Nature, or of fissures and cracks, but the proper shaped hole is always formed with great energy and intelligence. The mandibles are the only instruments and tools, though the legs with their spines act as rakes to scrape out and eject the portions of earth detached with the jaws, but the industry and perseverance of the insects are so constant, and their patience is so wonderful, that a small gallery is soon dug out, and an oval shaped cell is made at the bottom of it; sometimes several cells are excavated, but every species has its peculiar arrangement as regards the number of the chambers.

The chamber being formed, the *Sphex*, the *Pompilus*, the *Crabro*, or the *Odyncrus*, as the case may be, must immediately undertake to store up provisions within it, and therefore other insects are hunted, each species having its particular prey. Some like caterpillars, others the larvae of *Coleoptera*, and a few even attack spiders; and the *Sphex* is gifted with that peculiar instinct which has already been noticed in the *Ichneumons*, for if it obtains a larva belonging to an insect which grows to a considerable size, it only places one in each chamber; but if the larvae are small, several are introduced, and if they are minute it hunts down a crowd of them. When it seizes it prey, the fossorial *Hymenoptera* pricks it with its sting, the venom of which produces a lethargy, which lasts a long time, and, in fact, the wounded insect never recovers perfectly. There is thus a most interesting series of phenomena of a most remarkable character. The larvae of these *Hymenoptera* imperatively require nourishment, which can only be derived from living tissues; they would die close to a dead body, or if they were placed upon an animal which was decomposing, or which had become dry. Thus these larvae which have no offensive weapons, and which cannot move from place to place, come at last to gnaw and feed upon living insects which cannot do them the least harm. The difficulties which are required to be overcome in order that all this can be consummated would be insurmountable were it not for the bountiful providence of Nature, which causes them to disappear as if by enchantment. These larvae have nothing to fear from their victims, which have been rendered inert by a proper dose of the venom of the female insect; they cannot offer the least resistance to any sort of attack, and, condemned to be slowly eaten, they nevertheless live until a large portion of their substance has been consumed. Death can hardly be said to take place until the larva has had enough, and has reached its full growth. The poison of the sting seems to act upon the tissues as a preservative, and caterpillars and the larvae of beetles which have been stung by *Hymenoptera* may be kept for several months in pill boxes without showing any alteration, and without undergoing death or decomposition. It is to be hoped that the same Providence that feeds the larvae in this extraordinary manner gives the venom a power of subduing pain; if not, Povidence is certainly on the side of the *Hymenoptera*, and a more dreadful fate than that of the living death of the poisoned insects can hardly be imagined.

As soon as one of these fossorial *Hymenoptera* has collected a stock a provisions within its cell, it lays an egg inside and closes the opening with a portion of the rubbish of the excavation, and alters the outside appearance of the work so completely, and makes it look so much like the neighbouring surfaces, that it is very difficult even for intelligent collectors to find out the nests. The female constructs a great number of these cellular nests; sometimes they are close to each other, and at others they are widely apart, but it provisions them all in the same manner, and deposits a single egg in each; then its work is done, and it soon dies. Thus this mother has taken infinite pains, has shown a wonderful foresight, and has laboured without rest for the sake of a progeny which it never can see, and which has habits totally unlike its own. This is the result of instinct, whatever that may be, and it is evidently not the result of reason. Some interesting details bearing upon this subject will be noticed further on, when *Bembex vidua* is described.

There are some species in all the families of nest-making insects which have the usual organisation and the same method of development, but which have no structures with which they can labour, and which have no maternal instincts for the benefit of the larvae. We have noticed that some ants make slaves of the individuals of other species who can render proper assistance to the young, and do such work of which they themselves are incapable; and amongst the fossorial

Hymenoptera some may be found which cannot chase prey and enclose it in cells, but hey manage to introduce their eggs into the galleries and nests which have been well furnished with victims by hunting *Hymenoptera.* Thus the females which cannot hunt manage that their larvae live like comrades at the expense of those of the fierce and industrious kinds. They introduce their eggs, and the larvae are soon hatched, and crawl out of the egg sooner than those to whom the nest really belongs; the consequence is, these parasites—the offspring of lazy parents—eat up a large portion of the provisions laid up in store, to the great deteriment of the others. The female *Hymenoptera* which act in this very human manner have neither mandibles proper to excavate earth, nor legs so spined that they can act as rakes; so not having a mechanical power, they use their wits at the expense of the goods of those insects which are endowed with the means of labouring.

Fig. 5.16. *Mutilla Europaea.*

The *Mutillidae* were formerly taken for ants, because the females, having no wings, can only run over the ground, and thus resemble the neuter or worker ants. They are solitary insects, and consist only of males and females, the former having wings. They are found in all parts of the globe, and principally in the tropics. The genus *Mutilla* is found in Europe and in North America. *Mutilla Europaea* is by no means an uncommon insect; the male is of a deep blue colour, has a red thorax, with dusky wings, the margins of the first segments of the abdomen being ornamented with silky hairs of a silver grey. The female is black, its thorax is red, and it has three grey bands on the first three rings of the body. The females are found in woods, walking, slowly upon the earth, and they may be seen entering holes in the earth, but nothing is satisfactorily known concerning what they do there.

In the engraving the male is represented flying, and the females, which are without wings, are crawling over the ground, coming out of a hole, and crawling upon a leaf.

One author has satisfied himself that the females visit the nests of large humbel bees, and this assertion has caused them to be considered the mothers of parasites, but several naturalists have seen them attacking insects, and others have taken them in their holes, where the remains of grasshoppers and of flies were found, and these relics have been considered the leavings of the larvae of the *Mutillae*. The females have strong and spiny legs, fitted for excavating, and they also have dentated mandibles, so that the implements for forming cells are present, and therefore, they are sure to be employed. The females of the North American species have powerful stings and a very acrid venom.

The Australian kinds are arranged under the genus *Thynnus*, and one of them has a most extraordinary history.

Verreaux states that the males have long bodies, and straight antennae, but that the females, which are much smaller, have no wings, and twisted antennae. The male flies about, and carries of the female with him, paying her the greatest attention, and placing her on flowers, so that she can obtain her nourishment.

Frequently other males, which have not the happiness of possessing a wingless companion, come near and appear enhanted with her company. Of course, they all become jealous, and a fight ensues, and should her protector be unable to conquer the others, in order to disappoint, them he eats her up.

The species of the great genus *Scolia* have females with strong and arched mandibles, and very spiny legs; some of them include insects of very large size, and the commonest is often noticed in gardens in Central France, Italy, Spain and North Africa. A Florentine naturalist, Carlo Passerini, studied the natural history of *Scolia flavifrons* in 1840. If old rotten trunks of oaks, or masses of tan, such as are used in gardens and hot-houses, are examined, very large larvae of a beetle are often found; these larvae belong to the so-called Rhinoceros Beetle *(Oryctes nasicornis)*. They are usually very quiet, and live in spaces which they have formed within the tan, and they are particularly subject to the attacks of the females of this *Scolia*.

The *Scolia* is searching for food for its future larva, and it finds out that the beetle larva is within reach, so it digs down to it, and stings it, so as to produce the curious stupid condition already noticed as invariably following such an attack. As soon as this is done a small egg is laid on the skin of the torpid creature, and is glued to it by a special secretion. A young larva is born from the egg at the end of a few days, and immediately commences to gnaw the skin of its victim with its mandibles; it gradually forces its head into the wound, and begins to suck the animal juices. The young *Scolia* grows rapidly, and day by day the beetle larva becomes smaller and smaller, in consequence of being continually sucked. By the time that it has become completely exhausted, nothing else but the skin remaining, the larva of the *Scolia* has attained its full growth.

It then spins a thick cocoon of a brown colour inside the skin, which is all that is left of the beetle larva; so that this unfortunate creature affords food to the *Scolia* in the first part of its life, and a certain amount of protection and covering during metamorphosis. The *Scolia* passes the winter in its cocoon, and undergoes its transformation into a nymph during the spring, but the perfect insect does not break forth until the commencement of June.

In the engraving a male and female *Scolia* will be noticed, the one on the leaf and the other in flight. Immediately underground in a cell is a larva of the beetle *Oryctes nasicornis* on its back. The small larva of the *Scolia* has its head forced into the body of the beetle larva. Below, and on the right hand, is a cocoon which will be seen to rest inside what is left of the beetle larva; and on the left the nymph of the *Scolia* is shown within its cocoon.

In the Seychelle Islands, and in Madagascar, there are some species of *Oryctes* the individuals of which attain a great size; they do a great deal of harm to the cocoa-nut trees, and thus cut off one of the most important sources of food in tropical countries. There are two kinds of larvae of the genus *Scolia* which eat and destroy the larvae of these beetles, and which are metamorphosed exactly in the same manner as the European *Scolia*.

The *Sphex* is a well-known insect, and belongs to a genus which is closely allied to *Scolia,* and is commonly called the Sand Wasp. They are generally of considerable size, and their colour is of a brilliant violet blue, more or less deep, and some of them are marked with bands of yellow. Species are found scattered over all parts of the world, especially in warm climates, but only one is found in Great Britain. These *Hymenoptera* are remarkable for their activity and also for the elegance of their shapes. There is one which has yellowish wings and a red abdomen *(Sphex flavipennis)*, and it makes nests in the sand, in which it lays an egg; then it seeks the larvae of grasshoppers,

Fig. 5.17. The Metamorphoses of *Scolia flavifrons*.

and places them in contact with the egg, after having stung them. Another kind, which is found in Languedoc, attacks a very large grasshopper; and it is indeed surprising what very large insects the *Sphegidae* can carry and introduce into their nests. The females of the English species may often be seen in the eastern countries coming out of holes in the banks, and if they are watched, the process of their excavations is not difficult to understand. When the little hole is dug sufficiently deep in the soil, a tiny egg is laid within, the female comes out and stops up the opening with a piece of dust, and flies off in search of a victim. It pounces upon some small caterpillar or larva, or even small flies, stings one of them, and carries it off bodily to its nest. The female can always discover the hole it has made, and, after rolling away the earth that closes it up, the future food of the young *Sphex* is pushed into the cavity, which is then closed up from the outside. The activity, energy, and excitement of the insect whilst all this is going on are very great and it soon dies after the provisioning of the nest is completed.

Fig. 5.18. *Ammophila sabulosa.*

The genus *Ammophila* contains one species, the true Sand Wasp, which is very abundant in Europe. It is a slender, elongated insect, with a red band across the lower half of the third, the whole of the fourth, and the upper part of the fifth segment of the abdomen. It has habits like those of the *Sphex* just described, but it invariably chooses the caterpillars of certain moths for the prey of its larvae.

There are some *Sphegidae* in the tropics which have

Fig. 5.19. *Pelopaes spirifex* and its Nest.

enormously long waists and very arched mandibles, without teeth on their internal edge. The insects are black, with yellow marks upon them. They are perfect masons, and work in a most intelligent and persevering manner. A species common in Central France *(Pelopaeus spirifex)* builds its nest in walls, in corners of buildings, and sometimes within barns and houses, but almost always at the height of several yards from the ground. The insect is very slender, and does not seem fitted to carry burdens, but nevertheless it seeks clayey spots, and beats up small portions with its mandibles and carries them off in order to construct its nest.

It begins by forming one chamber, which is hollowed out in the clay brought from a distance and stuck on to a wall, and as soon as it is finished the female goes a hunting, and, curiously enough, selects the most dangerous, bold, and well-armed creatures there are for its prey. It does not hesitate to attack spiders, in fact, it rather enjoys the fight with them. Doubtless, the strong sting of *Pelopaeus* is a terrible weapon, and if it touches the spider there is an end of the combat, but if not, this last insect has its web at hand, threads of which it can throw with great rapidity and exactness.

The *Pelopaeus* is both bold and prudent, and approaches the web flying carefully, and usually it manoeuvres so well that the spider is stung before it can do anything; but it sometimes does happen that the spider is prepared for the attack, and the *Pelopaeus* finds its movements paralysed by some fine threads being thrown over it. Then the spider encircles its enemy with layers of web, and devours it.

But it does not often happen that the *Pelopaeus* is captured, on the contrary, it usually manages to bring one, two, or three spiders to its nest, according to their size. They are introduced and an egg is deposited close to them, and then more clay is brought, and all is closed. Thus the first chamber is constructed and victualled, and then the female builds a second one on the side of the first and upon the same horizontal line, and fills it in the same manner, then the third and the fourth cells are added, and sometimes there are six or eight of them. The external surface of the nest is marked with long depressions, which correspond with the intervals between the cells. The lower part of the nest is thinner than the rest, and when the larvae have become full grown and have undergone their metamorphoses the adult insects come out from this part and leave open holes.

If a nest be detached from a wall before the second metamorphosis has commenced—and this is by no means an easy operation, for the masonry of the *Hymenoptera* is good—the cells will be noticed to be occupied either by larvae or by cocoons, which have been made by them after the attainment of their full growth. The cocoons are made out of a paper-like tissue, and are soft and shining, and of a brown tint.

A very beautiful *Hymenoptera* renders immense service to the inhabitants of the Islands Maurice and Bourbon, and they only regret that the insect is rather uncommon. This *Chlorion compressum* constructs its rest very much like the *Pelopaeus,* but it does not hunt the same prey. All the seaport towns of those colonies are infested with cockroaches, which do an infinitude of mischief. Fortunately, they have many enemies, and the most terrible of all is the insect just mentioned. It is by no means an unusual thing for a struggle to be witnessed between one of these agile *Sphegidae* and a cumbersome *Blatta* or cockroach, but it appears wonderful what use the small insect can possibly make of the larger, and how it can manage to drag such a bulky body into its small nest.

Fig. 5.20. The nest of *Pelopaeus* Detached and seen from within.

The *Chlorion* may be seen hunting about here and there, and up and down the roadside and gardens, searching for its favourite prey. It spies out a cockroach, which appears to know intuitively that there is danger at hand, for it shows symptoms of great fright, and seems so confused that it cannot run away. The *Chlorion* pounces upon the insect, clasps it with its mandibles between the head and the corselet, and stabs it in the body with the sting. Then it flies off for a little distance, and awaits the effects of the poison thus introduced: and when the convulsions of the victim have ceased, the clever little insect seizes it stupefied prey, and drags the heavy burden with great efforts to its nest. Usually the opening of the cavity is so narrow that the cockroach cannot be got in, for its legs and wings stick out and prevent its introduction. But the *Chlorion* sets to work and cuts off the legs and the wings, and having thus lessened the difficulty, it strives hard to push the body into the hole, but as this plan usually fails, the *Hymenoptera* enters first of all, seizes the cockroach with its mandibles, and drags it in with all its force. As the integuments of the *Blatta* are more or less soft and flexible, the great insect is at last forced into the gallery, where it never could have been expected to have entered. Such proceedings on the part of the *Chlorion* almost verge upon the domain of reason; and it is difficult to explain them by the notion of that very indefinite quality called instinct, for the manoeuvres vary according to circumstances, and there appears to be an intelligent method of overcoming every difficulty.

The species of the genus *Pompilus* are fossorial Hymenòptera, which work with the same degree of perfection as the *Sphegidae,* and which hunt with the same bravery and with the same cunning. Those of this country are very small insects, and make their nests in sand or in old trees, for they seem to know that the rotten wood will relieve them from much trouble in the matter of hole-digging. They all provide spiders for the nourishment of their larvae. Most of them hunt those wandering spiders which scamper about and never make webs, but a few do not hesitate to attack and carry off the great garden spider, even when it is located in the middle of its beautiful web. Some *Pompili,* which have legs without spines, are incapable of digging, and therefore they act like some of the other insects already mentioned, and which have corresponding structural deficiencies, for they lay their eggs in the nests of other fossorial *Hymenoptera,* and their larvae act like parasites.

The *Crabronidae* appear at first sight to resemble the *Sphegidae,* but on examining any of them, it will be observed that the body is only moderately long, their antennae are straight, the upper lip is hardly projecting, their hind legs are not much longer than the others, and the fore legs end in a broad point. They are common insects all over the world, and their shapes vary with their habits and methods of life. There is a sub-family of them called the *Bembecidae,* and they are insects which have a large body, a very distinct lip, and pointed mandibles which have one tooth on their inner margin. The species of the genus *Bembex* are remarkable for the great length of their jaws, and their lower lip is formed into a sort of trunk, and they may be seen sucking up honey with it from flowers. *Bembex rostrata* is the commonest of them, and it digs deep, cell-like holes in the sand, and victuals them with flies and *Diptera* of all kinds; it then lays its eggs, and closes the orifice of its cell with that great care which has already been noticed amongst the *Sphegidae.* The examples we have offered showing the habits of the fossorial *Hymenoptera* have a certain sameness, for in every instance the female builds the nest, fills each cell with victims for the future larva, lays an egg close by them, and shuts up the habitation, and then dies without ever seeing its progeny.

But M. Fabre, of Avignon, has described the habits of *Bembex vidua,* which are certainly most remarkable and suggestive, and probably very rare, in the history of the *Hymenoptera.* In this species the female does not close up the cell, but penetrates into it every day, carrying in a fresh victim for the larva; and it always chooses a fly. Here is a case of a female insect caring for its larva which it sees, and which it notices to eat and care for food, so that the daily visit becomes a pleasure and a duty, according to the usual law of maternity.

Of course, the larvae of this species run great risks, for their cells remain unclosed, and carnivorous insects may enter in and destroy them. Moreover, the mother may be taken and killed herself, and then, as no food would be forthcoming, they would die from starvation. There is no doubt that the habits of this species cast a light upon those of the insects which only provide one store of provisions, and then close their nests; for it is not difficult to imagine that if the egg of a former *Bembex vidua*, the predecessor of all these, should not happen to have hatched at the second visit of the mother, she would have closed the hole and left it uncared for, not seeing the use of troubling herself to no purpose.

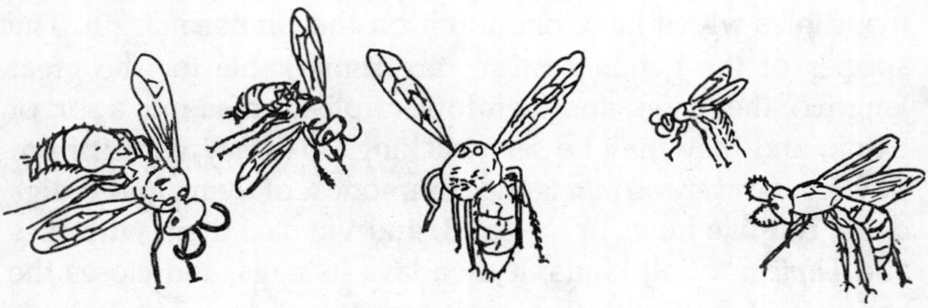

Fig. 5.21. *Cerceris arenaria.*

Females in different attitudes, those shaded darkly being magnified. A male natural size and magnified.

Cerceris arenaria is a species of the tribe of the *Crabronidae*, its members are very common in Europe, and have been carefully studied. These insects are very fond of dry ground and sandy soils which are exposed to the noonday sun. They dig deep holes, and it is very remarkable that they should provision them with beetles of the weevil tribe alone. It does not matter what may be the species of the weevil, so that it is one of the family; in fact, ten or twelve species belonging to different general have been found in the nests of a *Cerceris*, which certainly appeared to choose the toughest and strongest-skinned beetle it can for the use of its very soft-skinned larvae. But the *Hymenoptera* has the instinct to select weevils which have been lately born, and whose integuments, therefore, are not very firm; and, on the other hand, the larvae appear to have no difficulty in sucking the juices of their prey between the segments of its body.

Cerceris *quadricincta*, which has been studied by M. Fabre, has a great fancy for weevils, but it does not choose large ones, and selects those of the genus *Apion*, the smallest of the family, so that it has to collect at least thirty individuals to provision one of its cells. Another *Cerceris* even attacks those beautiful beetles called *Buprestis*, many of which glitter like diamonds; and some kinds chase other *hymenoptera*, and one beautiful *Cerceris* seizes honey bees, whilst an allied kind selects *Ichneumons* as its prey.

The genus *Philanthus* is well-known, in consequence of the peculiar habits of one of its species, which is a great bee killer. *Philanthus apivorus* or *triangulum* has a black head and body spotted with yellow, and its abdomen is yellow, but there is a triangular black spot on each of its segments, so that it is an insect readily distinguished. It digs galleries in paths in sandy soils and makes them vertical at first, and afterwards horizontal, and carries them on to some depth and distance; indeed, some of the galleries are said to penetrate downwards to the depth of a foot. It is a most audacious insect, and it settles upon flowers, and pulls them about, and plays with them, as if it did not want anything to eat, and had no particular cares; but it is on the look out for its prey, and a little patient watching will be rewarded by a very curious scene. Presently a bee comes to the flower to collect honey and pollen in its usual quiet and business-like manner; the sly *Philanthus* looks at it, and examines it, to see if it is in good condition and if it is to its taste, rushes upon it with great impetuosity, seizes the bee between the head and the corselet, and stings it in the body. The bee, being a brave insect, offers great resistance, but the *Philanthus* rarely misses its aim, and usually succeeds in imparting some of its venom to the unfortunate victim. The bee becomes convulsed, and thrusts out its sting here and there, but it is of no use, and insensibility soon comes on. Then the attacking insect seizes the bee with its mandibles and legs, and flies off with its burden. As the *Philanthus* approaches its nest it may often be noticed to stop, as if it feared some hidden danger, but sooner or later it introduces its prey, lays an egg, and closes the gallery.

Fig. 5.22. The Metamorphoses of *Philnthus opivorus.*

In the engraving two females may be observed carrying bees which they have stupefied. In the earth, on the right-hand side, a bee has just been introduced into a cell; in the next gallery a large larva has consumed the greater part of its provisions, and in the three other cells the larvae may be observed in their cocoons, which have transparent walls.

The *Philanthi* often frequent particular places in great numbers, and many hundreds of them may be seen working simultaneously, but without taking any notice of each other. Not only do they display indefatigable industry, but very great boldness also, for every now and then they do not hesiate to come near beehives, and to seek their prey in such very dangerous neighbourhoods, for certainly they might easily be destroyed by the efforts of a number of their victims. The larva is rather compact in shape, and attains its full growth when it has eaten its bee; it then constructs a silken cocoon with very thin walls, which are almost transparent. The cocoon is of a very peculiar shape, for it resembles a small long bottle; its bottom is round, and the neck is very visible, and appears to be closed with a black wax; in fact, the cocoons of *Philanthus* resemble the tiny bottles in which homoeo-paths carry about their remedies. Water renders the cocoons quite transparent.

Fig. 5.23. *Cemonus lugubris* and its Nests.

The Great Crabro *(Crabro cephalotes)* makes holes and cells in rotten wood, with the aid of its mandibles, which have a bifid termination, and it victuals its nest with flies.

The small *Cemonus lugubris* constructs regular cells in the stems of briars, ad the accompanying engraving represents the curious appearance of the cells in a twig of a wild rose.

The species of the genus *Odynerus*, or the False Wasps, are pretty little black insects banded with yellow tints, and they are most interesting on account of their agility, graceful movements, and singular powers of nest constructing. *Odynerus parietum* is a well-known species, and some years since our attention was directed to a very curious spectacle, which was observable at a short distance from Denain, in the Department du Nord, early in June. The roadside was bordered by a sandy bank of about seven feet in height, which encircled a great field of lucern. The hedge or bank was formed of a rather dense soil, and was exposed to the full power of the sun; thousands and hundreds of thousands of this *Odynerus* were flying about the field and hunting over and upon the plants with wonderful perseverance. They brought little green larvae between their mandibles, dug holes in the earth, made little galleries, introduced their prey, and closed up the cells, each one pursuing its particular duty without being in the least disquieted by the thousands that were doing the same thing all around. The liveliness and animation of the scene was beyond description. The work was constantly going on, and old and young individuals were occupied either in just commencing to hollow out the soil or in closing up and finishing their labours. In some spots the insects were working hard at victualling their cells, and had completed a curious gallery which projected in front of the soil for more than an inch, like a tube. This tube was slightly curved, and was made up of little flat pieces or cylindrical morsels of earth placed in circles one over and in front of the others; small spaces being left in the walls of the tube, which gave it a lace-like look, but without admitting the rain. These vestibules are excessively fragile, and fall to pieces when they are touched, but the insects crawl into them without doing any damage. These singular entrances to the galleries and cells last as long as the victualling goes on, but when everything is finished inside, and the eggs are laid, the *Odynerus* breaks up its fragile vestibule and thus hides the entrance to its nest most perfectly.

We broke away the earth which formed the bank here and there, and exposed several cells which were just below the soil, and many of them were more or less perfect. In one place a gallery was noticed to form the entry of two, three, or four cells, and in another of only one cell; but all the little chambers were provisioned with the same sort of insect, namely; the green larva of a weevil *Phytonus variabilis)*, and there were fifteen or sixteen of them in each cell. The larvae were intact in most of the cells, but then only the egg of the *Odynerus* could be found; while in others the provisions had been partly consumed, for the larva had been hatched, and had increased in size at the expense of its victims. The shape of the larva is short and rather oblong.

In the engraving some of the vestibular tubes will be noticed on the right hand, and an *Odynerus* will be seen crawling into one; other perfect insects may be observed in different positions. On the left hand, there are two cells exposed in section, and in each there is a larva with a crowd of its victims upon it. All the *Odyneri* finish their particular duty by the end of June, and die; and such a bank as we have described presents no external traces of the wonderful insect work that has gone on within; but inside the larvae are living, eating, and growing, down in the dark and far away from danger. When they have consumed all their provisions they will form a silken cocoon, and hybernate in it until the next spring; then they will turn into nymphs, and, after two or three weeks have elapsed, they will fly as perfect insects.

Fig. 5.24. The Metamorphoses of *Odynerus parietum.*

The *Odynerus* which attacks the briar chooses dry twigs which have a slanting direction, or which are horizontal. The insect sets to work to hollow out a twig by getting rid of the pith, and when this is completed it goes off to seek earth and tiny stones, which it forms into a sort of clay. It then completes a cell and victuals it, and afterwards constructs a second and perhaps a third and fourth. Very often six or ten are to be found. The insect which occupies the cell nearest the opening becomes hatched before any of the others, although this cell was made last of all, so that the youngest egg hatches before the others. This is very wonderful, and by thus hatching sooner than the others this youngest *Odynerus* makes room for the free passage of the insect in the second cell, and so on. The cells open through one passage; and if the insect in the most distant cell, and which was filled first of all, was hatched first, it would force its way through the other chambers to gain the open air, to the great determent of the included larvae.

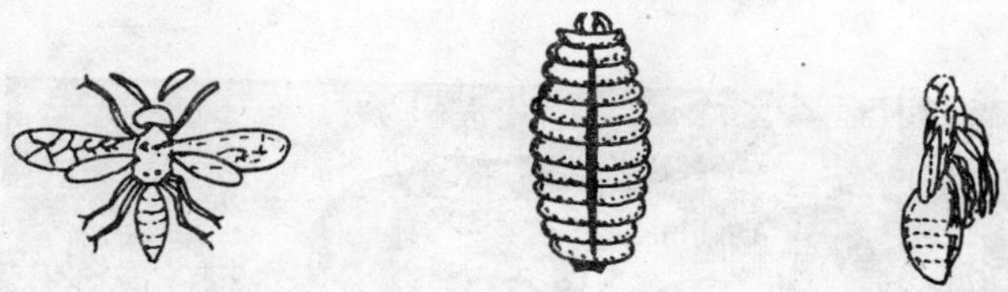

Fig. 5.25. A Species of *Odynerus*, Larvae of the *Odynerus* Pupa of he *Odynerus*.

Eumenes pomiformis is a black insect with yellow rings, which builds nests upon walls, and it does not hollow out the earth, but constructs little round capsules out of clay, sticking them to bricks and stones, and making one cell in each. In the engraving on these little nests are shown, and the mature insects are represented upon the wall.

Hymenoptera which Make their Nests of Papyraceous Substances (Vespidae)

The *Hymenoptera* which make their nests with papery-looking substances are the Wasps, the true social wasps, and not the solitary wasps which belong to the genera we have just described. These true wasps are arranged in one large family, the *Vespidae,* which is subdivided into three groups, containing several genera. There are three kinds of individuals amongst the wasps, as there are amongst the ants, namely, males, females, and neuters or workers, and these last are really females which never lay eggs, but which are provided with wings, and yet differ much from the perfect females. These insects live in societies, the individuals of which are more or less numerous according to the species. Usually the assemblage only lasts a year, and it breaks up at the commencement of cold weather; by that time the males are all dead, then the workers begin to die off, and only the true females manage to hybernate by seeking refuge in the holes of walls, cavities of old trees, or wherever they can hide themselves easily and get out of the way of the cold. In the early spring a solitary female, a mother, begins to build a nest; it is small at first, the cells of it are few in number, and an egg is laid in each.

Fig. 5.26. The Nests and Adult Insects of *Eumenes pomiformis*. The capsules with holes in them have been opened by the perfect insects after their metamorphoses, but the others still contain larvae or nymphs.

As soon as the larvae are born the wasp wanders to and fro fro food for them, and spares no pains in order to provide its progeny with sufficient nourishment. The larvae soon become full-grown, and spin a silken cocoon, which fills up the little cell, and thus enclosed they are transformed into nymphs, the metamorphosis into the adult form occurring shortly. The new wasps thus produced are all workers, or rather workeresses, for they are sterile females, and they are created to take the part of nurses to a progeny which does not belong to them. These workers begin their duties as soon as they become perfect, and from that moment the mother of the nest, who has been so laborious, and who was so industrious when all alone, begins to lead a quiet life, and to do little or nothing, for she has plenty of nurses to take care of the little ones. The workers increase the extent of the nest, and prepare cells for the larvae, and, when this is done, the mother completes another great batch of egg-laying, and deposits an egg in each cell. The larvae which come from these eggs do not turn into workers only, but into male insects and females which are not sterile. It is not known how many egg-layings take place during the year, but they probably very with the species. Very often the combs of the nests are found to contain larvae late in the season, and when cold weather suddenly sets in. In this case the wasps appear to understand that the fruits will soon be all gone, and that the means of subsistence will be cut off; but, in order that the larvae shall not suffer from prolonged starvation, these intelligent creatures kill them at once, so that the nest which was a few days before full of life, activity, and animation, becomes a dreary solitude. The wasps construct their nests out of vegetable matters, such as woody fibres and dead leaves, and they make a sort of paper from these substances. They triturate their building materials between their mandibles, moisten them with their saliva, and mould them into a homogeneous paste, admirably suited for the construction of the cells and the walls of the nests. The nests are made upon different plans, all of which appear to have a distinct reference to some peculiar conformation of the insect. But, although there are many minor differences, still, in the main, all wasps' nests resemble each other. The essential and most important parts of the nests are the cells of the comb, and they are always built upon the same plan. In the commencement the cell is a kind of cylindrical cup, and others of the same form are built close together on a horizontal floor, and they are all closed at their summits. The cells of many wasps' nests are always cylindrical in shape, but usually they are hexagonal in outline. When a round cell is in contact with six other cellules which surround it, the walls flatten regularly on account of the equal pressure of the surrounding structures, and a geometrical shape is perfected. The result is that the wall of one cell becomes really a portion or a layer of a partition which is common to two cells, and very soon the double wall is no longer seen to exist, for it is not wanted. By building hexagonal cells there is a great economy of materials, and this is a very important matter when the substances of which some very large nests are composed are scarce.

Everybody respects the wasp, and is aware that the mothers and workers are armed with a very strong sting. The larvae are feeble creatures; nevertheless, the head of the wasp larva is stronger and larger than that of the other *Hymenoptera;* even their mouth-pieces are stouter, and this enables them to receive bits of fruit or fragments of insects as nourishment, besides the usual fluid or very soft food. It is supposed by some naturalists that the sterility of the females is more or less due to a difference in the food given to them by the workers; and there is some reason for believing that those larvae which eventually become mothers are fed upon an animal diet; and the others, which become sterile, and turn into workers, are fed upon vegetable nourishment. There may be some truth in this opinion, but another will be noticed further on.

One of the most common wasps is the Wood or Bush Wasp *(Vespa sylvestris)*. It is a little smaller than the common wasp, and it attaches its nest to the branches of trees and bushes, hangs it under roofs, or attaches it to the corners of walls. The little delicate round nests of this species may often be met with in the spring, and a very slight examination of one of them will prove that the covering, or envelope, is made up of a smooth, grey paper, which is slightly shiny and flexible, and perfectly impervious to water. A thick column is found in the middle of the inside of the nest, sustaining a single comb, which is composed of eight, ten, or twelve cells; it is the work of the mother wasp, which has passed through the winter in a condition of sleepy stupidity. After having built her nest, she deposits an egg in each cell; and when the larvae are hatched, they become her workers and nurses. They enlarge the first comb by adding new cells around it; then they construct a second comb, and attach it to the first by two or three pillars ; then a third is made, and fixed to the second; and so on for a series which may include a fourth, fifth, and a sixth comb. The number of the combs varies according to the increase of the population of the nest. The whole has three papery envelopes, one placed over the other, and they serve to protect the combs, to which, however, they are never attached. The largest nests are sometimes a foot in height; but whether they are large or small, they look very pretty amongst the foliage of beautiful bushes. The opening is situated at the lowest part of the nest, and underneath, and is always sufficiently large to give free ingress and egress to the wasps.

In the accompanying engraving there is a figure of one of these nests, which is suspended from a small branch.

The common wasp *(Vespa vulgaris)* makes its home in the earth, and after having found a small cavity somewhere, it brings constructing materials to it, and makes it nest in the dark retreat. When the population of the nest increases, and when the number of the sets of cells has to be augmented, the wasps are obliged to undertake a great deal of labour in order to enlarge the area of their homes. The nests of the common wasps have several papery envelopes, like those of the wasps just mentioned, and they often assume very great size, and contain thousands of individuals. To all appearance, the societies of the common wasp do not perish during the winter; and if

Fig. 5.27. *Vespa sylvestris* and Nest.

one of the localities where wasps are often seen is examined, it will become evident that the wasps never leave it year after year, unless disturbed.

The Hornet *(Vespa crabro)* is always feared, on account of its great size and fierce appearance; and it may be easily distinguished from all other wasps by the russet colour of the head and front part of the body. Hornets usually make their nests in old trunks of trees which contain large hollow spaces, for they want much room for their great habitation. They attach their first comb to the upper part of the cavity by means of a large mass of wax made up in the form of a supporting column. A second comb is added, when required, beneath the first, to which it is fixed by more or less numerous pillars or little columns, and all the combs are attached in this way, one under the other consecutively. When the nest is completed inside some cavity or other, the walls of which protect the whole sufficiently, the insects only make coverings to those parts which are more or less exposed, but under other circumstances they take great pains to cover up their nest with many envelopes. Some hornets' nests, which are found under the eaves of barns or granaries, attain a very considerable size, and are really very pretty and extraordinary objects. They are wonderfully fragile, and not strongly built, like the nests of the other wasps. Ordinary wasps scrape off woody fibres from living trees, and form a strong tenacious paper, but the hornets content themselves with rotten wood, out of which they manufacture a yellowish or russet-coloured paper, which although very pretty to the eye, is very friable, and does not possess any lasting properties. The larvae of these large wasps are fleshy grubs, and are destitute of feet. They have to be fed by the workers, and undergo metamorphoses similar to those of the other wasps.

The wasps which have elongated bodies and the first segment of the abdomen formed into a long pedicle constitute the group of the *Polistites*. Linnaeus, without much regard to the geographical distribution of one of the species, called a very common wasp of this kind *Polistes gallica.* The French *Polistes* is a black insect decorated with yellow tints, which are also observed on the antennae. It frequents open spaces in woods, and there are few prettier sights than that presented in the spring-time by one of the females when it is building its little nest, or is attending to its larvae. It is not difficult to observe all this, for the *Polistes* attach their nests to low plants and bushes. The brooms especially furnish them with straight and narrow twigs, which are very convenient for their articular method of nest building. The mother, after having hybernated during the winter, begins to work earnestly and with great perseverance early in May, and commences to construct her nest with materials which resemble those used by the bush wasp. The fibres of bark are reduced into a homogeneous paste, which is converted into a strong grey paper. The *Polistes* first uses a large quantity of this paper to form a solid foundation for the nest, and a strong peduncle, which has to maintain the comb. The comb is very small, and there are not often more than five, six, or eight cells in it. It is never covered up with envelopes, and although it is increased in size subsequently by the addition of new cells, it is never enclosed and protected from the weather. Any one interested in the habits of these industrious *Hymenoptera,* and who will visit the neighbourhood of their nests daily, can study all the phases of the life of the larvae, for these pretty nests without envelopes, and unsheltered, offer every opportunity for the investigation; but although apparently so liable to be injured by rain and wind, they are so made that the larvae are perfectly safe inside. The nests are always placed obliquely, so that the rain falls off them without entering into the cells, or remaining upon them in the least; moreover, they are turned towards the east, and as the bad weather, cold rain, and strong winds usually come from the opposite direction during the summer and autumn months, the inhabitants of them have not much chance of suffering from causes acting from without. The little *Polistes* having taken care of its tiny comb, the larvae are soon transformed into nymphs, and before this occurs they close their cells with

Fig. 5.28. *Polistes gallica* **and nests.**

a little lid, by forming a silken cocoon. Some observers state that as soon as the adults fly it becomes evident that they are all workers, and that they begin to extend the nest, which, however, never reaches a great size. Fifty or sixty cells constitute a large nest, but it happens sometimes that a second comb is attached to the first by little pillars, as is the case in the nest of wasps.

In the engraving are representations of one large and two small nests of *Polistes gallica.* That on the left-hand side projects from a peduncle which is attached to a stem. On the right-hand side there is a front view of a small nest, and a *Polistes* is building the cells. The large nest shows many open cells, a few of which contain larvae, whose heads are placed outwards, so as to receive food from the workers; there are other cells which appear to be closed in consequence of the larvae having spun their cocoons within them.

Professor C.T. von Siebold has made most careful examination of the habits and peculiarities of these wasps, and he notices that isolated female wasps originate different nests, or colonies, in the spring-time. He has proved that these females were born in the previous summer, and that they lived in the virgin state until the autumn, when they were fertilised just before their sinking into the profound sleep of hybernation. He states that when the female thus fertilised awakens in the spring-time, she lays eggs which turn to females, but not to males. The progeny consists at first of small females, and not of workers or neuters, as stated above. Von Siebold has dissected many of them, and has found that they are not workers—that is to say, females whose reproductive organs are dwarfed and arrested in their growth—but that they are perfectly formed females, full of eggs. This progeny, consisting of virgin females full of eggs, assists the original mother, not in working, but in filling the cells, for the eggs they lay are fertile, although they have never been fecundated. The increase of the nest takes place rapidly, and the larvae, receiving abundance of nourishment, are transformed into wasps as large as the mother. Towards the end of June or the beginning of July the comb presents a large surface, and is composed of a very great number of cells. At this period some males individuals may be remarked for the first time among the numerous large and small females. Their number soon increases considerably. The observations of these facts suggested to Von Siebold that there might exist in *Polistes* a division of physiological labour—in this sense, that the fecundated females of the preceding year produce only female eggs, whilst the virgins of the new generation produce male eggs parthenogenetically. Experiment has confirmed this hypothesis in the most striking manner. Von Siebold selected a certain number of nests in the spring, at a period when the mothers had already reared one or two assistants. He removed the mothers from the nests and dissected them, in order to ascertain the condition of their egg-bearing organs. He always found evidences of fertilisation which must have occurred the year before. At the same time he entirely emptied all the cells of those nests which contained eggs, or any small larvae, preserving only the larvae of large size. Notwithstanding the disappearance of the mothers, the little virgins continued to take care of the larvae which had preserved, and consequently the colonies did not perish. He took the precaution to mark, in each of the nests experimented upon, occupied and empty cells. In a few days he perceived that some of the latter contained eggs. Careful examination even enabled him to surprise some of the little virgin wasps at the movement when they were depositing their eggs at the bottom of a cell. These individuals were at once sacrificed and examined. There were eggs within them in abundance, but no evidences of fertilisation by the male. During this time, thanks to the assiduous care of the young virgins, new female individuals—produced from the large larvae which had not been sacrificed—arrived at their complete development, and at once took part in the labours of the society. The nests were consequently enlarged by new cells, which were speedily occupied by eggs laid by the virgins. All these eggs—and this is an important fact—were developed

Fig. 5.29. *Polybia palmarum* and Nest

notwithstanding the absence of fecundation, and gave birth to young larvae, which prospered under the care of the virgin society. *All these larvae, at their final transformation, furnished males,* in oppotion to the larvae which had been previously produced by the original mother, and which had furnished *females* only. It is therefore evident that in *Polistes gallica* the male individuals originate by parthenogenesis, from unfecundated eggs.

There are some other wasps, which have very slim bodies, that constitute the genus *Polybia.* They are not known in Europe, but are very common in tropical America, where they form either enormous nests or tiny and beautifully-formed habitations. One of the species *(Polybia liliacea)* makes a most wonderful nest, which is often nearly four feet in height, and which contains thousands of cells. It is attached to branches of trees, and its covering, which is almost like cardboard, is rough, and is even attached to the internal combs. Mr. White described a huge *Polybia* nest from Central America, and noticed that its very thick envelope was covered with large pointed tubercles. On the other hand, some of the small species make their nests on the lower surface of leaves, and one which is here represented is common in Guatemala.

Polybia palmarum is a very small wasp, about a quarter of an inch long, and it constructs its nest on the lower surfaces of palm leaves. The nest does not occupy much space, and the breadth of one leaf suffices. There is only one oval-shaped comb in the nest, and it is clothed in an envelope made of a rather fragile reddish-coloured paper, in which there is a small inferior aperture. The size of the cells of this comb is very minute, and does not reach to more than from the twenty-fifth to the twelfth of an inch. The regularity and the perfection of the construction of the cells are very wonderful, but the insects build so well, that although the palm leaves move with the wind, and sometimes very strongly so, the larvae are all safe within the delicate little nest.

Some foreign species of wasps belonging to the genus *Chartergus,* make their nests of a substance like pasteboard; hence they are called Pasteboard Wasps. The nest of *Chartergus nidulans,* a native of Brazil, has a beautifully polished white appearance, and is so solid as to be impenerable to the weather. Reaumur showed some of the substances of which these nests are composed to a cardboard manufacture, who declared it to be most likely the produce of a particular manufactory at Orleans.

Another wasp, a native of Brazil, which is a honey-making insect, is said to form its nest out of very different materials, for it collects the dried dung of a species of tapir, and moulds it into a rough sort of paper.

A black wasp, which has smoky-coloured wings and a very large head, lives in Guiana, and is called the Armadillo Wasp *(Patua Morio).* Its nest is perhaps one of the most wonderful examples of intelligent design in insects that can be put forward. These wasps choose the straight and upright branches of a tree which has no lateral twigs, and they make it the axis or the support of the nest. The combs, which are composed only of a few cells, are fixed to the branch by means of a very solid mass of wax. They are separated by a considerable interval, are placed one over the other, and sometimes there are ten of them. In the accompanying engraving their numbers and mutual relations are very well shown. The walls of the nest are formed by a fusi-form envelope, made up of a woody paper, and marked with transverse tubings, which are, as it were, goffered. Moreover, this paper, the fibres of which are arranged with wonderful regularity, is tinted with longitudinal bands of different colours. The envelope is attached closely to the branch above and below the extreme combs, and is perforated low down by a small opening for the insects. Inside there is plenty of room between the combs themselves, and the envelope is not attached to them, so that the wasps can pass up from the bottom to the top of the next with ease.

Fig. 5.30. *Tatua morio* and Nest.

Bees

The nest-making *Hymenoptera,* which collect honey and pollen, are more interesting, so far as their organisation is concerned, than any others; and if it were not for the ants, they might also lay claim to possessing greater intelligence and higher instincts than any insects. All the insects which form the family of the *Apides*—of which the bee is the commonest type—have in fact a higher organisation than any others. The concentration of their nervous system is greater than in any other *Hymenoptera;* their respiratory apparatus is unusually developed, the tracheae of their abdomen being transformed into large sacs; and the appendages of the head are especially adapted for particular purposes, their structures being very elaborate. The mandibles are turned into pincers fit to cut wood, or into instruments for digging and working earth, and they are formed to enable the insects to knead cement in moulds to pound their wax in, according to the habits of different species. Among most of the bees the jaws and the lower lip, which are long and delicate, are used as a trunk to suck up honey from the nectaries of flowers; and the hind legs, especially the first joints of the tarsi, are transformed so as to act as pollen collectors.

The bees are vegetarians during the whole of their lives; the adults fed upon honey; and the larvae, which require more substantial aliment, are supplied with a sort of cake, which is composed of honey and pollen. The larvae of all these *Hymenoptera,* like those of the wasps and the fossorial kinds, are incapable of moving about, and are destined to live in the narrow cells of the hive, where they are nursed by particular bees. Some bees are solitary, and others live in colonies, which are either permanent or only last a year. The solitary spectes lead of life like that of the fossorial *Hymenoptera;* and a single female constructs a nest, makes the cells for its larvae, and victuals them. The social bees have three kinds of individuals—namely, males, fertile females, and neuters, which are also called workers or sterile females. In those colonies which only last a year some solitry females lays the foundation of the hive, and the workers are gradually born and increase in number; but the fertile females which are found in permanent colonies never do any work, and, indeed, they are incapable of it, for the necessary structures are wanting in them. There are certain species amongst the bees—whose habits are like those of some insects which have already been noticed in treating of the *Sphegidae*— which have the instinct to enter the nests of industrious kinds, and to deposit their eggs in the cells that have been partly filled with food intended for the larvae.

The bees we shall first notice differ from all others by not having their hind legs made to collect pollen, and the genus *Osmia* contains the most important of them. These *Hymenoptera* build very elaborate structures, and take almost of fantastic care about the different materials which they use for their cells, each species choosing particular substances. They collect pollen, like other nest-making *Hymenoptera,* and make a cake of it for their larvae, but their legs do not assist them in collecting it. But there are some stiff and crowded hairs which cover the lower part of the abdomen of these bees, and they answer the purpose of pollen collectors just as well as the hind legs of the other bees. The *Osmia* rubs itself against the stamens of flowers, and detaches the pollen, then it collects some grains against the hairy structure, which retains them, and then the insect returns to its nest carrying a greater of less quantity of this important food. Of course, the first care of the bee should be to get rid of its burden, but it does not appear, at first sight, how this sticky pollen is to be moved from off the hairs; but a structure is at hand by which this difficulty can be overcome at once, for the *Osmia* is provided with a kind of brush, which is placed under the first joint of the hind legs, and as these extremities can be moved in all directions every morsel of pollen is speedily detached from the whole body. One of the commonest *Osmiac* is the mason bee of Reaumur *(Chalicodoma muraria).* This species has a massive body and strong mandibles, which have four small teeth upon them, and its name, which signifies "stone house," is by no means badly chosen. The female mason is of a velvety black colour, and the male, which is always smaller, is clothed with a russet pubescence, the end of the abdomen being black.

Fig. 5.31. *Chalicodoma muraria*. A male in flight. A female working at her nest. A female seen in front. A completed nest.

Spots of earth may often be seen on walls which have a sunny aspect; they are more or less egg shaped, and look like lumps of mud which have been thrown against the wall and become dried up. They are the constructions of this *Hymenoptera*, and when they are examined closely they will be found to be as hard as Roman cement, and to be made up not with mud, but with a mixture of gravel and earth. The lumps are more or less arched, and it is very evident that their rounded state is not a matter of chance. This mixture of earth and gravel sticks to the wall with the greatest tenacity, and the chisel and hammer are requisite to detach it. Now, these ugly and dirty looking masses have been carefully built up, with a wonderful amount of art, by a single *Hymenoptera,* which has been the architect and workman at the same time. The labours of the *Chalicodoma* begin in the month of May, and soon after it is born. The female explores and goes over a wall and chooses a particular spot, and having settled that essential point, she goes off to collect building materials, and if followed, will be noticed to alight upon a sandy and gravelly soil. The insect takes up little masses of sand and small stones, of a certain size, with its mandibles, then it disgorges a little saliva and sticks them to some grains of earth, and thus agglutinates the earth and the sand so as to form the mortar which it is going to use in building. When it has worked up a little piece it flies off with its burden, and returns to its wall to fix on this first quantity of cement. The same trouble is taken over and over again, and then, the mass of mortar appearing sufficient to permit other building operations to be commenced, to be sets to work to mix all the earth, and usually she labours so well as to complete a little cell in one day. This cell is open to a certain extent, and the bee enters it several times to make the inside walls smooth. Then something else besides building has to be done, for the cell has to be victualled, and the *Chalicodoma* flies off to collect honey and pollen from the flowers. It makes up a sweet cake from them, which will be the future nourishment of its larvae, and then it stuffs the cell nearly full of this food, and lays an egg in the midst. The *Hymenoptera* walls up the cells and sets to work to construct another close to it, then a third, and sometimes as many as twelve or more. These cells are placed rather irregularly, and are not found to be of the same number in every nest. All the cells are built, filled with provisions, and closed perfectly, but yet the labour is not over, for the bee builds up a sort of roof or general covering to the whole, and uses larger sand grains than those which enter into the composition of the cells, and thus the external walls of the nests are found to have great thickness and wonderful hardness. The larvae will have plenty to live upon, and certainly every care has been taken to place them out of danger. When they are full grown they add to their imprisonment by making a cocoon of a paper-like tissue, which looks as if it were varnished. Their metamorphosis is completed within, and the adult insects fly. But how do they manage to get out of their strong and prisonlike home? Do they contrive to pierce the cement, which is as hard as stone, and which resists the blow of a hammer? Every care has been taken not to give too much trouble to the young bees in their first attempt at coming to the light. When the roof and general wall of the nest was constructed, a slit was left low down near each cell, and it forms a kind of door, hidden very carefully by rather soft sand or cement.

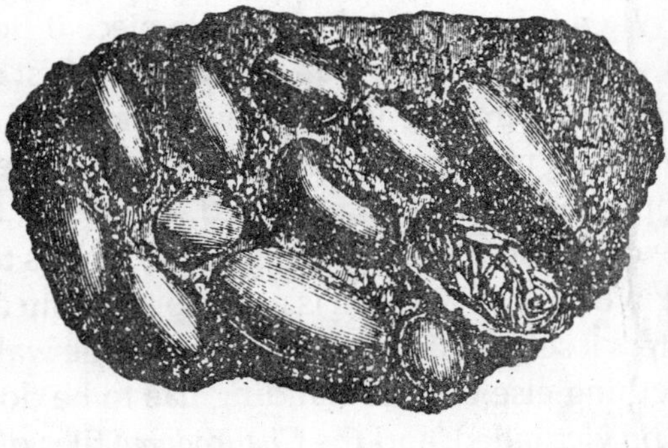

Fig. 5.32. A nest of *Chalicodoma muraria* Detached and seen from within.

It is very remarkable that this insect should not always build up its nest in the same manner, and that it should occasionally use the ruins of last year's nests for the foundation and walls of the new. Old and more or less broken down nests, which contain vacant cells and the skins of the nymphs, frequently remain attached to walls, and the *Chalicodoma*, when exploring, appears to keep a sharp look out for them, and takes care to spare itself all the trouble possible by repairing the broken down cells and making them as good as new. The female cleans out the inside, removes the debris of the cocoons, and brushes out the skins of the larvae and nymphs, and all uncleanly matters. Then the holes are repaired, the openings are closed up, and the old nest begins to resemble one of the year. Sometimes the mason bee becomes waxy, and then she penetrates into the nest which some other one is making, and, if she likes the position endeavours to maintain it by force. Reaumur says that when one of these bees was gone to collect some cement wherewith to finish up the cell, M. du Hamel saw another one enter in without any ceremony, crawl all about, and turn everything over, just as if the nest belonged to it. That this was not done out of curiosity was gleaned from the fact that when the true mistress of the nest arrived, laden with building materials, the other one would not allow her to enter, and she was obliged to fight the usurping bee in order to obtain her rights, and the new comer was quite ready for the combat. An *Osmia* which lives in Algiers has the wit to choose snail shells for its nest, and forms cells within them with a mixture of earth and cow-dung. Another well-known insect *(Osmia aurulenta)* seeks for the dried twigs of brambles, clears them out, and constructs cells inside, one after the other, with a small quantity of mortar. Its work is comparatively easy and simple, especially when compared with the labours of the *Chalicodoma* and other *Apides*.

In the engraving the perfect insects are seen in flight, and one is crawling upon a leaf. The cells within the bramble are occupied by cocoons, one of which is opened to show a nymph.

Formerly most villages and country houses were covered with thatch, and a stunted vegetation grew over the roofs, in which many insects found very comfortable homes. Many may recollect the pretty villages with their thatched cottages which were so common before slates began to be used, and many naturalists were aware that nearly every straw of those simple roofs was the cradle of many a numerous family of

Fig. 5.33. The *Osmia aurulenta* and its nest.

little bees. Now-a-days granaries and barns tell the same story, and little *Osmiae,* which have long and notched mandibles, make comfortable galleries in the old straw. These insects are called *Chelostomes,* and they are so small that the inside of a straw forms a spacious gallery for them. Their strong mandibles are admirably adapted for cutting into the straw, and when this is done the inside is properly cleaned out, and cells are made within with the aid of a little mortar. It happened, therefore, in the olden time that the labourers inside the cottages protected and gave lodging unintentionally to hundreds of industrious and hard-working little bees.

Some *Hymenoptera* which belong to the bees are leaf cutters, and have very long mandibles with four teeth upon them, and they are the instruments for perpetrating this clever piece of work. These bees are called *Megachilidae,* and the commonest species, which is well-known in every garden, and much dreaded in green-houses, is *Megachile centuncularis.*

It is not a remarkable looking insect; in fact no one would believe it to be gifted with so much intelligence by observing its very humble exterior. These bees are smaller than hive bees, are black, and have some red hairs on the thorax, and a white down upon the head. A corresponding down ornaments the first three segments of the abdomen in the females, and all the segments of the male insect.

The *Megachile* makes a perpendicular hole in tolerably solid earth, works it downwards for some inches, and then enlarges it into a horizontal gallery of considerable length. This is to be the future nest, and cells will be made in the side. There is nothing very extraordinary in this, for it is the procedure of most of the *sphegidae, Crabronidae,* and the solitary bees, but something else is about to be done. The insect flies about the rose bushes, and finally selects a nice leaf, upon which it rests either above or below; it then cuts out a large piece with its mandibles, just as if it had been done by a pair of scissors. No time is lost, and the insect cuts quickly and soon detaches the piece of leaf, which does not fall, however, to the ground, heavy as it appears to be in relation to the small insect, for it is taken up between the legs and the mandibles, whilst the wings are strongly agitated, and the bee flies off with it to the nest. The *Megachile* will thus cut ten or twelve pieces of leaf of different shapes, and will carry them to the bottom of its gallery, and there she twists the different pieces and folds one within the other, so as to form a sort of funnel-shaped cone, the end of which is rather narrower than the orifice. The whole looks something like a long sleeve with folds upon it. The pieces of leaf are not sewn or attached to each other, but they are fitted together so perfectly that they retain their shape even when they become dry. This pretty funnel-shaped cell, of from a quarter to a third of an inch long, is destined to be the residence of a larva of the leaf cutter, for it proceeds to make up a cake of honey and pollen, places it inside, and then lays

Fig. 5.34. Leaf Cutting Bees and Nests.

an egg. Nothing more is required but to wall up the nest, and the *Megachile* goes off to the rose bushes again, where it snips off a circular piece of leaf, of a diameter exactly fitting the opening; it pushes this leaf so that its sides fit into the walls of the funnel-shaped roll of leaves which compose the lining of the cell, and endeavours to close it completely. But this intelligent little worker is not satisfied, for it would appear to be aware that the honey which she has laid up with the egg might flow out through some narrow slit; so she goes off and cuts two more circular pieces, and uses them as she did the first, and thus the little cells are closed by three layers of leafy covering, and sometimes by a fourth. A second cell is constructed in the same manner, and its end fits into the opening of the first, and thus a series of eight or ten of them is usually found, looking like a long tubular sleeve. When all this is completed, the eggs being laid and the cells victualled, the *Megachile* closes the hole of its perpendicular shaft with the earth which she dug out in the first instance, and this is done so well that there is no trace of it left.

In the engraving the leaf-cutting bees are depicted either in flight, on the leaves, at work carrying the leaf to their hole, or upon the ground, and one is completing a cell underground, on the left hand side.

There are some other bees which have three teeth to their mandibles, and which, perhaps, are more particular in making up their cells than the leaf cutters. They are called the flower cutters, and one of them, which uses the poppy as its material, is not very uncommon. *Anthocopa papaveris*, which selects the common scarlet poppy, is a small insect of a velvety black colour, with a white down on the edges of the segments of the abdomen. The female digs perpendicular holes in dry and sandy soils, and pounds their sides so as to make them hard and lasting; then it search for poppies, and cuts off pieces of their petals, choosing the freshest, youngest, and the most beautiful flowers. The insect flies off with the piece of scarlet stuff, introduces it into the hole, and pushes it down. Of

Fig. 5.35. *Xylocopa violacea.*

course, the delicate tissue is ruffled during this proceeding, but the bee presses it against the sides of the hole, and works away until every fold is removed and the gaudy tissue becomes perfectly smooth. Then three or four pieces are cut off consecutively, and introduced into the hole and made perfectly smooth, and, finally, the honey-cake and egg are placed inside this beautiful little piece of workmanship. Before closing the hole the insect appears to remember the peculiar nature of the soil in which its cell, and in order to prevent any grains of sand from falling upon the honey, the bee folds in the free

Fig. 5.36. *Amthoropa papaveris* and nests.

ends of the lining of the cell, just as a man would fold in the top of a sack; this being done, the opening in the earth is filled up, and no trace of it is left.

These bees are represented in the engraving, some flying, some cutting pieces of red poppy, and others are either upon the ground or entering a nest.

The Carpenter Bee *(Xylocopa violacea)*, which is not uncommon, is larger than the largest humble bee, and may be known by its velvety black body and violet-coloured wings. This large bee, when about to construct its nest searches for pieces of wood which it can pierce without much difficulty; then it sets to work and gnaws away with its mandibles and makes a hole, working industriously and with no small difficulty. This bit of carpentry may take several weeks, as the hole sometimes reaches to a depth of a foot or more. But if the bee should happen to find an old gallery in a tree, or any tubular holes, she takes possession of them, establishes cells within, and makes their party-walls with sawdust, which she agglutinates with her saliva. The diameter of the holes is often half an inch, and after the sawdust is formed into a paste it is arranged so as to make the floor of the succeeding cell, in which an egg is deposited with the usual supply of pollen paste; this is covered with a layer of the sawdust, and thus the floor of another cell is made; and about a dozen are thus constructed one above the other.

There are some very common bees, which abound in the hot districts of Europe and Africa, which resemble hive bees, but are larger and more hairy. They are solitary bees, and belong to the genus *Anthophora*. One of them is a black insect, clothed with grey hairs, and having the last segments of the abdomen russet coloured. It digs holes between the stones of old walls, and collects all the sand dug out, agglutinates it with viscid saliva, and fixes it in front of the hole, arranging it in the form of small cylinders. This little vestibule, like that of an insect already described, is destroyed after the nest has been provisioned, and its materials are used to wall up the entrance. Other *Anthophora,* which make nests in rubbish, do not collect the substances which they have dug out, for they have plenty of materials at hand.

Anthophora personata has fawn-coloured legs, and makes a nest in loose earth, and builds five or six cells which open into a common gallery. When the cells are opened, large larvae, or nymphs, as the case may be, are found. The larva of this *Anthophora* remains upon its side, and is a stout and large insect, especially as regards its abdomen, and the creature seems to be content always to remain in this curious position, with the body rather bent in front. In the engraving several of these solitary bees are drawn in flight, and a section of the cells and gallery is drawn on the right hand. Two of the cells contain nymphs, and two others, larvae.

The bees which act as parasites upon others have been collected in one group called the *Nomadinae.* Some of these are parasites of species of *Osmia,* and others of *Anthophora;* and these different insects have great resemblances to the kinds they rob. But there are certain peculiarities of structure which distinguish all the parasites, for they invariably have simple hind legs and no masses of hair which enable them to collect pollen readily, for they are not destined to work, and therefore they are not supplied with the necessary implements. Their jaws and lip are very short, and they have no necessity to gorge themselves with honey, for they will never have to supply their young with it. The true *Nomadinae* are slender-shaped *Hymenoptera,* are almost smooth, and resemble little wasps on account of their black and brown colours variegated with yellow. One species *(Prosopis signata)* introduces its eggs into the nests of other solitary bees; and the *Melectidae,* which are called Cuckoo Bees, and which have a large body not unlike that of the *Anthophora,* are the parasites of these last-mentioned insects.

Everybody knows the Humble Bees, the large velvety *Hymenoptera,* that hum about the garden and make so much noise when they are caught in a room. They form colonies or societies, which commence, enlarge, and end like those of the wasps. A fertile female takes upon herself the duties of a worker early in the spring, after having passed a long hybernation, and lays the foundation of the future swarm. She soon finds herself surrounded by workers, which enlarge the nest. Thus one humble bee *Bombus muscorum)* builds nests which are not very easily found, for the female establishes them in little cavities in the middle of fields, and takes care that they are in the immediate neighbourihood of plenty of flowering herbs. Mowers often cut into these nests, and if they are examined they will be found to be composed of a vaulted roof formed of species of moss cleverly entangled and fastened together. The opening of the nest, however, is from below. If the covering of moss be removed, a second will be noticed, but it is formed of a thick layer of greyish-yellow wax. The moss, although collected in considerable quantities, would permit water to enter into the nest during heavy rains, but the lower roof of wax keeps everything beneath it water-tight. After having removed the waxy covering, the interior of the nest comes into view, and a mass of little round pellets crowded one over the other is observed, and these little things form the paste which is destined to nourish the future larvae. Mixed with them, and encircled with wax, there are either eggs, larvae, or cocoons, and at certain times of the year cups of wax, which contain a very pure honey, may be seen on the sides of the nest. The larvae resemble those of other bees, and they spin themselves cocoons, which are egg-shaped, and which are usually found pressed one against the other. They are of different dimensions—the smallest, always the most numerous, are the cocoons of workers; those of medium size belong to males; and the largest contain fertile females. Before getting out of the cocoons they have to be cut round at one end so that a portion comes off resembling a lid.

Fig. 5.37. The Metamorphoses of *Anthophora personata*.

In the accompanying engraving there are many workers of this bee, and a large female is represented on the ground on the right hand side. The cocoons are exposed to view.

Another humble bee *(Bombus terrestris)* is a very large kind, and forms its nests in the earth at a considerable depth. These bees do not use mosses, but make a lining to the roof of their dwelling with layersm of wax.

Fig. 5.38. *Psithyrus vestalis.*

Formerly certain *Hymenoptera* were classed by all naturalists amongst the humble bees, and, indeed, a superficial examination of them permitted this to be done, for they have the same shape size, and colours, and look like them; but Lepeletier de Saint-Fargeau recognised them as parasites, and collected them under the genus *Psithyrus.* These insects which are so like the humble bees have only two kinds of individuals amongst them, for none of them are workers. There are males and females, and these have as might be expected, no implements for working. Every species of humble bee has its parasite, its particular *Psithyrus;* and curiously enough the parasite has the peculiar colouring of the insect it robs, and usually there is no difficulty in telling the habits of a *Psithyrus* by its shape and tints. There is something in this resemblance more than chance, for it will be found that some flies, not wasps or bees, but *Diptera* of the genus *Volucella,* which are parasites in the nests of humble bees and wasps, resemble their unwilling hosts in their colours. One of the commonest species of the genus *Psithyrus (Psithyrus vestalis)* resembles the common humble bee of our gardens. It enters the subterranean nests of this species, lays its eggs, and then returns to the fields.

The habits of the common hive bees, and their methods of cell building, are so well-known, that it is not necessary to describe them here. It is a curious fact that many virgin bees lay eggs whence male larvae are always derived, and that this female after fecundation lays other eggs which turn into workers and fertile females. This faculty forms part of the phenomena of parthenogenesis, which have been so carefully studied of late years. Hive bees when wild establish themselves in cavities, such as those inside old trunks of trees. They make their hive, and their society consists of a crowd of workers and a single female or queen. Cell making goes on, the queen lays, and larvae are born, and a new generation in thus produced; and thus some hundreds of males may be counted in a hive at one particular period of the year, besides the queen and the workers.

These three kinds of individuals present striking differences. The fertile females and workers of wasps and humble bees are very much alike; the fertile females are only slightly larger

Fig. 5.39. *Bombus muscorum*

than the others—but were they not workers at one period of their existence? The queen bee is incapable of working; the workers or neuters are smaller than she is, and are not so large as the drones or males, and it is not necessary to describe them, for they are well-known to every one.

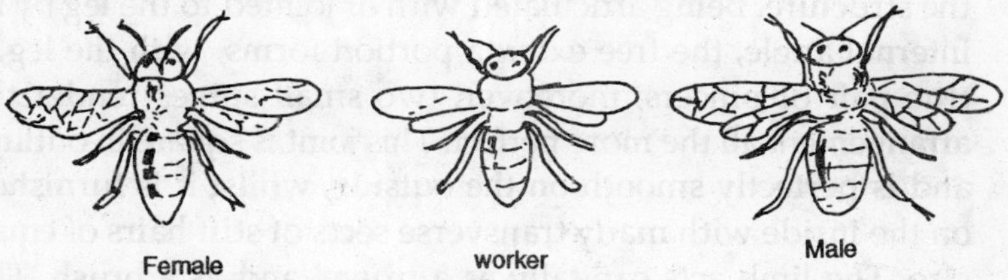

Fig. 5.40. Domestic Bees (*Apis mellifica*).

The production of wax is one of the most remarkable physiological phenomena of the organisation of these *Hymenoptera*. It was generally thought, formerly, that the bees disgorged their wax from the mouth, and Reaumur certainly held this opinion; but John Hunter discovered the manner in which the wax was formed; and it is now evident that the bees carry within themselves this important building material. The segments of the abdomen of bees overlap from before backwards, but then the margin of one is lifted up, two broad and smooth surfaces will be noticed on the uncovered surface of the next wing; these surfaces maintain during one part of the year two thin, white, and almost transparent laminae, which are really composed of wax. The wax is really secreted by some small glands which are within the abdomen, and it transudes through the soft and smooth integument between the rings or segments. It would appear that the sugary matters which are sucked and digested by the bees are to a great extent transformed into wax, which is to all intents and purposes a sot of fat.

The hind legs of the bees are so important to them in everything relating to the larvae that it is necessary to notice them briefly. They are very wonderful implements. The leg is much enlarged near its farther end, and resembles a long triangle in outline; a set of sharp points or many strong spines, which are regularly distributed, form a kind of rake towards the extremity, and this implement enables the bee to seize the waxy plates that are between its abdominal segments. The first joint of the tarsus or foot succeeds the leg and attains a great size, when compared with that of the other members of the structure. Being articulated with or jointed to the leg by its internal angle, the free externa portion forms, with the leg, a true pair of pincers; moreover, two small spines render the arrangement all the more perfect. This joint is square in outline, and is perfectly smooth on the outside, whilst it is furnished on the inside with many transverse sects of stiff hairs of equal size. The limb acts capitally as a trowel and as a brush. The fertile female, which never works, has the traces of this arrangement, and the males also; but neither of them has the pinching and brushing structures. These are restricted to the workers. The fore legs have a deep notch in the first joint of the foot, which is very useful in building, and there are some long and close hairs which form a nice little brush, which can often be placed in contact with the brush behind.

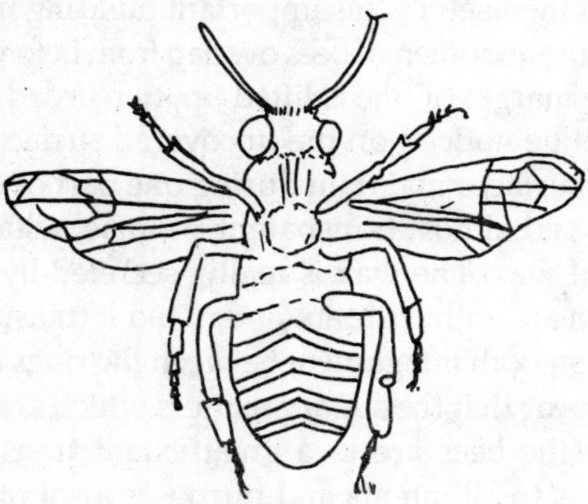

Fig. 5.41. Under surface of a Bee, showing the wax Between the segments.

The mouth has some remarkable structures in it. The mandibles enlarge towards their ends, and, as they terminate in a plane surface with a projecting border, they only close together sufficiently to permit the proper quantity of wax to rest between them that should be worked at one time. The jaws and lower lip are long and flexible, and constitute a most perfect appartus for sucking up honey.

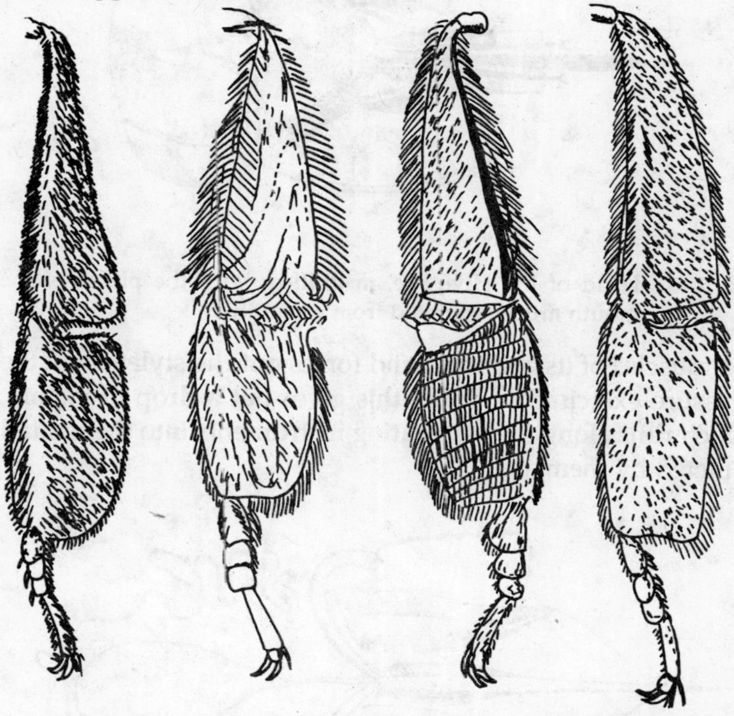

Of a fertile female. Of a worker, seen from outside. Seen from the inside. The male.

Fig. 5.42. Hind Legs of the Bee.

The females and workers of nearly all the nest-making *Hymenoptera* are armed with defensive weapons, which very with the type of the insect. The sting of the bee, or rather its effects, are well-known to most of us. It is made up of two very sharp stylets, which are mounted upon scales attached to the last segment of the abdomen, there being two valve-like sheaths which encase and protect them. The poison glands are formed by two twisted tubes which lead into a large reservoir, the small opening of which is in a tube that emerges between the piercing stylets. When the bee is about to sting, it contracts

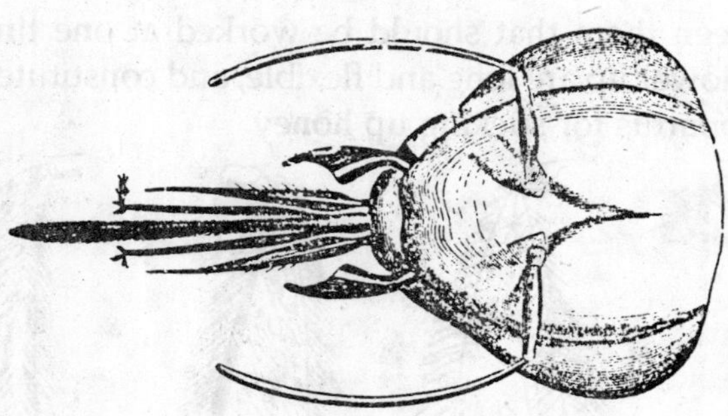

Fig. 5.43. Head of the Hive bee, magnified, with the pieces of the mouth much separated from each other.

the muscles of its abdomen and forces out the stylets, and the pressure exercised in doing this gives out a drop of venom, which runs along the perforating instruments into the wound inflicted by them.

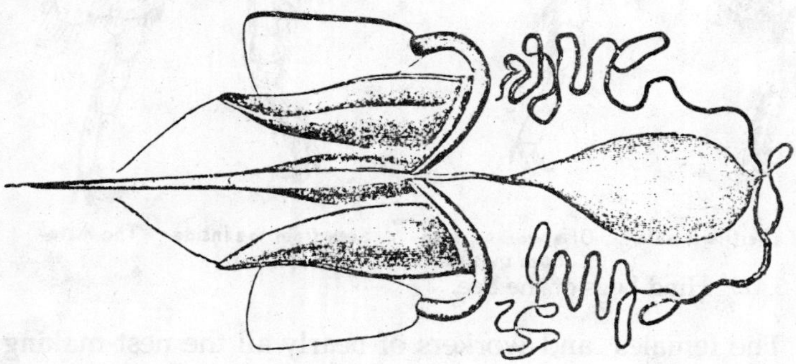

Fig. 5.44. The sting and Venom Glands of the bee.

When a swarm of bees has taken possession of a cavity—of a hive, for instance—the workers immediately close all cracks and interstices by using a resinous matter, which is admirably adapted for the purpose, and which is obtained from the buds of poplar and other trees. This substance is called *propolis*, and is usually found in large quantities in the upper part of the roof of the cavity, where the comb is to be fixed eventually. This is the first part of the labour, and, when it is completed, the bees unite in groups, so as to form the cells of the comb. Each bee seizes a lamella of wax from its own body with the pincer on its hind legs, takes it to its mandibles with the fore-legs, and kneads the mass until a small round piece of proper consistence is formed. This is struck on to the vault of the hive, and, when a number of such pieces has been collected and placed in their proper position, the cells are hollowed out by the workers, which excavate them side by side, and form them into the regular hexagonal shape by pressure.

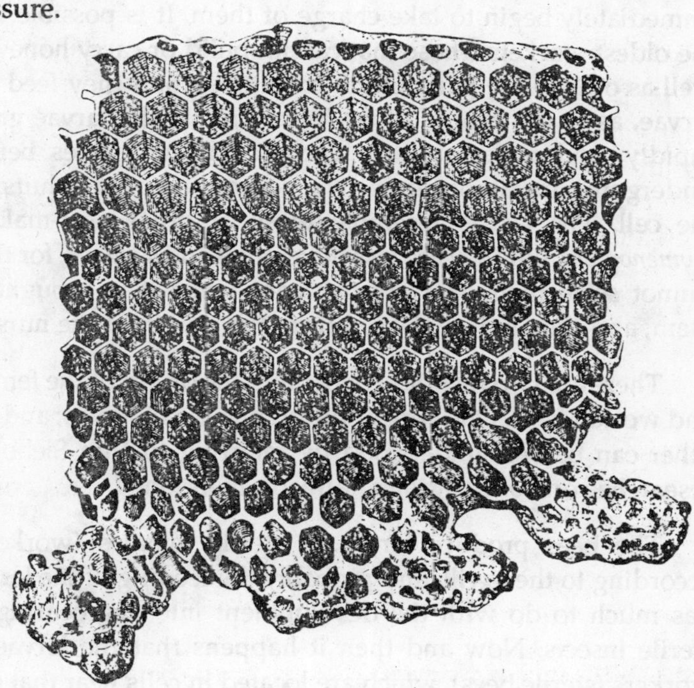

Fig. 5.45. A Portion of Honeycomb.

The size of the cells varies, but the majority of them are small, some are slightly large, and a few are of considerable size, and are usually at the ends o the combs. The smallest are for the larvae of workers; the next size will be filled by those of males, and the largest cells will contain queens or fertile females.

As soon as the cells are funished the queen or fertile female of the colony, or swarm, runs over the surface of the combs and lays an egg in each cavity, and she is attended by a host of workers, who take care that only one egg is dropped into each cell. Should two fall in, one is pulled out and destroyed. When the laying is finished the work of the queen is at an end, and she does not concern herself in any way about the future larvae, or the attentions of the workers to them.

The larvae are hatched in three days, and the workers immediately begin to take charge of them. It is possible that the oldest workers (those that cannot build or carry honey so well as others) are particularly the nurses, and they feed the larvae, as in the case of other *Hymenoptera*. The larvae grow rapidly, and make a silken cocoon for themselves before undergoing metamorphosis. The cocoon, as usual, shuts up the cell. The larvae resemble those of other nest-making *Hymenoptera*, but they are less developed and weaker, for they cannot seize their own nourishment when it is brought to them, and they have to be fed from the mouth of the nurse.

There are two theories concerning the origin of the female and worker bees. One very doubtful one is as follows; and the other can be best understood by referring to Von Siebold's researches on *Polistes gallica*, already mentioned.

The bees produce fertile or sterile females (workers), according to their will, and probably the feeding of the larvae has much to do with the development into egg-bearing or sterile insects. Now and then it happens that the larvae of workers (sterile bees), which are located in cells near that of a fertile female larva (a future queen) receive, more or less accidentally, some of the particular nourishment intended for their important neighbour. The small quantity of royal food alters the condition of the development of the reproductive organs in the workers, and they become more or less able to lay. When a hive has lost its queen the bees produce others by feeding up the larvae which should be workers.

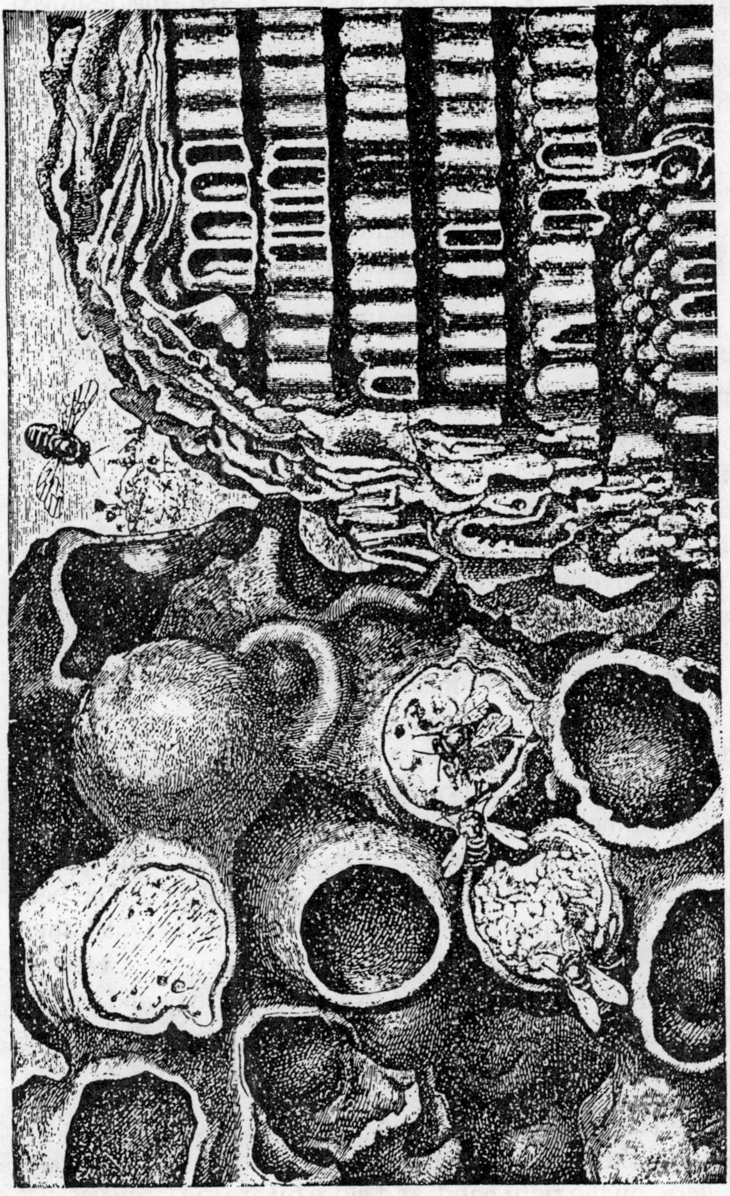

Fig. 5.46. The nest of *Melipona scutellaris*.

A larva is chosen for this important position, and its cell is enlarged, the neighbouring insects and their cells being destroyed to make room. Then the larva is fed up royally, and the development of its reproductive system increases, so that finally a change into a fertile females is perfected.

It is perfectly certain that a maiden female can produce a progeny, and that when there are several fertile females associated with males they fight until one is the conqueror, the workers dragging out the dead, and witnessing the combat.

The influence of parasites upon the larvae is considered elsewhere, as are also the destructive habits of the death's head moth.

There are some short and squat little bees with long legs and very varied colours, which have no sting, and several fertile females are said to live together in harmony amongst them. There may be some doubt about this last statement, but the species lives in South America, where naturalists are rather scarce. There are many species of these *Meliponae*, and *Melipona scutellaris* is one of the largest, so far as the bulk of its insects is concerned. These bees make their nests in hollow trees, but are not slow to occupy a box or a basket, and they are soon domesticated. The engraving represents part of a nest, only, for these structures are very large, and the *Meliponae* are just as prodigal of their wax as the other bees are careful and sparing of it. The centre of the nest is occupied by the cells for the larvae in the form of a comb of one set of alveoli. The bees hoard up their honey in cells exactly like those in which their larvae are to live, but the *Meliponae* make very large vases of wax, in which they collect and store away honey and pollen, and locate them close to the larvae cells, and the walls of these great store-houses are very thick and strong.

The single rows of cells in combs placed one over the other are shown in the engraving of the nest of *Melipona scutellaris* on the right hand, and close by are the large vesicles of wax, which are being filled by the workers.

All the *Hymenoptera* we have noticed are terrestrial, and the perfect insects either fly in the air or crawl on dry land. Immersion in water is fatal to them after a time. The larvae of many *Ichneumons* appear to exist surrounded by the juices of their victims, and they must respire either by means of the air passing through the air tubes of the insects they live in, or by the more or less aerated fluids which form their nourishment. But the majority of the hymenopterous larvae respire by spiracles, and cannot live under water.

A discovery of Sir John Lubbock's is, therefore, very interesting. He observed many specimens of a small hymenopterous insect swimming in pond water with their wings, and quite at their ease. The fly belonged to a genus the larvae of which are eminently parasitic in the eggs of some *Lepidoptera;* and the accomplished naturalist just mentioned considers it highly probable that the slow-flying swimming insect was in search of some aquatic victims for its progeny. He found a second species, which used its legs as oars, but did to swim with its wings; and he noticed that both kinds of these *Polynemae* were fond of creeping along the sides of the vessel in when they were placed, or on the leaves and stems of aquatic plants. They frequently swam boldly out into open water. The history of the metamorphosis of these aquatic *Hymenoptera* has yet to be written; and it is evident that its comparison with that of the *Hydrocampidae* amongst the *Lepidoptera* will form a most interesting inquiry.

The metamorphoses of the *Hymenoptera* differ from those of the *Lepidoptera,* in consequence of the larva remaining for a greater or less period inside a cocoon before transforming into a pupa. Instances of retrograde metamorphosis appear to occur in the genus *cynips,* and the wingless condition of the female members of the species of *Cynips terminalis,* and the absence of wings in *Cynips aptera,* are usually stated as examples. But the metamorphosis can hardly be said to be retrograde in the strict sense of the term, for the perfect female of *Cynips terminlis* and the adults of both sexes of *Cynips aptera* are more fully develop than their larvae and nymphs, and possess additional structures. In no sense can they be said to be lower in the insect scale.

The transformations of these insects are not then retrograde, and they are not incomplete in the usual sense of the term, for there is a marked distinction between the three stages of growth. The nymphs have not the same powers and structures as the adults. The *Cynips aptera* never has wings; the transformation is imperfect, and the evolution of the insect does not reach the point usually attained by the other *Hymenoptera.* Wings would be of infinite use to the *Cynips,* and it does not appear that there is anything relating to the natural history of the species which would account for the loss of wings by disuse during many succeeding generations. The wingless condition and the imperfect metamorphosis probably have a deep signification, and these exceptional *Hymenoptera* may be the oldest of their order; for it will be shown, in succeeding chapters, that there is great reason for believing that wings have been superadded to the earliest insects during the long geological periods.

Nothing can be more diverse than the development and instinctive powers of the larvae of the saw flies, the parasitic *Hymenoptera,* and of the ants and bees; and it is remarkable that the very imperfect larvae of the last two kinds should be transformed into much more intelligent adults than the others. The examination of the structural changes which occur in the nervous system during the transformations of the bee, will suggest that the concentration of the nervous masses in the adult has much to do with the development of great power of instinct. The occurrence of parthenogenesis in *Cynips,* and in some of the wasps and bees, has been proved; and it has been noticed that the workers or wingless adults are probably the result of this process. They are without the power of reproduction, and thus their metamorphosis is a stage more imperfect than that of *Cynips aptera.*

Very little of the intense activity and instinct of the *Hymenoptera* appears to be squandered upon the pleasures of the adults, and every unusual gifts and habit they may have, with nearly all their restless industry, are employed in one direction—in perpetuating the species and in preserving the young during their transformations. All the honeycomb making, the excavation nests, and the building-up of subterranean edifices, are for the preservation of the feeble larvae and still more helpless pupae.

6

Metamorphosis in the Coleoptera— Beetles

The *Coleoptera* are those insects which are commonly called beetles. There are great numbers of them, and perhaps they have been more carefully studied, so far as their specific differences are concerned, than any other insects. More pains have been taken to collect them than have been given to the discovery of any other orders, and special collections have been formed by a great number of amateur naturalists, who have searched the whole world over for these objects, which appear to be so very interesting to them. Books which contain simple descriptions of the *Coleoptera* appear year by year in the different countries of Europe and America, and some treat of one genus of beetles, whilst others refer to the species of a district. The beetles of France are not yet all described, and a distinguished naturalist, M. Mulsant, of Lyons, has been at work at them for more than twenty-five years. This may give us an idea that the species of beetles are tolerably numerous; but really their numbers are legion, and it may be asserted, with a tolerable degree of truth, that 100,000 species have been collected and placed in museums. The greater number of them have been described in large works, in periodicals, and in pamphlets, and it will readily be understood that it is not very easy to find the description of any particular beetle, in consequence of the scattered nature of the literature of the subject. Fortunately, Professor Lacordaire has published a work enumerating the beetles, and giving the names of books which contain their descriptions and drawnings.

Considering what charms the *Coleoptera* have for so large a number of naturalists, one might think that there was something unusually interesting about them; but this is not the case. The *Coleoptera* have no industrial occupations, and their instincts are very limited. In examining into their physiology we do not find any of those remarkable gifts and instincts, almost amounting to reason, that characterise the *Hymenoptera*, and which are amongst the most instructive and wonderful phenomena of Nature. It would appear that the charms of the *Coleoptera,* and the interest that they have excited amongst naturalists and collectors, are referable to the beauty of some of them, and to the ease with which they may be preserved, in the consequence of the density of their structures and the nature of their wings. There is no difficulty in distinguishing any one of the beetles from all other insects. The most distinctive character is in the wings, which, so far as the beetles are concerned, are four in number, and in two pairs. The fore wings are always of the same hard coriaceous texture as the other parts of the body; they never cross one over the other when the insect is at rest, and they cover the body more or less perfectly. They are sheaths, and act as coverings to the hind pair. The name *Coleoptera* is taken from the peculiarity κολεος, sheath; πτερα wings), and the sheaths are usually called "elytra." These elytra, when the insect is at rest, are joined to each other at their inner margins, and thus form a sort of middle line upon the top of the body. They open at right angles with the body and project straight out from it, and remain fixed in that position when the beetle is flying, and they do not beat the air or vibrate so as to produce motion. The hind wings, which are almost always much larger than the fore wings, are membranous, and are marked with many branching nervures, and these wings, when the insect has done flying, fold themselves up and are hidden underneath the elytra. The membranous wings are strongly vibrated during flight, and do not remain at right angles to the body like the others. The structures of the mouth are usually well developed and strong, and are fashioned for the purposes of mastication.

The *Coleoptera* undergo complete metamorphoses. They are born as larvae, like the *Lepidoptera* and the *Hymenoptera*, and the larvae are transformed into immovable or quiet nymphs, which are, like those of the *Hymenoptera* the exact casts or models of the perfect insects. The beetle or perfect insect usually passes a short existence.

There is great diversity amongst the *Coleoptera*, both as regards their anatomy, physiology, development, and the manner in which they live. Many appear to be inferior or lower in the insect scale than others; whilst some are so elaborately constructed that they must be placed in the most perfect class. The most decided anatomical differences, so far as the internal organs are concerned, coincide with striking distinctions in the external conformation of the insect; and in those beetles whose limbs denote superior powers the nervous system is very concentrated; and it is less so in those species the structures of which are more simple. The food of the *Coleoptera* is of all kinds, so, of course, the modifications of the structures of the digestive organs are very great. Many of the larvae are leaf-eaters, and certain kinds do a great deal of harm to all sorts of crops and trees. There is much interest attached to the early conditions of the *coleoptera*, and there is often a striking resemblance between the larvae of those families, the adult forms of which differ in the strongest possible manner. In some families of beetles the development of the larvae is very slight, and they have no legs, so that they look like worms. In other families the larvae have structures which are intended to act as legs, but they are feebly developed, and locomotion is therefore very imperfect. On the other hand, many larvae are born so fully formed, that they are very much like the perfect insects, and are more or less active in their habits; and, finally, some on leaving the egg possess the habits and the methods of movement which characterise the perfect forms. They cannot fly, however. All these insects, more or less well organised, must undergo their principal metamorphosis during the period of rest and quietude when they are nymphs.

The metamorphoses, the habits, and the instincts of many of the *Coleoptera* are well-known, and those of a considerable number of species in each of the great groups into which these insects may be divided have been carefully studied.

There are at least twenty families of the *Coleoptera* which can be readily distinguished on account of their members being very characteristically formed; but it is impossible, from what we know at present, to collect them into groups, and to distinguish one from the other by attributing to certain to certain forms higher or lower positions in the insect scale.

Formerly it was attempted, and it was supposed very satisfactorily, to divide all the *Coleoptera* into four great primary divisions, according to the particular number of the joints of the joints of the tarsi. Geoffry, a great entomologist of the last century, invented this method of classification, which, although open to many objections, is very useful, and is generally employed by modern authors.

The divisions are as follow — I. *Pentamera,* in which all the tarsi are five-jointed; 2.*Heteromera,* in which the four anterior tarsi are five-jointed, and the two posterior four-jointed ; 3. The *Tetremera,* those which have four joints to all the legs; 4. The *Trimera,* which have never more than three joints to all the tarsi.

It must be understood that this classification is very artificial, for there are many beetles which are very closely allied so far as their general structures are concerned, which have great differences in the numbers of the joints of their of legs; and, on the contrary, those whose joints are of the same number, are occasionally noticed to be very different in their other characteristics; so that the number of the joints does not always coincide with the natural affinities of the insects. This ready method must, therefore, give way to one which groups the *Coleoptera* by considering the general characters.

The family of the *Scarabaeidae* is, perhaps, the most

interesting as well as the most numerous of the *coleoptera,* and its members present remarkable distinctions in their habits, foot, methods of life, and in their general conformation. Nevertheless, although they vary so very much amongst themselves, these beetles constitute a very natural and very characteristic assemblage. They possess antennae which are short and inserted into a cavity underneath the lateral edges of the head, and which end in a club-shaped mass made up of several leaflets. The antennae enable us to distinguish most of the *Scarabaeidae* at once.

Dumeril, who was a good naturalist, called these *Coleoptera* The *Lamellicornes,* from the structure of their antennae; but it should be noticed that all of these organs are not composed of leaflets or lamellae. Some are in the form of a cone reversed, or are nearly globular towards their ends, the joints being box-like. The tarsi are invariably made up of six pieces; the body is always more of less heavy, thick, and stout, and in many of the members of the family the structures of the mouth, and even the mandibles, either remain partly or entirely membranous. This feeble condition of the mouth-pieces is generally accounted for by the nature of the diet of the insect.

The *Scarabaeidae* are usually large, and some of them attain the greatest size amongst insects. The principal forms of the family are more or less known to everybody, and they are the *Scarabaeidae* proper, the Dung Beetles, the Cockchafers, the Rose Beetles, &c. During the adult stage, the *Scarabaeidae* present very notable differences in their external shapes and in their habits, but they resemble each other in most wonderful manner when they'are larvae. These are the great " white worms," as they are usually called, and they may be the larvae of any of the family, although they are almost always attributed to the cockchafer Every one has seen these white fat things, whose bodies are thick and cylindrical, and usually curled up, and whose skin is very this and furnished here and there with stiff hairs, which give the insect considerable sensibility of touch and may assist in its movements. The larvae have a rounded head, which is covered with fawn-coloured or brownish and very hard skin, and they have strong mandibles, land antennae that have four of five joints.

Fig. 6.1. The Metamorphoses of the Rose Bettle *(Cetoni aurata)*.

The larvae of the *Scarabaeidae* were first carefully studied by a Dutch naturalist, De Haan, in 1836, and since then other observations have been made upon them. The larvae are always hidden up either in the ground, in the roots of plants, or in the middle of vegetable or other rubbish. They never come to the light; they are colourless and rather soft, and as they always live in darkness they have no eyes. They are always eating and consuming a large quantity of nourishment, and therefore their digestive organs are very voluminous. They walk with great difficulty, especially as the curvature of their body necessitates their resting on one of their sides. When first born they are rather active, but they soon get lazy and dull, and spend the greater part of their lives in destroying everything in their immediate neighbourhood, and eating it.

After a longer or shorter life, according to their peculiar habits,they form a rounded or egg-shaped cocoon, and undergo their metamorphosis into nymphs. In order to make the cocoon, they bring together either grains of the soil of pieces of wood and rubbish, and unite them together with their saliva. The principal types of the family of the *Scarabaeidae*, when arranged in different tribes, are the *Cetoniidae*, the *Glaphyridae* the *Melolonthidae* the *Scarabaeidae proper,the Geotrupidae*, the *Coprinae*, and the *Passalinae*,.

The *Cetoniidae* are very beautiful insects, which like the honey they obtain from flowers, and which are distinguished from the other tribes of the family by their membranous mouthpieces and the ten joints which compose the antennae. There are two groups of them; and the first , which consists of the rose Beetle tribe, properly so-called, is will-known by its containing the genus *Cetonia*, the commonest species of which, *cetonia aurata*, is so abundant everywhere in Europe. It is to be seen during the whole of the summer and especially in the months of May and June, flying about gardens or resting upon the flowers. It is a beautiful beetle whose golden-green tints and delicate linear and white ornamentation strike everybody with admiration, for when it moves in the sunshine, and flies slowly or quickly, the whole of its body looks like a mass of glowing gold-like metal. It is very remarkable that these rose beetles should not fly like most of the others, by extending their elytra at the same time as the membranous wings, for they only lift up the elytra slightly, and never separate them; and this peculiarity is shown in the engraving on the previous page.

On the left-hand side of the engraving one of these beetles may be seen flying downwards with its membranous wings extended, but elytra still cover the whole of the body as they do in the insect which may be seen at rest on the upper part of the picture and on the right hand. Low down on the right hand is a nymph in its cocoon, and around it are larvae of different sizes.

This beautiful beetle chooses the most elegant and elaborately-coloured flowers as its house and foot; it likes to dig deep into the hearts of roses and honeysuckles, the petals of which it eats, and, moreover, it sucks their honey. The females usually lay their eggs in decayed wood at the foot of trees or in their trunks, and some have the extraordinary instinct of depositing them in such curious places as ants' nests. They look like the larvae of cockchafers, but their heads are much smaller, their legs are shorter, and the last segment of their body has a little projection upon it which isn't observed in those insects with which we are now comparing them. The larvae live together in some numbers. At the approach of winter they grub downwards into the earth, even to the distance of one or two feet, and return again in the spring. Probably they live three years in this manner, and then they construct oval-shaped cocoons out of chips of wood and pieces of rubbish, all of which are glued together by the secretion of the salivary glands. They undergo their metamorphoses in the cocoon, and they proceed rather rapidly, for the beetle forces itself out in the early summer. Dr. Baird says that a very pretty species is found in the Philippine Islands, which is of a fine brilliant metallic green colour. The ladies of Manilla keep them as pets in small bamboo cages. The beetles of this family of the *Cetoniidae* are widely dispersed, but most of them are found in tropical climates, and they vary much in size and in ornamentation in every country.

There are some beetles of a Brazilian species which rest in the morning time in the plantations of maize, where they are found sitting under the leaves; they fly during the day round the lofty trees, upon the leaves of which they feed. They are styled the princes of the beetle tribe, but are inferior in size to the Goliaths *(Goliathus)*, which are peculiar to tropical Africa. The same author informs us that the great *Goliathus caciucs* is roasted and eaten by the natives. Several species used to fetch very high prices amongst collectors, and as much and £50 have been given for a single specimen. The larva of this species makes a cocoon of mud, and undergoes its metamorphoses in it.

The *Trichiitae*, which differ from the *Cetoniidae* in not having a particular portion of the mesothorax prominent, have a species which is small and black, and which is often seen running about the ground *(Valgus hemipterus)*. It has five spines on its legs, and the female has a long and straight ovipositor, with which it introduces its eggs into the cracks and holes of decaying wood.

The *Melolonthidae* have a very short lower lip and strong mandibles, which enable them to browse leaves, and the hooks of the legs are constructed in different manners, according to their methods of life. The cockchafers, or May Bugs, are representative forms of this family. These beetles appear in enormous numbers in some years, and are just as scarce in others; and this depends upon the long duration of their development. If the spring is warm they appear in the month of April, but the greatest numbers fly in the month of May; towards the end of June they are almost all gone. They eat the leaves of many trees—such as the oak, the beeches, the maples, and the poplars, and sometimes the elms. Occasionally the trees are completely spoiled and killed by the May bags. They are almost nocturnal in their habits; they dislike the sunshine, and seek the shade during the day under the branches of bushy trees. They wake up in the morning, and are moderately lively then; but during the evening they are particularly active, and fly here and there, making a sonorous buzzing noise. They are heavy and very clumsy in their flight, as most of us know, for when we are walking out in the warm spring evenings there is often some difficulty in escaping these dull and stupid-looking beetles; in fact, they prefer banging against everything, rather than guiding themselves away from obstacles which are hard enough to do them injury.

The May bugs do much mischief when they are beetles, but the ravages they commit in the larva state are of most serious consequence to agriculturalists. The female beetles dig into decaying matter, or light soils, and hide themselves whilst they lay about forty eggs. The larvae are hatched at the end of four or five weeks, and by this time the tender vegetation of early summer is progressing rapidly. Then these white worms or cockchafer grubs, as they are often called, find themselves surrounded by abundance of food, and, unfortunately, of that kind which is very precious to man. They do not eat the leaves, but the delicate roots which come in their way as they crawl underground ; and thus a vast number of plants is destroyed. The larvae grow very slowly, and do not undergo any metamorphosis until their third year; and when they are transformed into nymphs during the months of March of April. Some weeks afterwards the beetles escape and fly. The long duration of their growth and evolution explains how it is that these beetles usually visit us in great numbers every third year ; but it would appear that unusual heat, or seasons of great cold, have something to do with the acceleration or retardation of their development. It is not uncommon for larvae to undergo their metamorphosis towards the end of their second year, when the summer has been unusually hot; but even then the cold of the autumn may prevent the perfect insect from leaving its cell—so that it does not fly until the spring of the third year. The larvae are shaped like those of the *Scarabaeidae* generally, and they are remarkable for their strong mandibles, which are furnished with a diamond-shaped tooth that enables them to cut across rootlets very easily.

It would appear that the larvae of the May bugs, found as they are over the whole of Europe, were not formerly such pests to farmers. They were certainly not so common in the olden time, but they appear to have increased their numbers with the spread of agriculture. Farming operations favour these creatures, for they cannot live in undisturbed ground; they cannot move about solid earth or in old rocks, but they can do so in alluvial soils, or in those which the process of tillage renders light and suitable for vegetation; and, formerly, the acreage of cultivated ground where the May bug larvae can live was, of course, much less than it is now, and consequently the beetles were not so common. The large must by dug up to be destroyed, and this is a very awkward operation whilst ploughing and spade husbandry are going on. It is much easier to collect the beetles themselves, and if they were will hunted down everywhere simultaneously, there would be no doubt about a rapid diminution in the numbers of the larvae; but,unfortunately, all the efforts of men who are sufficiently enlightened to advise the farmer rarely meet with success. Thirty years ago, a Prefect in the Department De la Sarthe—M. de Romieu—took great pains, and persevered in urging upon agriculturists to destroy the real originators of the mischief they suffered from ; and of late years M. Riset has been studying the amount of mischief these creatures really commit. He found that in two arrondissements in the Department of Seine Infferieure, 3,60,000 lbs., weight of cockchafer grubs were found; and as a grub only weights about thirty-one grain, there must have been 81,290,322 of them.

With this calculation before us, which refers only to two small districts, it is not difficult to understand the enormous amount of injury that is done to agriculture by these grubs.

Very fortunately rooks, jays, magpies, crows, and other birds, eat immense quantities of them, and this fact along ought to make us careful of preventing the wanton destruction of the birds that deep down these insect pests.

The engraving describes the metamorphoses of the cockchafer. The larvae are seen in different stages, and there is a nymph underground also, whilst the perfect insects are crawling above.

There was at one time a speculation by which people hoped to make money by extracting oil from cockchafers, which was supposed to have extraordinary virtues, and to be useful in many ways; but, unfortunately, the expense of manufacturing the oil was too great, and effort came to nothing.

The species which most resemble the common cockchafer are not very widely distributed in Nature, and the mischief they perpetrate is not much feared or cared for. One of them may be noticed on account of its great size, and list favourite locality. *Melolontha fullo* is a fine beetle whose general colour is brownish, and it has little white spots scattered here and there over the whole of its surface, especially on the elytra, on which they produce curious designs. The males of this kind are always admired on account of the great size of the lamellae of their antennae, which resemble beautiful feathers. These cockchafers are not found inland, but choose the seaside for their residence, and they may be found on the dunes or sand-hills of Dunkirk land the environs of Ostend, and they have been noticed on different spots by the sides of the Baltic Sea and the neighbouring ocean, as well as all round the Mediterranean. They fly with great velocity, and make much buzzing; and they appear so fond of noise, that they constantly produce a sharp sound when they are on the ground or amongst the leaves, by rubbing the elytra against the body.

In the engraving the males may be seen with their great antennae, and the female on the left-hand side of the picture shows the white markings on the elytra very distinctly.

Fig. 6.2. The Metamorphoses of the Cockchafer. *(Melolontha vulgaris.)*

The *Rutelidae* are beautiful beetles, and are intermediate between the *Melolonthidae* and the *Scarabaeidae*. The species of some of the genera of this tribe have magnificent colours, which hint at their existence where the sun shines in all its glory, and where the climate is most genial. Some of them live in the hottest countries of the world, others in tropical America, and some in Australia. There are no *Coleoptera* in Europe as brilliant as these. The prettiest of these brilliant creatures belong to the genus *Euchlora* ; they are small *Colcoptera* very much like the cockchafers in shape, and are usually of a beautiful green colour, and their antennae are made up of nine joints. One of the commonest species is the vine *Euchlora;* it is of a beautiful metallic green colour, and its elytra are striated and covered with points. It often does great mischief to vineyards. Another European kind also resembles the cockchafers, and is about half an inch long ; it is of a dark green colour, and is marked with light brown stripes. It is called *Phyllopertha horticola* and every one must have noticed it on trees and bushes during the summer time.

The *Scarabaeidae* comprise the giants of the order of the *Coleoptera*, and are well-known by their heavy and clumsy bodies, which are solidly protected by an integument which acts like a cuirass, by their rudimentary lips, their strong mandibles, and by their head and prothorax being almost always furnished, in the male insects, with prolongations which look like horns. The species of the genus *Scarabaeus* have their jaws furnished with teeth, and the general structure of their mouths indicates that they live upon hard leaves and level upon woody tissues. The horns which the males carry give them a very curious aspect. These horn-like prolongations present great diversities of shape in different species, and the endeavour to find out their uses does not meet with much success. There is nothing which leads to the suspicion that they have any particular function, when the peculiar habits and method of life of these insects are examined ; so that, if we take a limited view of the question, we may consider them as decorations of ornaments.

Fig. 6.3. Beetles of *Melolontha fwlo*.

The larger species of the *Scardaeidae* live exclusively in those countries where Nature produces the most luxuriant vegetation, such as the Antilles, South America and the Moluccas.

The larvae of these enormous insects live inside the trunks of old trees; and any one can readily believe the mischief they must do, and what a large quantity of vegetable matter they must consume during their growth. The most remarkable of the larger *Scarabaeidae* is *Scarabaeus Hercules* ; it has a black body and olive elytra, which are very brilliant and spotted with black. The males have the forehead and the prothorax each armed with a prodigiously long horn. Another *Scarabaeus* is as large as this one, it lives in New Granada, and is black in colour ; another which is covered with a fine layer of hairs, is found in the Brazils ; and the great *Scarabaeus Atlas*, from the island of Amboyna, has a brilliant bronze tint. The *oryctes* are the *Scarabaeidae* of our part of the world, and may be distinguished from the members of the genus *Scabaeus* by not having teeth in their jaws. The finest specimen of these insects is *Oryctes nasicornis* ; it is of a brilliant brown colour, and has a horn upon its forehead like a rhinoceros, whence it has derived its title of "nose-hots." The larva is larger than that of the cockchafer, but has smaller legs, and it lives in the trunks of old and rotten oaks, but it is more readily discovered in those places where tan has been allowed to collect and to remain for sometime.

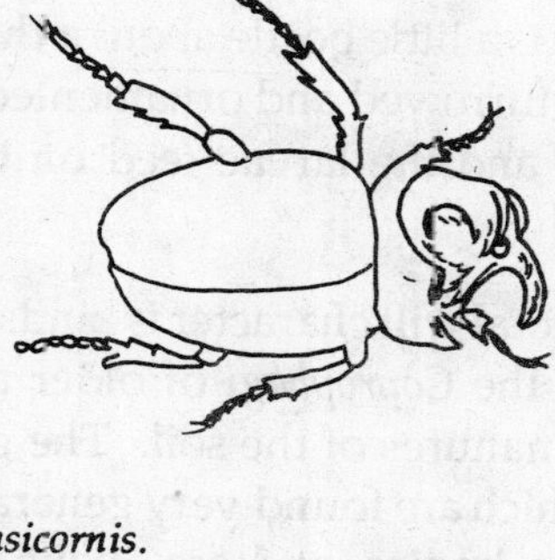

Fig. 6.4. *Oryetes nasicornis.*

The *Geotrupidae* and the *Coprinae* form other tribes of the *Scarabaeidae.* They are essentially the *Coleptera* of dung-hills, manure, and of stercoraceous matters in general. The *Geotrupidae* have weak mandibles and jaws, the first being of a leathery consistency, and the antennae of ten or eleven joints. The *Coprinae* have antennae which are formed of nine joints, and the mandibles and jaws, which are small and membranous, are suited for the liquid diet in which these insects indulge. The principal genus of the first tribel is *Geotrupes,* and its members are commonly called dung beetles, as are also the *Coprinae.*

A large beetle, which is of a sting black colour, and whose tints pass from green to blue or to violet , and whose elytra are strongly striated, flies and makes great buzzing in the fine evenings towards the end of summer and the beginning of autumn, all about the roads and near walls, or wherever filth has been dropped. The female of this beetle takes particular care about laying its eggs. She makes a deepose hole it the earth and a comfortable chamber at the bottom of it ; in this retreat she lays an egg and then collects and packs up a quantity of filthy dung, which she stores up there, so that when the larva is hatched it will have its natural foot at once. The species of another group form the genus *Trox,* and they have projecting mandibles. These species are small, but are readily distinguished on account of their grey colour and the roughness of their elytra; they like the dust and they are exactly of its colour, and one of them may often be seen on dusty roads *(Trox subleases).* It is a little beetle about a third of an inch long, having the elytra furrowed and ornamented with little tufts of hair. The adults and the larvae feed on the dead bodies of animals.

There are several will character is kinds of insects amongst the *Coprinae* (or the *Coprophagi* of older authors), and their duty is to act as manures of the soil. The genus *Aphodius* has many species, which are found very generally over the world. They have oblong bodies, and are about the smallest of the *Scarabaida. Aplodius fimetarius* is to be found on all dung, and its lustrous black colour and red elytra are very marked. *Aphodius fuser* is found in the same places, and is known by its perfect black colour and larger size.

These dung beetles keep their skins grassy and beautifully shining by secreting an oily fluid, which prevents all the nasty things they live amongst collecting upon them, or staining their skins. This is particularly observable in the sacred beetle of the Egyptians, *Ateuchus sacer,*which belongs to the group of the *Coprinae,* formerly called *Pilulares.* The group to which this genus belongs is world-wide, and the American forms are representative of the European. The genus just mentioned, and which may be considered the most important, contains large black *Coleoptera,* which are particularly numerous around the Mediterranean. Their great broad and flat bodies attract attention at once.

The sacred beetles have the fore legs enlarged, and furnished with several very strong tooth-like spines, but there is no tarsus at their ends, and they are evidently designed for a special purpose. The heads of the beetles clearly point to the same design, and are admirably adapted for the peculiar habits of the species. The Egyptians represented this beetle—which is now common in Provence and in eastern and northern Europe, as well as in North Africa and Egypt—on their oldest monuments; they figured it on amulets, and cut its shape deeply on their tombs. They worshipped the beetle, knowing its habits, for they were an agricultural people ; and the title of "sacred" has descended to the present day.

The *Atcuchus sacer* is quite black, has two tubercles on its head, the elytra are finely striated, there are four spines on the fore legs, and its scarcely an inch in length.

The instincts of this beetle are very remarkable. It may be found buried in dung, or seen rolling a pill of like substances, which it drags about with its hind legs slung the roads and paths. It does not do all this, or select the offensive matters and roll them up and wheel them off in its peculiar fashion to please mankind, or to fertilize the soil for our benefit. The care for the future offspring, which it will never see, is at the bottom of the whole proceeding, and it desires to place its

Fig. 6.5. The Sacred Beetle (*Ateuchus sacer*).

eggs in a safe place. One of the females lays an egg, but instead of being satisfied with allowing it to remain, like other *Coprophagi*, in the midst of a mass of manure, or in a hole in a dung-heap, it rolls it up in a little mass of refuse in the shape of a pill. The beetle rolls the mass about over the soil with its hind legs, and soon makes it into a round ball. The ball is not left to shift for itself, for the insect in most particular about the place where it is to be put by ; it seeks a position in which the future larva can live comfortably and safely, and, in consequence, often has to carry and push the burden for considerable distances. The beetle takes its time, perseveres, and after using much ingenuity, usually succeeds in its endeavors. Should the surface of the soil be very irregular, the beetle places the bundle on its head, the flat shape and tubercles of which are admirably fitted to receive and hold the weight, and carries it over everything until easier ground is reached. Sometimes an impediment is so great that this maneuver is of no use, and the beetle drops the ball, and apparently gives up the labour in disgust at the unexpected difficulty ; but this is not really so. The *Ateuchus* flies off, and soon returns with several others; and it is said that patience will always reward observers with a wonderful sight. The beetles set to work, and push and pull this way and that, and crawl over all sorts of places, mutually assisting each other. When several of them are thus seen rolling a ball, it is a suggestive sight, and very in comprehensible to those who are ignorant of natural history.

As soon as the *Ateuchus* has rolled the ball to the place where it wishes, the insect begins to dig a hole, and in this it is assisted by its front legs, the large spines of which enable them to dig rapidly and easily. The tarsus is deficient, for it would have been in the way when the legs were digging into hard soil. Having such capital tools, the insect excavates, the hole rapidly, and when it is sufficiently spacious and deep, the ball is rolled down to the bottom; then the beetle uses its long hind legs, which are furnished with a sort of brush, to fill in the cavity and smooth the earth lover it.

Many amulets and Egyptian pictures represent an *Ateuchus*, which resembles the shape of *Ateuchus sacer*, but it has a brilliant golden colour, which was supposed to have been an invention of the artists of the days of the Pyramids, but this is not the case; and the coloured *Ateuchus* was found, in 1819, by M. Collide, and lit is not rears in sinner.

There are some *Scarabaeidae* which live in Europe, and are smaller than the sacred beetles, but which have the same habits. One kind, which is very common in France, is named after the pill of dung it carries about, and is called *Gymnopleurus pilularius*. Another species, which has a thick body and long hind legs, is named after Sisyphus, and is sometimes found in the environs of Paris.

The last tribe of the family of the *Scarabaeidae* we have to notice comprehends the genus *Passalus,* and contains very large beetles, which have long flat bodies, the prothorax separated from the trunk by a waist, antennae bowed, and spined fore legs. All have a brilliant black colour, and inhabit the American continent the East Indies and Australia. They live in the trunks of old trees, and hide beneath the bark, and the flattening of their bodies enables them to do this. The larvae, which live upon decaying wood, whilst they resemble for the most part those of the *Scarabaeidae,* have a peculiar construction, which proves that their development is not far advanced. Their first and second pairs of legs are as long as usual, but those of the third pair only exist as little tubercles.

The *Lucanidae,* or the Stag Beetle family, were formerly included amongst the *Scarabaeidae,* but their longer antennae and peculiar form give them a different aspect. There are also some differences in the larvae; and it is now proved that the nervous system, which is very centralized in the *Scarabaeidae,* is much less so in the *Lucanidae.* The two families are therefore separated by structural differences.

The *Lucanidae* are usually very large insects, whose males are well-known on account of the enormous size of their mandibles. These structures, which belong to the mouth, are differently curved and toothed, according to the species, and they look like large pincers, and give the beetles a very formidable appearance. It is impossible, from our present knowledge, to account for the use of these great mandibles, the growth and size of which differ in the individuals of the same species. The female beetles do not have these large mandibles, and the structures are of ordinary proportions in them, being furnished with teeth which enable them to be used as leaf-cutters. The *Lucanidae* are found in many parts of the world, but their species are most abundant in the East Indies, the islands of the Straits of Sunda, and the Moluccas, whence they are occasionally imported into Europe. These insects, which inhabit woods, and only fly in the evening. are difficult to collect in those countries where no one cares to spend a night in the middle of forests ; so that these *Coleoptera,* which are very beautiful and whose shapes are very remarkable are much prized by collectors.

The genus *Lucanus* has been greatly sub-divided, but it is not necessary to describe it, because the type of it is so well known under the form of the Great Stag Beetle, *Lucanus cervus,* our largest beetle. When in the larval condition it lives in the trunks of old decaying oaks, and it is very common in some districts. The beetles fly in the hot evenings of the summer months, and their large mandibles render them very visible. The accompanying engraving of the metamorphoses of the stag beetle represents a large male upon the trunk of a tree; the mandibles are separated, and their denotations are very visible. On the right hand, female beetle is upon a leaf, and her small mandibles make her look like another species especially as the head is so much narrower than that of the male. Low down in the picture, on the left hand side, there is a nymph in a sort of cocoon hollowed out of the wood. It is a male, and the future mandibles can readily be seen. It is something like the perfect insect, and at the same time resembles the larva. The larva is figured to the right, and its shrp, trenchant jaws are srikingly suited for their purpose. The development of the fore legs is worthy of notice, and so is that of the spiracles which are so very visible on the sides of the body. It is said that these insects are six years in coming to their full growth. Some very natural affinities connect the *Lucanidae* with the *Histeridae,* which are short insects, with broad bodies and antennae ended by a little solid club-shaped mass ; their nervous system is rather less centralized than in the *Scarabaeidae.* One division of them the *Histeridae* proper includes little shining beetles , with truncated elytra which do not cover the whole of the abdomen, angular antennae and short legs. The species of the genus *Hastier* are black in colour, the tint being very brilliant, and sometimes spotted with red. They work in manure-heaps and amongst dung, which they eat; and they even enjoy the decomposing flesh of dead animals. Their larvae, which live in the same way, are soft and white, and have an elongated and compressed shape, and are provided with short legs. Moreover, the end of the abdomen has two double-jointed structures upon it, and a long tubercle also, which has something to do with the progression of the insect. The genus *Saurians* contains beetles which are usually of a bronze colour, which may be greenish or bluish in tint ; and the heads of the insects are remarkable, for they appear to be sunk into the prothorax. Others very closely allied to them, live upon decaying wood, and exist generally under the bark. The insects of the other division of the *Histeridae* have not small contractile legs nor angular antennae. They live upon dead bodies, in mushrooms, in decaying wood, and some frequent flowers. They are usually very small insects, and little is known about their metamorphosis. The commonest species is of a green bronze colour, and is about the twelfth of an inch in length. It is common on flowers.

There are two genera which are will worthy of attention, and they belong to the family of the *Silphidae*. The *Silphidae* have the tarsus formed of six joints, the antennae ending in a club shaped mass divided into leaflets, the mandibles and biles are projecting. They are found principally in Europe and North America.

The members of the genus *Necrophorus*, the Sexton Beetles, have a thick body, stout legs, and their hind thighs are more or less enlarged. One spices is particularly interesting, on account of its remarkable habits.

Necrophorus vespillo is common throughout the greater part of Europe, and may be readily recognized by its colour. It is black, and is ornamented with yellow hairs on the side of the body; the club end of the antennae are reddish, and the elytra are crossed by two broad bands of a light red tint. When the dead body of a mole or a field-mouse is left in the fields, the sexton beetles soon find it out, and begin to collect around in considerable numbers. They're not going to eat the corpse, but they intend to lay their eggs in it. If the body is allowed to remain exposed to the air, it will dry up, or be eaten by other animals, so that if any larvae were therein they souls be destroyed, or they would perish forwat of food. But the sextons have an instinct which prevents their young from being exposed to such dangers, and they set to work and bury the body, so that their larvae can nourish themselves appoint it without being disturbed. They hollow out the ground beneath the body, and make a tolerably deep hole by throwing out the soil with their large legs ; then the animal sinks down, and they cover it up with the earth which had collected around the margin of the pit. These beetles are remarkable for the subtle sense of smell they possess. They fly swiftly, and are constantly hunting, with the aid of their acute sense of smelling, for the dead bodies of animals. In Russia, where dead bodies are often buried in shallow graves, the sexton beetles may be seen in hundreds in the churchyards. They dig a hole in the earth over the body, and take up their abode in the substance of the corpse. The generally manage, by dint of hard work, to bury a small animal in twenty-four hours; they then take up their dwelling in the body and feed upon it. The female speedily lays her eggs in the decomposing substances; these soon become larvae and feed, like their parents, upon the putrid matter until they acquire their full growth; they then get out of the body and bury themselves in the earth to the depth of several feet, and remain there for three or four weeks, and undergo their metamorphoses. The larvae make a cell, in which they undergo their transformations. They are oblong in shape are yellow wish in colour, and they have leathery plates on the the segments of the thorax and abdomen. They have very small antennae, power full mandibles, toothed like a saw, the jaws furnished with little points like those of a comb, and short legs, which enable them to dig. Although they live in darkness, they have two eyes, which are very distinct in young individuals, but which seem to become atrophied as the growth of the adults approaches its maturity.

Fig. 6.6. The Metamorphoses of the Sexton Beetle *(Necrophorus vespillo)*.

The engraving represents a small dead animal, which is beginning to disappear in consequence of the beetles having dug beneath it. The perfect insects are figured in different attitudes on the left, low down, there is a lava, and above it a nymph.

The carrion beetles of the genus *Silpha*—and which are commonly called Shield Beetles in France, on account of the flattened shaped of their body, and from the projection of the prothorax above the head—heave ten joints in the antennae, and the last four pieces form its club-like end. Their elytra are margined off, as it were, and they have long and thin feet.

These insects, which are rather small, run with great rapidity, and have very subtle powers of scent, for, like the *Nrrrophori*, they soon find out the dead body of any animal which may have been left on the soil. They have different methods of life; and the larvae and beetles differ in their habits. Some seek out and live in putrefying bodies ; others exist in living creatures; and a few feet upon vegetables matters. This genus of beetles presents us with a remarkable example of the existence of very different conditions of life in closely allied species.

Fig. 6.7. The Metamorphoses of *Silpha quadripunctata.*

The larvae of the carrion beetles are well advanced in development, and approach the adults in the perfection of their structures. They are very flat, and all the segments of the thorax and of the abdomen are covered with a hard integument.

They have six eyes on each side of the head, and long and strong legs, which are furnished with spines, and they move with great rapidity. These larvae are represented in the engraving as crawling upon the ground, whilst the full-grown beetles are upon the leaves.

Silpha obscura is the commonest species, and may often be seen running about the roads. It is less than an inch long,is of a black colour, and has three ridges upon the elytra. Its larva lives upon the dead bodies of animals, and is of a brownish colour, and in marked with black lines.

Silpha rugosa us as common as the species just mentioned, and its larvae, which are perfectly black, are often seen in little groups upon carcasses.

Silpha opaca is found in uncultivated lands during the spring, when its larva devours the shoots of young beet-roots. The *Silpha* represented in the engrving inhabits woods and forests, where it climbs about the oak trees, and chases the caterpillars which frequent the trees in a most active manner. This insect cannot be mistaken, for its body is black, and it has its elytra, which are of straw colour, ornamented with four black points. The larva remains upon the ground, and, like the perfect insect, is carnivorous, for it attacks any insects which it may come across under the fallen leaves or amongst the grass. Finally, the largest of the carrion beetles, *Silpha thoracica*,whose prothorax is of a bright red colour, is as carnivorous as the rest of them, and is found amongst underwood, especially in damp places.

The *Staphylinidae* are a family which contain a great number of beetles, some of them being of moderate size, but the generality are very small. They can usually be distinguished from all other *Coleoptera* by the shortness of the elytra, which only cover a portion of the abdomen. If the elytra of other beetles were compared to a long coat, those of the *Staphylinidae* would resemble a waistcoat ; but, nevertheless, the wings are as large as usual, and are hidden, when the insect is quiet, underneath their cases, being folded up once or twice , so as to be packed away safely. The antennae are rather long and thread-shaped, or more or less beaded.

They are very agile insects, and live under very different conditions of life. Some are absolutely carnivorous, and only feed upon the living prey they catch; others frequent dunghills and manure-heaps ; whilst a few are found upon dead bodies. Mushrooms of all kinds are the favourite diet of a number of species, and rotten wood and all sorts of vegetable and animal remains are enjoyed by a great number of them. The *staphylinidae,* with their plain legs and antennae, are evidetly more simply organized than the *Silphidae* and the *Scarabaeidae;* but their larvae are born much more fully developed than are those of this last family. The larvae of the *Staphylinidae* are partly covered with a hard integument, and are very active, and recall the adult forms, not only by their shapes but by their peculiar. The largest of the *Staphylinidae* have very short elytra, and the last segment of the abdomen is retractile. One of them is very well-known , and belongs to the genus *Staphylinus,* and it is usually called, in this country, the Devil's Coach-horse. It is a black insect, which is constantly seen on the roads, and emits a very disagreeable smell. It is a carnivorous beetle, and has astonishing impudence and boldness, like many other animals that live by the chase ; so that when it is disturbed it does not attempt to run away, but stops, and puts on a menacing attitude. It grasps the ground with its legs, and cocks up its head and tail, snaps its sharp and curved mandibles, and exhales a very disagreeable smell. The larva is quite as carnivorous as the beetle ; and its head and thoracic segments are hard, its legs are long, and the abdomen, which is narrow towards the end, is furnished with sort of tail, which assists in walking. It is very agil, and takes on the habits and methods of progression of the adult.

In the engraving a perfect beetle is represented with its wings expanded, and on the left hand, immediately beneath it, is an adult whose wings are shut up under neat the short

Fig. 6.8. The Metamorphosis of *Staphylinus olens*.

elytra, and which is in menacing attitude. Below it are the larvae, and under a stone a nymph. The larvae hide themselves underneath stones or in quiet places, and only go out by night. They hybernate, and when the fine weather returns they they make a cell in the ground, where they undergo their metamorphosis. The nymph is of a shining yellow colour, and has a crown of hair on the front of the thorax. Other members of the genus, much as *Staphylinus maxillous*, which has prodigious jaws, live upon carcasses; others, which have thin and almost linear bodies, are very common in damp woods, and are found underneath fallen leaves and decade dark,where they chase their prey. The larva of one species lives underneath the bark, and attacks others which feed upon decayed wood. There is a little group of the same tribe, the species of which live, in the larval and adult condition, in decaying vegetables, and especially in mushrooms. An allied form does not care about the mushrooms, but seeks manure heaps, where its larvae devour the maggots of flies; and another species lives in hornets' nests, where it probably eats the larvae of those great wasps.

Some *Staphylinidae,* which belong to this division, and which are found in the nests of the *Termites* of Brazil, are distinguished by the enormous development of the abdomen, which is carried in a most peculiar manner, for it is turned up and allowed to rest on the back of the insect. The enormous destention of this part of the body is due to the fact that the beetles do not lay eggs,but produce living larvae. This is a very curious and interesting discovery, and it was published by schiodte in 1864. They are the only beetles which do not lay eggs and whose larvae are born alive, and this is a most remarkable distinction. The insects are about the tenth of an inch long, and their abdomen is soft. It is supposed that the hairs which cover some parts of the abdomen are furnished with a peculiar secretion that is very much liked by the *Termites*

There are some other *Staphylinidae* which are very small beetles with short elytra, and they are remarkable on account of their antennae, which are swollen at their ends, and their long palps. *Iselaphus* and *Claviger*, genere to which these beetles belong, usually live in an heaps. The species of the first. genus are very active and hide under stones and about moss, and those of the second really require the attention of ants for their existence, but neither produce larvae viviparously.

The species of a small family of the *Coleoptera* do much mischief to manufactured animal substances. They are called *Dermestidae*. These insects are always small, and are of grey,

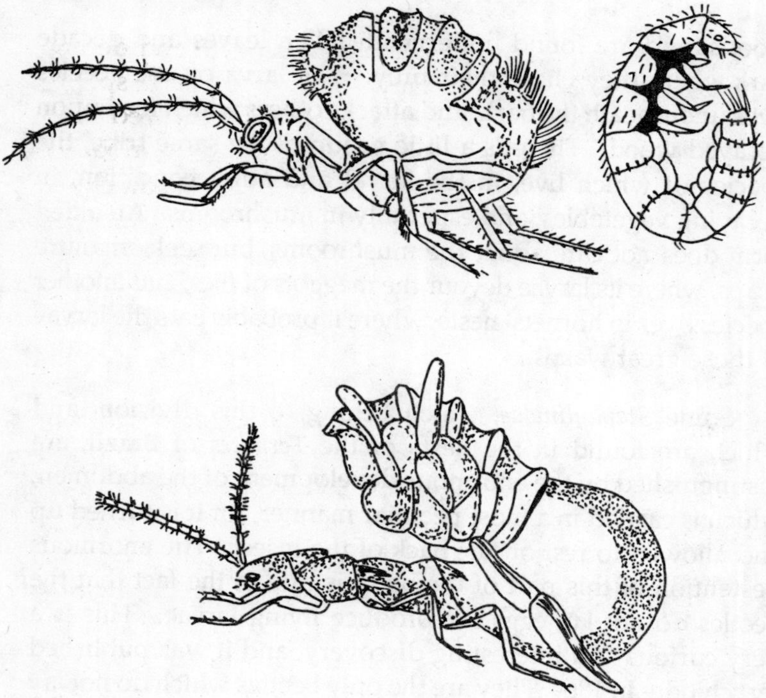

Fig. 6.9. The upper figure are those of Cortoca. The turning up of the hinder parts of the body is very evident in the engraving.

brown, and dusky colours. Their antennae are short, and terminated by a large, club-shaped end, and their mandibles, which are small, do not project. it is certain that these insects live upon animal matters in the larval state. They attack furs, skins, and leather, animals which are in museums, and the dried flesh of bodies, &c. Many of them have been transported by shipping from distant countries, and have become acclimatized. They have committed much mischief in nearly every part of the world and some of the foreign species appear to propagate very rapidly. When they are full grown these insects have slender, contractile feet, and when they are disturbed they fold them under the body and pretend to be dead. The larvae are clothed in an elastic and thin integument, which is furnished with hairs collected in bundles, in masses, and in slender brushes of great elegance of form. They are gifted with a certain amount of agility; they have little legs, the tarsus ending in a simple hook, and they have small heads, there being usually six eyes on leach side. The species of *Dermestes* are readily known by their one bodies and their antenna, which consist of tend joints, the last three king formed into a club. They are found in houses, and especially in warehouses. A remarkable species(*Dermestes lardarius*) is a black insect, and the bases of its elytra are fawn-coloured and marked with three dark points. It consumes very little nourishment in the adult state, but very much the reverse occurs when it is larva. Then it is a blackout brown insect, covered with long tufts of hairs, which are placed in regular series on the margin of each segment, and it is a perfect pest to furriers and leather-sellers. If it eats lard, as its name indicates, it does not hesitate to devour the skins of beasts and of birds that are placed in museums, and it is equally dreaded in shops where anything of the kind is sold.

The *Dermestes* which frequent hides and stuffed animals are by no means disinclined to enjoy bacon and ham. If they cannot get anything of this sort to eat they have been known to destroy cork, whole cargoes being almost completely ruined by them. They have also been found in lasbestos, perforating it in various directions, and undergoing their transformations there. Some other species than those we have noticed have been found in Egyptian mummies. The larvae of the *Dermestes* which is so destructive to museums have two strong mandibles, with which they commit their mischief, but usually the perfect insects of these pests do not do much harm.

Another species of the genus, *Dermestes vulpinus,* although not so common over the whole world as the last, is occasionally found in great numbers. The larval are found under the same circumstances as those just mentioned, and about twenty-five or thirty years since they caused such great damage in the furriers' warehouses and shops in London that a large reward was offered for a remedy which would destroy them. It is easy to kill them by evaporating benzine, or bisulphide of carbon, but it is necessary to do this several times, and especially when the adult insects have set undergone their metamorphosis, so as to prevent eggs being laid.

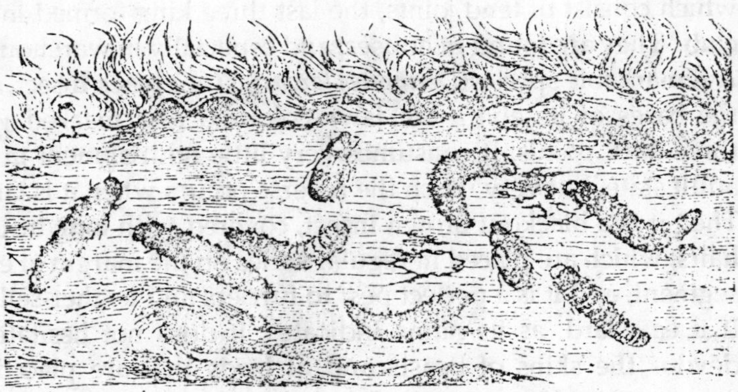

Fig. 6.10. The Metamorphoses of *Dermestes lardarius* and *Dermestes vulpinus.*

The engraving shows on the right hand the black beetle of *Dermestes vulpinus* and its larvae, and on the left hand Dermestes *lardarius* is represented, the beetle having paler elytra. A *Dermestes* which has an oval shape and antennae of eleven joints is readily known by its black colour and bye white spot on each elytra. It also destroys furs, and is frequently found in houses. *Megatoma pellio,* as it is called, has a reddish-brown larva, which is covered with long and brilliant hairs, and which has on its end a brush of them looking like a tail, eats all kinds of furs, and even feathers. Some very small kinds, which have short, thick-set, and almost globular bodies, frequent flowers in fine weather. One of them *(Anthrenus museorum)* is a beetle about the tenth of an inch long, and is the terror of all collectors of insects. The larvae do not hesitate to demolish any collection of insects amongst which they may be born, and they commit an infinity of damage in a short time. They get into the interior of the insects, which they eat, and are hardly to be seen except when they undergo their metamorphosis. The nymph is protected by the skin of the larva, underneath which it lives.

Another large fails of small insects, allied to those we have just described, are known as Mushroom Eaters. These little beetles have a broad body, and their elytra are often decorated with yellow aired points. Their larvae resemble those of *Dermestes*, but they have not so many hairs.

The *Hydrophilidae* are for the most part leaf-eating or herbivores insects, and they are developed so as to be able to swim and and even to lie in water. They have short antennae, which are inserted under the sides of the head, and they have very long pulps, which are thread-shaped and pendent. The remarkable size of these appendages to the mouth formerly gave to these beetles the name of *Palpicornes*. One of the most notable peculiarities of the structures of these beetles is observed in the females, which have some glands in the abdomen which produce a silky substance that covers their eggs with an impervious cocoon. This is quite unique amongst the *Coleptera;* and the female *Hydrophilidae*—the extremity of whose abdomens is furnished with two cylindrical appendages or thread drawers—set to work immediately after they have laid their eggs to enclose them in a cocoon, the silken threads of which afford them a safe and secure protection against the multitudes of other insects that are always on the look-out for animal matters. The eggs float in this cocoon until they are hatched. The larvae are generally oblong, and have rather a projecting head without any neck-like constriction. They have large mandibles, which are curved near the extremity, and are often dentated. They have cuirasses plates upon the body segments, and the last segment of the abdomen is very small; their legs are rather long, and theatres are usually unguiculate. Although they have not the agility of the larvae of *Dytiscus* (the true water beetle), they are very active, and care principally for a vegetable diet, but, nevertheless, they are sometimes carnivorous and very voracious, and eat the fry of fishes and small mollusca. Being almost amphibious they may be seen to leave the water

Fig. 6.11. The Metamorphosis of *Hydrophilus Piceus.*

and to crawl with some difficulty amongst the grass on the sides of ponds and streams. When about to undergo the first metamorphosis they bury themselves in the earth or in the mud, where they make a cavity, daubing over its inside with their saliva; when the cavity is finished they become nymphs. Although there are not many species in the family, there are some very characteristic forms amongst them.

The genus *Hydrophilus* contains large insects of an oval form having antennae with nine joints, the last four forming the club. These beetles have two long, very solid, and sharp pointed spurs, which, when they are laid hold incautiously, can inflict a sharp wound. They are found in many parts of the world, but only one kind, *Hydrophilus pieces,* common called *Hydrous pieces,* inhabits Europe, and it is the largest. It is a well-known beetle, and is an inch and a half in length , being of an olive-black colour. It swims with great facility, on account of its hind legs being flattened, ciliated, and able to be used as oars. It is an interesting inhabitant of fresh-water aquria. The manner in which it comes to the surface of the water, and the way in which it breathes, when it does so, always excite interest. The *Hydrophilus,* which is a big beetle, could not maintain itself horizontally at the surface of the water, but it pokes out. its head above water mark, and worlds its antennae, the club-shaped ends of which are partly hollowed out; and thus carries downwards a small bubble of air, which is pressed against the sides of the thorax and of the abdomen, and which then comes in contact with the respiratory orifices. The beetle is equally interesting when it is about to lay its eggs, and whilst it is forming its cocoon. Crawling over the aquatic plants head downwards, it lay its eggs and immediately drags its silk-drawers over and around them, and it envelopes them in a cocoon, which it fixes by itself. These cocoons are of a high grey colour are a short oval in shape, and have a long conical pedicle.

The beetle on the leaf close to the surface of the water, in the engraving on the previous page, is laying its eggs and forming its cocoon. Another beetle is on the ground on the left hand; a nymphs enclosed in its cell on the right, and above it there is a full-grown larva.

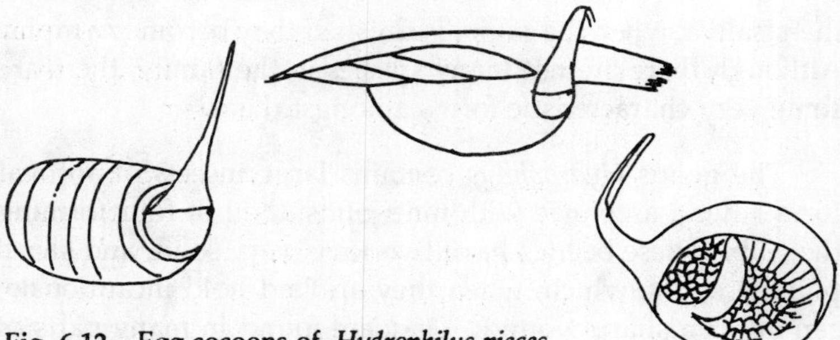

Fig. 6.12. Egg cocoons of *Hydrophilus pieces.*

As soon as the young larvae are hatched they begin to feed, and grow very rapidly ; and when they have attained their full size they are large insects of a dark grey colour, and the hard parts of the head and of the thorax are of a lustrous brown tint. Whilst they are quiet, and even during their progression, the head and the tail are turned up so that the body is always more or less curved. They swim easily with the assistance of ciliated legs and of flexible abdomen. The nymph has some very strong and thick hairs on the thorax and abdomen, but their use is unknown.

The central cocoon represented in the above engraving is attached to a leaf, but that on the left hand is isolated. On the right hand a cocoon has been opened so as to show the position of the eggs in it.

The *Dytiscidae* live in most stagnant waters, and in the little streams which roll slowly along on account of their water weed. These beetles are provided with large wings, with which they can fly and they are frequently noticed to crawl out of the water, and to fly from one pond to another. They thus appear to lead two lives : one, the ordinary sub-aerial existence of beetles; and the other the special life of water insects. They are made like land beetles; and they are therefore obliged to come to the surface every now and then in order to breathe, and a very simple arrangement enables them to obtain a fresh supply of air. The spiracles, or breathing organs, are situated on the upper part of the body, and are protected by the elytra, or wing cases, which fit the upper part and sides of the body, so that no water can get underneath them. When the insect wishes to breathe, it lets itself float to the top of the water, so that the upper part of the back projects above the surface; then the elytra are raised up, expiration takes place. and a supply of fresh air is inspired, and then the beetle shuts down the wing-cases, and includes beneath them a certain quantity of air, which it uses from time to time.

The *Dytiscidae* are carnivorous, and attack all kinds of aquatic insects, which they tear and injure with their strong mandibles. Their larvae live in the water, and have the same propensities at the adults ; and they are constantly chasing a number of insects, mollusca, and even young fish. They are long creatures, narrowed behind, and are clothed in such a hard integument that they may be said to be truly armoured. Their head is large and broad, and the mouthpieces project ; their mandibles are arched, and are perforated at the extremity, so that the insect can suck through them; and their legs are long. They are very active, and swim readily by means of a flexible abdomen, which is terminated by two appendages, which are often leaf - shaped.

These insects are provided with spiracles on the sides of all the abdominal segments; nevertheless, when they come to breathe on the surface of the water, they life up their tails where the least spiracles are, and apter to respire with them in particular. When about to undergo their metamorphosis, they come out of the water, and hide themselves in the mud or sand by its side, and make a cell according to their size, and are transformed into nymphs.

The commonest kind is *Dytiscus marginalis*, and it abounds in all the stagnant waters of Europe.

The *Gyrinidae* are water beetles, and are well-known by their peculiar habit of describing circles upon the water with wonderful rapidity. This extraordinary movement has given them their scientific name, which infers that they turn round, and we call them popularly Whirligigs, and the French *Tourniquests.* They are small beetles, the body is oval, and more or less convex, and the feet are perfectly organized for swimming; the middle and hind ones act as broad oars, and the front legs are used as rudders. They are generally of a very brilliant metallic bronze colour and when the sun shines on them they look like pearls dancing on the surface of the water. Sometimes they stop short, appear immovable, and look as if they were easy to be caught, when all at once they disappear with the rapidity of lightning, darting about from place to place, whirling round and round and finally plunging deep into the water. They pass the greater part of their existence on the surface, and watch for their prey below and look out for dangers above. Moreover, they occasionally snap up a passing fly; so Nature has given them four eyes, or rather their eyes on each side are divided and the insects appear to have two eyes that look upward and two that look downward. When the whirligigs are caught they exude from their body a milky fluid of an extremely strong and disagreeable ordour, which remains for a considerable time afterwards.

The female beetle lays her eggs upon the surface of the leaves of aquatic plants. The adults disappear from the surface of ponds and rivulets when the weather becomes cold, and hide themselves at the bottom, under stones, amongst water weeds, and evening the mud. The larvae are long, very thin and white, and the hard parts of their bodies are of a yellow tint. The head is small, and all the segments of the abdomen are furnished upon their sides with long solitude appendages, which act as swimming organs, and also as gills. The last segment has four movable hooks upon it, which enable the insect to jump. The larvae, when full grown, are very active and voracious, and they get out of the way of danger by rapid jumping. About this time they leave the water and take up their abode upon the leaves of aquatic if plants, spin a cocoon, and undergo their transformations.

Fig. 6.13. The Metamorphoses of *Dytiscus marginalis*.

The most ferocious insects of the order of the *Coleoptera* are the *Carbidae,* and they are essentially carnivorous, being furnished with powerful offensive weapons in the form of cutting mandibles, and being able to run very rapidly. Their bodies are almost always elongated in form; the antennae are long and cylindrical, like threads, and the legs are admiral adapted for running, but not particularly so for climbing over leaves. But it is the conformation of the jaws that furnishes the most remarkable characteristic of the family. The mouth-pieces have two pulpi, the external lobe of the jaw being fashioned into a true palp, which is divided into two or three joints, as is the case in the *Dytiscidae,* which are aquatic carnivorous beetles. This character not found in other families, and thus becomes important. The males have the front tarsi enlarged, and they hold the females by them when they do not wish to dispense with their company. The larvae of these *Coleoptera* are as carnivorous as the adults, and most of them hunt their prey and run it down, so they are encased in more or less solid armour, the head and the segments of the thorax and abdomen being clothed with coriaceous integuments. Their mandibles, which are arched like sickles, have a large tooth et their base, and in some species there are other long and sharp teeth,which give these structures a very terrible appearance. Their be-lobed jaws have a single pulp of four joints; their legs, which are long, usually end in a tarsus, which consists of two joints; and the end of the abdomen is generally furnished with two jointed appendages and a tubercle, which enable them try walk.

There are two principal forms or types in this family. The *Carabidae* and the *Cicindelidae,* and the genus *Caratus* is an example of the first. They are for the most part large, brilliantly coloured beetles; and their peculiar shaped heads, their long legs, and their beautifully armoured bodies not only are very elegant, but give them a fierce aspect. They are usually found concealed under stones or amongst moss at the foot of trees; and the common *Carabus* is constantly seen running about the

Fig. 6.14. *Carabus auratus.*

roads and fields in search of prey. It destroys a great number of insects, which do much mischief to agriculture, but of course country people crush them whenever they have an opportunity, instead of preserving them. They are of a beautiful golden green colour, the antennae and legs are of russet tint, and the elytra are decorated with three rounded, sculptured ribs. This *Carabus,* which the French call the Gardener, and, for some extraordinary reason, the Sempstress, eats caterillars, snails, and even attacks the May bugs, in spite of their weight. Sometimes in the spring a very curious massacre may be witnessed, when a May bug bumps against something in its careless flight, and tumbles down on the ground on its back.

The *Cherubs* on the look-out for such eventualities, rushes from a hiding-place, and before the May bug can arise, seizes it with its sharp mandibles and opens its body, tearing its intestines and eating them. The poor May bug wriggles about, and crawls off at great pace; but to no purpose, for the enemy never leaves off the attack; and, finally, it rolls over on its back and becomes an easy prey.

In the engrving a May bug is seen on its back, and a *Carabus* tearing away at it. Behind the beetle a larva is just pouncing on an ant; and above the May bug a larva is under the shade of a stone, upon which another beetle is about to attack an insect.

The larvae of *Carabus* are of a shining black colour, and do not come out in the daytime; they hide themselves up under stones, in holes, or under clods of earth. They chase their prey like the adults. There are nearly two hundred species of this genus, an they have the wing-cases more or less soldered together; for as the beetles run in search of their prey, their wings are not always required, and, therefore, are imperfect.

The *Calosoma syscophanta*, which we have already noticed in treating of the *Lepidoptera*, belongs to this family, and is, perhaps, the most beautiful beetle in Europe. It has wings, although it runs quickly and well ; and it has a large body, which is of a deep violet colour, the elytra, which are striated and punctucated, being of a green and gold tint. Its decorations are superb, and its habits are most attractive; and it runs about the trunks of large oaks with incredible rapidity. It attacks by preference the processional caterpillars; and the adult insect and the larva are often seen destroying them together, caring little for the webs, or the dangerous hairs. If the engraving of the metamorphosis of *Bombyx processionea* is examined a *Calosoma* beetle will be seen on the trunk of the tree, about to attack the leader of the caterillars; and low down on the trunk a *Calosoma* may be noticed. The larva resembles those of the genus *Carabus*, and is larger than they are. When full grown, it grubs into the earth, makes a sort of cell, and is transformed into the nymph, which is decorated with small tufts of hairs, that are very prettily arranged. There are some beetles belonging to this great family whose habits are very different to those of the species just mentioned. The *Carabus* and the *Cherubs* enjoy the full sunshine, and pursue their prey like hunters, and scramble over the trunks of trees, and do not like water; but there are other beetles which live in caverns, where not light ever penetrates, and they are consequently colourless almost transparent, and blind; and there are also species which live in those parts of grottoes and caves where the gloom is less dense, and there is some light, and they have imperfect eyes *Blennis areolatus*— which was studied by Victor Audouin on the coasts of the island of Noirmontier— passes the greater part of its life at a very considerable depth below the surface of the water, and it is only found at low spring tides, when the sea has retired more than usual. It hides under stones, and in the crevices where a certain quantity of air is apt to remain; and it appears that the long hairs which cover the beetle keep a quantity of air-bubbles between them and the body, for the purposes of respiration.

These beetles are mentioned with the hope that researches will be made at sometime or other concerning their metamorphoses.

Another group is that of the *Brachinidae*. They hide themselves under stones and congregate together. They are well-known on account of their peculiar manner of repulsing their enemies, for when they are disturbed they elevate their tails and eject a quantity of a vaprorous fluid with a considerable noise. There are few things more curious than what happens when a group of these Bombardier beetles is disturbed by the stones which cover then being suddenly removed. Every beetle hastens to get out of the way, and begins by discharging its weapon, and as this goes on the smoky, cloud-like gas which follows the noise is sufficiently remarkable. The fluid which they eject is pungent, acrid, and volatile, so that it becomes a bluish vapour in contact with atmospheric air. Chemical tests prove that it is a strong acid, and it is sufficiently acrid to produce the sensation of burning upon the skin. The apparatus and the secretion are of course super added during metamorphosis to the usual structures of the larva.

The *Cicindelidae*, or the Tiger Beetle family, is not so numerous in its genera as that of the *Carabidae*. They are very elegant beetles, and their tints are so varied that they are much sought after by naturalists. They have large jaws, long pulpi, sharp and toothed mandibles, long and thin legs, so that like the family just mentioned they are admirably adapted to act as carnivorous creatures. The common green tiger beetle, *Cicindela compositors*, is well-known in Europe, North Africa, and in Asia Minor. It is of a beautiful sea-green colour, with copper tints upon the head, the body, and the margins of the elytra, which assume a fiery look when in the full sunlight. These insects are very interesting when they are larvae, and their habits and artifices resemble those of some very different genera. The larvae live in the same localities as the adults, and these last are so perfectly covered in by their thick integument, and are so active, that they have no fear of owunds, and can readily escape from too active an enemy. The larvae have all the desire for slaughter evinced by parents, but their delicate skins, long bodies, and short legs, not only prevent them from chasing prey, but from attempting a struggle with an insect of any size; nevertheless, these imperfectly armed creatures manage to obtain their prey without exposing themselves to much risk. They have short, thick, and spiny legs, which enable them to dig holes in the ground and they have a flat head, with which they cast forth the pieces of earth they have detached, and furnished with these instruments they construct a vertical tunnel, which curves at a certain depth and becomes a horizontal gallery. The larva has a cuirasses plate on each of the segments of its body, and there are two fleshy tubercles upon the fifth ring of the abdomen, which is larger and more swollen that the other, and which are provided with hooks that curve forwards. The insect crawls in lets tunnel with ease, and if it wishes to remain set fast it sticks the back of its body against the sides, and rests safely with the aid of its hooks. In this position it can poke its head out of the ground, and it closes the entrance of its tunnel and waits until some little ant or other insect passes over. The top of the larva's head forms the floor of the cavity down which the insects are expected to fall, and when one of them touches it the larva descends at once and with great precipitation. The prey feels the soil giving way beneath its feet, and falls down the hole, and is forthwith eaten by the larva, which is soon ready for another. When full grown it closes the orifice of its hole, and undergoes the metamorphoses.

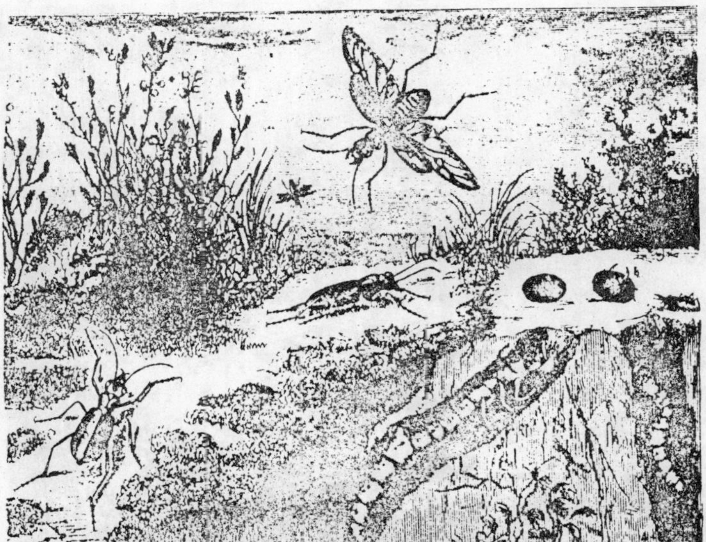

Fig. 6.15. The Metamorphoses of *Cicindella campestris.*

The engraving shows a green tiger beetle flying; beneath it there are two on the ground, and to the right hand there are two galleries or tunnels each of which contains a larva, looked into the side by peculiar structure of the fifth segment of the abdomen. The habits of this larva resemble those of the ant lion.

The Meal Worm Beetle whose larva is commonly called the Meal worm, and which is very injurious to flour and bran, and which does a great deal of mischief to ships' biscuits, belongs to the family of the *Pimelidae.* It frequents granaries and mills, and especially bakeries, for it likes an elevated temperature. The long, cylindrical, and shining yellow larva passes its life enveloped in the flour, and even makes a cell there, where it can undergo its transformation into a nymph. Its only virtue is that it makes capital bait for fishermen and a good food for pet birds.

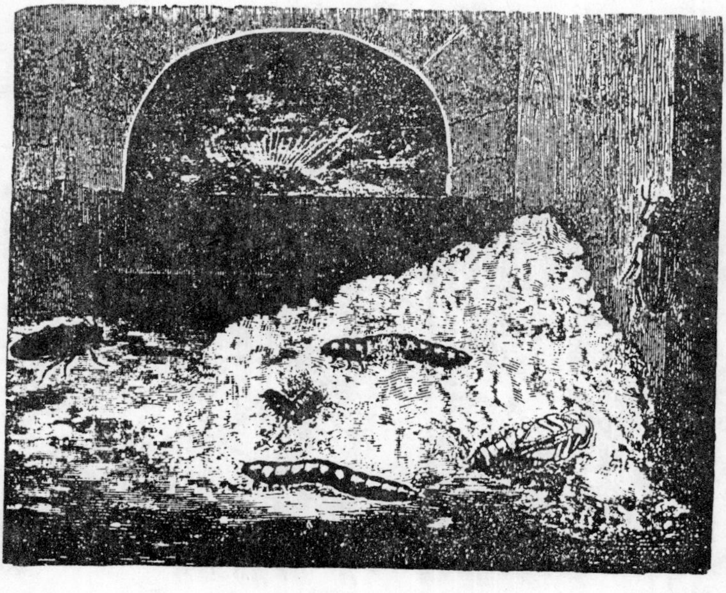

Fig. 6.16. The Metamorphoses of *Tenebrio molitor.* (The Meal Worm.)

The larvae are represented in the midst of the flour in the engraving, where there is also a nymph on the right hand.

The beetles may be seen on the right and on the left side of the engraving.

The beetles of the genus *Blaps* are closely allied to those just mentioned. *Blaps mortisaga*, or the Churchyard Beetle, is the best known species in this country. It hates the light, is of a deep black colour, and has the power of producing a very offensive smell. It lives upon rubbish of all kinds, and keeps in damp places, and the darkest and most retired spots of caves and graters. Almost all country people look unfavourably upon these beetles, and consider them of evil omen. They are very tenacious of life, and there are instances recorded of their having been kept in spirits of wine for many hours without any ill effect. Curiously enough, the larvae have been ejected in great numbers from the human stomach, but how they got there we cannot portent to say.

The *Cantharidae* are well-known as the blistering flies, and many species of the family have been used in the healing art every since the time of Hippocrates and Aerates. The integuments of these insects contain the blistering substance, and as this is so much used a great deal of trouble has been taken to investigate their habits and metamorphoses but without much success.

The oils beetles belong to this family, although they are apparently very differently formed to the brilliant blistering fly. These oil beetles, which belong to the genus *Meloe,* emit an oily liquor like those just mentioned, and it has vesicating properties. The insects do not fly, they have no wings, and and their elytra are very short, their bodies being clumsy, heavy, and soft. They crawl over low plants and grasses, and feed upon wild buttercups, and they are apt to be eaten by cattle, whose mouths soon become much affected. The metamorphoses of some of the *Cantharidae* have been very ably investigated by George Newport and other naturalists. Thus, in the year 1700, Goedart collected the eggs of some species of *Meloe,*and obtained some young larvae. Later De

Gear did the same, but in both cases the larvae perished immediately after their birth, and the observations of the two authors were forgotten. In 1802, Kirby, a celebrated English entomologist, found a small insect which resembled the larvae described by goedart and De Geer upon some *Hymenoptera* related to the bees, and belonging to the genus *Anderna* or solitary bees. He thought they were parasites. In 1828 Leon Dufour found an insect which probably belonged to a parasite of the same kind as those just noticed. He considered it to be a new species, and he called it by the name of *Triongulinus.* Many observers managed to hatch the eggs of *Meloe,* Spanish flies, and of species of the genus Sitaris, which belongs to the same family , but does not possess blistering properties, but they did not get further than the point obtained by Goedart in the year 1700. After a time, however, it became evident that young larvae hooked themselves on, whenever they had the opportunity, to other insects provided with wings, and especially to the nest-making *Hymenoptera;* and Victor Audouin found some full grown and adult individuals of species of *Sitaris* in the nests of the solitary bees. There appeared, then, some evidence that these beetles lived during their early age at the expense of the bees. In 1845 George Newport threw a vast amount of light upon the question, which he had examined assiduously for may years. He studied two kinds of *Meloe, Meloe proscribes* and *Meloe cicatricosus,* and he satisfied himself that the young larvae of these beetles attached themselves to the hairs of the bees, which carried them to their nests. There the larvae underwent their transformations. They nourished themselves with the honey cake of the bee, and when they were full grown they formed for themselves a kind of cell or covering, in which they remained perfectly inactive for sometime. They did not become transformed into nymphs at once, but remained as larvae in this quiet condition. Considering how many accidents might happen to the larva of one of these beetles before it can reach the position which appears to be necessary for its development, it is really wonderful that any one should come to the adult condition. But Nature compensates

Fig. 6.17. *Meloe cicatricosus* and *Sitaris humerals.*

by giving the females an enormous number of eggs, and newport counted 4,218 in one of them. The meloe beetles lay their eggs just below the surface of the ground, and when the little larvae are hatched they climb upon the plants in their neighbourhood, and should any insects with wings alight near them they do their best to hook themselves on. They do not appear to select bees in particular, although they alone can afford the growing larvae their nourishment, but choose the first fly that comes, and thus vast numbers are carried about which soon die from starvation.

"According to M. Fabre—the only naturalist who has yet observed these changes in *Sitaris*—its life-history has eleven events :—1. The egg is laid in the galleries of the *Anthophora*. 2. A little larva is hatched with six serviceable legs, which remains in list birthplace without food or change of size from October to April. 3. In April it becomes active, and attaches itself to the males of *Anthophora* when they emerge from their cells, which they do before the females. 4. From the male it passes to the female *Anthophora*. 5. From the female it passes to its egg, on which it springs as the female lays it on the surface of the honey on which it was to feed, where it sits as on a raft. 6. It then eats up the egg, which takes it eight days. 7. Seated on the empty shell, floating on the honey, it now undergoes a metamorphosis from its previous slim, active, carnivorous form, to a white fleshy grub which is no longer carnivorous, but feeds on honey, and is so organized as to float on surface of the honey, mouth below, spiracles above. 8. After the honey in the cell is done, it changes its skin, and passes into something under a corneous envelope, like the pupa of a fly. 9. It moults again, and once more resembles its second stage (I presume the grub state ; but as it is only said to resemble it, not to revert to it, there may be some difference intended : nothing is said as to its food or feeding in this stage. 10. It then passes into the ordinary pupa state. 11th and lastly. It becomes the perfect insect."

These *Cantharidae* thus change their habits of life in the larval condition, and have a succession of alterations of form which has few analogues. Fiber calls the whole proceeding hyper-metamorphosis.

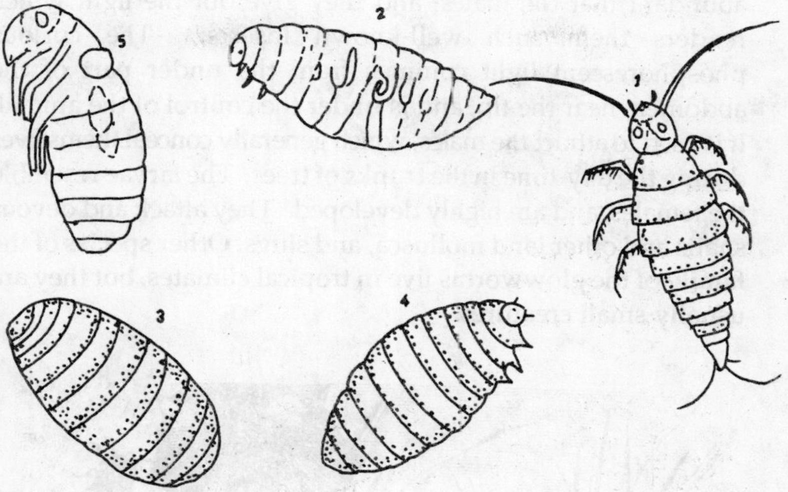

Fig. 6.18. Metamorphoses of *Sitaris humerals*.

1. Larva in its earliest form. 2. Larva in the second form. 3. False pupa, or quiet larva 4. Larve in the fourth stage. 5. Nymph.

In the engraving, the first figure is that of a very small larva (magnified) which is ready to attach itself to the bee, and to devour the egg of its unwilling host. The second figure represents the larva after its first change of skin, when it has ceased being carnivorous and has become a honey eater.

Glowworms, which are frequently seen in many parts of England during the summer nights, are very common in some parts of the Continent, and they sperkle like little gems amongst the grass the dense foliage up the hill sides in Southern Europe. They are beetles, and belong to the genus *Lampyris*, and they resemble somewhat, but cannot be confounded with, the *Cantharidae*. Their chest pieces, which form a broad mass, project more or less over the head, like a buckler. The males have large wings and long elytra, but the females have only rudimentary wings, and they therefore cannot fly. *Lamphyris splendidula).* The females resemble the larvae ; they are more abundant that the males, and they give out the light which renders them such well-known insects. The curious phosphorescent light commes from the under part of the abdomen, near the tip, and is under the control of the animal. It is used to attract the males, which generally conceal themselves during the day-time in the trunks of trees. The larvae resemble the females, and are highly developed. They attack and devour snails and other land mollusca, and slugs. Other species of the family of the glowworms live in tropical climates, but they are usually small creatures.

Fig. 6.19. Metamorphoses of the Glowworm *(Lampyris splendidula).*

There is a male glowworm represented flying in the engraving and there is one on the tree on the right hand of the picture. A female is shown beneath the flying male, and there is another on the ground to the right hand. The segmented condition of the body and the want of wings causes the female to resemble the larvae, and these are represented to the left.

Fig. 6.20. Metamorphoses of *Alaus oculatus.*

There are some beetles which are very generally known as snaps, or Springing Beetles, or Clip Beetles. They belong to the family of the *Elateridae*, and they are very easily distinguished. They have solid integuments, the antennae either cone-shaped or toothed, and the tarsi have five joints, the first being more or less enlarged, and furnished underneath with flexible plates, which enable the insects to walk easily over plants. But the most characteristic structure is a projection of the first piece of the under part of the body, which can be fixed by the will of the insect in a cavity of the second portion. Generally speaking, these insects are of an elliptical or oval form ; sometimes they are nearly linear, and are usually flattened. They fly very well, but they cannot run fast. When placed on their back they are able to spring up to a height of ten or twelve times their own length, and this leaping motion is accompanied by a snapping noise, which has given the insects their popular name. Their legs are so short, that when they are on their backs they cannot turn over, but be contracting the body vigorously, the projection of the under part of the body, which we have just noticed, is placed within the cavity in the second port; it remains there, the insect exercising great force to withdraw it. Suddenly, by a sort of trigger motion, the point gives way, the curved body becomes straight, and of course bounds into the air; and should the insect not have the good fortune to come down on its feet, the same process of gymnastics is tried over and over again , until the desired result ensues. One of these beetles is well-known in meadows and corn-fields. When in the larval condition it is called the Wire Worm; and as it attacks the roots of grass, it does an infinity of mischief.

Allows oculatus belongs to this family, and is found in Louisiana. In the engraving the beetle, with curious eye-shaped markings on the thorax, is upon a leaf; there are two larvae below it, and the nymph is represented in its underground cell.

The *Cleridae* are very pretty little beetles, which are ornamented with bright and varied colours, and which resemble somewhat the Clicks, and moreover, they have some likeness to the *Dermestidae*.) They are of an elongated, nearly cylindrical form, and have club-shaped antennae, the head and thorax roughened with hairs, and the elytra adorned with lively and marked colours, disposed in transverse bands. Their integuments are moderately strong, but they are not quite so flexible as those of the glowworms. The adults are very fond of flowers, and run about the trunks of trees. The larvae are elongated, have small legs, and two projections at the end of the last segment of the abdoen. In this stage of their existence they are very carnivorous. One species of the family *Trichodes alvearius* has very curious habits. The females get into the nests of bees, and lay their eggs there. The larvae, as soon as they are able, begin to eat the larvae of the bees, and very likely they enjoy some of the comb also. The beetle has bluish black elytra, marked with three red bands, and may be noticed to enter the nests of solitary bees. Another species selects the hive bee, and lays its eggs in the cells with the larvae of the bees. The larvae of beetles, when they have destroyed those of the bees in one part of the hive, search out the others , and proceed from cell to cell devouring the inhabitants of each.

Fig. 6.21. The Metamorphoses of *Trichodes alvearius*.

In the engraving the nests of the bee are seen in section, and the larva of the beetle is drawn between two adults.

Another species (*Clerks formicarius)* is a little red insect, which is rather common, and its larvae live under the bark of old trees, where they chase and devour many wood-eating insects, especially larvae. Others belonging to the genus *Necrobia*, feed, both in the perfect and larval state, upon carrion and dead animal matter, such as dried skins and old bones. They are very common in houses, and one species frequents prisons and other gloomy places. Some of the larvae of the genus *Anobium* live in old wood, which they eat, and they undergo their metamorphosis in a small cell, which they from within planks and beams in old furniture, and when they become beetles they eat their way out by drilling small round holes, which every one has been in old houses. The beetles are nocturnal in their habits, and call each other by striking their mandibles against wood, and they answer each other in the same manner. The blow produces a little tick and it is repeated at intervals. The beetle being invisible, superstitious people have a nervous dread of it, and many old women, especially nurses, call it the death tick, and prophesy a fatal issue of the complaints of the sufferers they and watching if the beetle calls it love in this very disagreeable manner.

Elm trees are very often noticed to be scored beneath the bark buy radiating lines, which are arranged in a very regular manner. This is the work of a most destructive beetle, which is called *Skillets destructor*. It is worthy of its name, for many large trees and whole forests finally succumb to its attacks. They have small and projecting heads, club-shaped antennae, and very strong mandibles. In the adult from they nibble the bark of trees, and suck the sap, and the females lay their eggs on the trees, the sap of which does no circulate very actively. The female penetrates underneath the bark, and screws herself in between it and the woody tissue. She makes a gallery, into which her body fits exactly, and bores little side alleys on either side and at regular distances, and lays an egg in each.

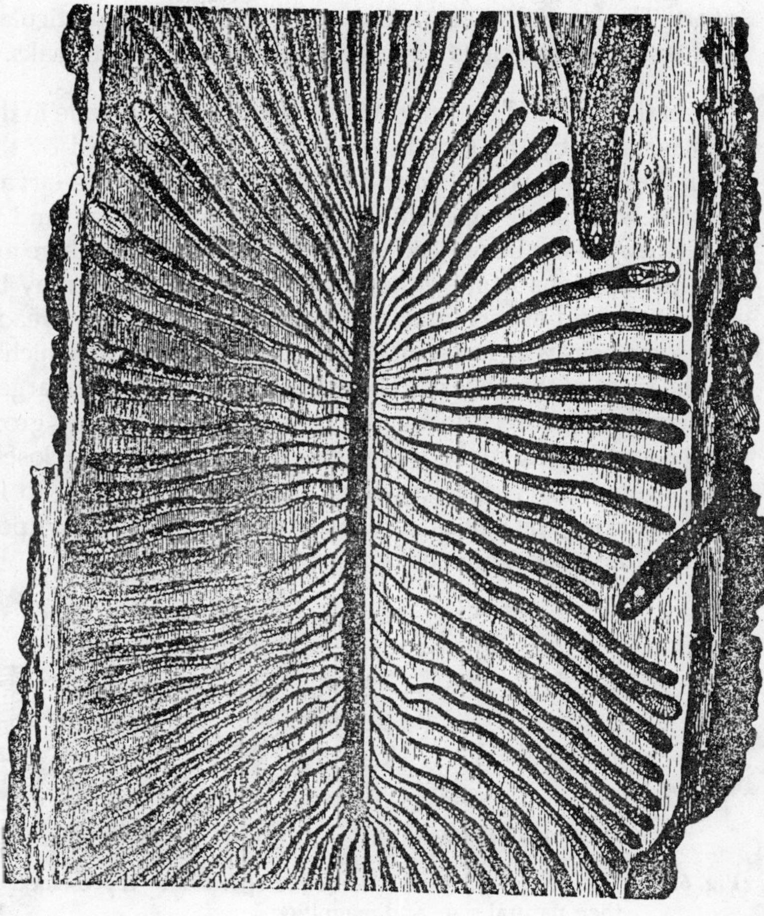

Fig. 6.22. The Bark of an ElM Pierced and Marked by *Skillets destructor.)*

As soon as the larvae are born they set to work to gnaw the wood, doing this so as to enlarge and lengthen their side alleys, and they always eat away in determinate direction. The bark and the wood of the trees which have been attacked by the *Scolytus* are marked with regular designs, which are peculiar to certain species, and some of them resemble tatooings when they are done on hard wood, and really look like works of art. The large elms are destroyed by this beetle in particular, and a smaller one does a great amount of mischief to oaks.

In the center of the engraving the long gallery made in the first instance by the female is seen to be surrounded by the radiating alleys, which have been gnawed out by the larvae. At the end of some of them nymphs may be observed, and in others larvae my be noticed. On the right hand side there are two perforations from without, and a female beetle may be seen at the end of each making its gallery. They have already laid some eggs which resemble white points within little niches on the sides of the gallery. These little niches will be enlarged in a radiating direction by the larvae, which as they grow larger and larger, make wider and longer alleys. A closely allied genus has a species which does much mischief to fir trees, marking the just as if plans or maps had been cut upon them. It is called Tomicus *thpographus*.

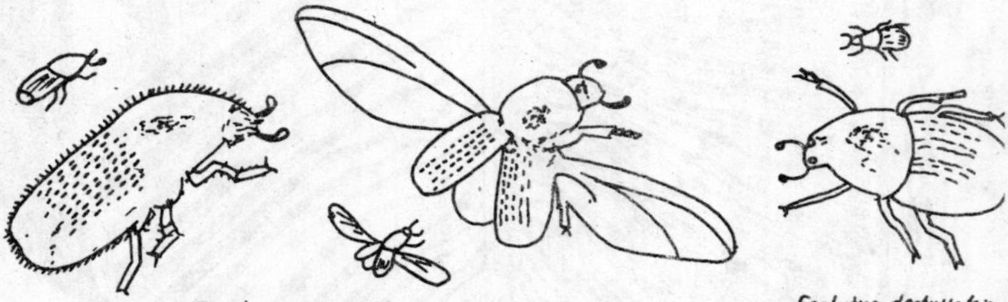

Fig. 6.23. The beetles of these destructive species are represented in their natural size, and magnified.

The most numerous family of the order of the *Coleoptera* is that the *Curculionidae*, and many species are known as Weevils and Hog Beetles.

The larvae of the *Curculionidae*, are almost always thick, massive, and slightly curled, like the larvae of the *Scarabaeidae* they are colourless, white or yellow, and they have flexible skins. These insects live hidden up in the trunks of trees, and in twigs or in grains. They have no legs, or they only possess the rudiments of them, in the form of small tubercles, which may be distinguished in a great number of species. The head is strong, and is clothed with a leathery integument; their antennae are very small , their mandibles are strong, and often dentated; and—as a rule, there being a few exceptions to it—they have no eyes.

The larvae are, therefore, very embryonic in appearance, and can be contrasted with those of many other *Coleoptera* which are much more developed.

Three principal forms may be distinguished in this great family—the *Bruchidae,* whose beaks are short and broad; the *Attelabinae,* whose beaks are long and almost cylindrical; and the *Curculionidae,* which have beaks of greater or less length , or whose antennae are bent after the first joint.

The *Bruchidae* are well-known, on account of the mischief committed by the commonest genus, *Broaches.* Almost every one has seen dried peas pierced by a perfectly circular little hole; and some may have noticed a small beetle, about the eighth of an inch long, coming out of them; it is a blackout insect, marked with white lines or spots. It is called *Broaches pisi,* and is a very common beetle. The mature insect lays its eggs when the pass begin to ripen. As soon as the larvae are hatched, they make their way into the substance of the pea from without, and devour the inside. When the larva is about to undergo its metamorphosis, it gnaws in such a manner as to come so close to the outside skin that it only leaves a delicate layer between it and the world. When the adult has undergone its transformations it has only to break through this think partition in order to come into the light. In some instances the larvae remain within beans and peas during the whole of the winter, and undergo their transformations before it is time to sow the peas next year. Their ravages were at one time so universal in North America , as to put an end in some places to the cultivation of peas altogether.

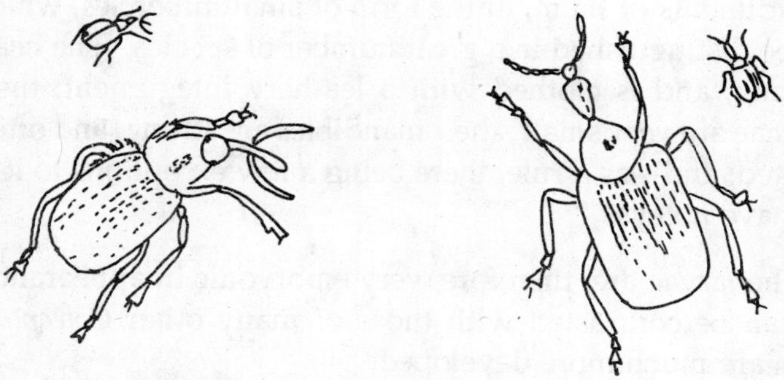

Fig. 6.24. *Rhynchites Bacchus.* *Apoderus coryli,*

Brucus granarius is another species which is very destructive to the same plants in this country ; and in tropical countries other species attack plants with seeds like peas, and some are even found in the cocoa-uut. Dr. Baird writes that the species of *Anthribus* which belong to this family are most frequently found amongst old wood, and some of them appear to live as parasites upon the *Cocci* ; one species, which lives which lives through the winter in the pine-forests of Germany, destroys the *Cocci* in great numbers. The genus *Attelabus* has a very pretty species, with red elytra and corselet ; its habits have been very carefully observed by M. Goureau. The female deposits an egg at the end of an oak-leaf, then she splits the large median nervure across several times close together, then she folds the leaf and rolls it up, and thus makes a safe home for her larva, which she never sees, and of whose habits she can form no conception. The larvae live in leaves and flowers; they only eat the tissue between the outside skin. They change their skin several times before they attain their full growth ; and having spun a cocoon of silk, or of glutinous matter, they shut themselves up and undergo their transformations. The larvae of other species live either kin the stems of plants or in fruits, which serve them as a shelter and as food. They do great mischief to the plants upon which they live, but they are perfectly harmless as beetles. The species of the genus *Apoderus* and *Rhynchites* — the last being characterized by a long and thin beak—have the instinct of cutting the stems of leaves or the ends of twigs, and of laying their eggs just inside ; and their larvae, which can only live upon fading leaves, have thus to thank their parent for giving them life, and their peculiar means of subsistence also.

In the engraving above the real and magnified sizes of two common beetles, whose long beaks very visible, are shown.

Many of the species are very hurtful to fruit trees, and it appears that the male insect, in some instances, helps the female find a shelter for the egg. One kind attacks the vine and the pear tree. The female insect selects a leaf, cuts the stalk with her beak almost half through, and then rolls the leaf together, and in doing this she is assisted by the male. Having made a symmetrical roll, she cuts into and pierces it with the beak and lays an egg in the opening, and then she pushes in the egg in such a manner that it remains on the inner side of the leaf. Five or six eggs are thus introduced, and the leaf is rolled up in such a way that it is impossible to discover from the outward appearance how the eggs were deposited.

The *Apoderus* in the engraving is a pretty red insect with a black head, and it rolls up the leaves of nut trees. The *Rhynchites* in the same engraving has a violet red tint, and looks every silky, and it twists up vine leaves.

The little *Apion* is scarcely a line in length, lives in its larval condition in seeds and grains, and is very injurious. It is a true weevil in spite of list size.

The *Larini* frequent the shores of the Mediterranean. They are well-known by their elytra, which are covered with sort of efforescence, which imitates spots and cloudy markings. The ir larvae live in the fleshy part of the receptacle or eye of composite flower. The engraving represents the larva of *Larinus maculosus* in the center of the receptacle of *Echinops ritro*. The adults are on the ground and on a leaf.

Fig. 6.25. Beetles and Larvae *Learinus maculosus*.

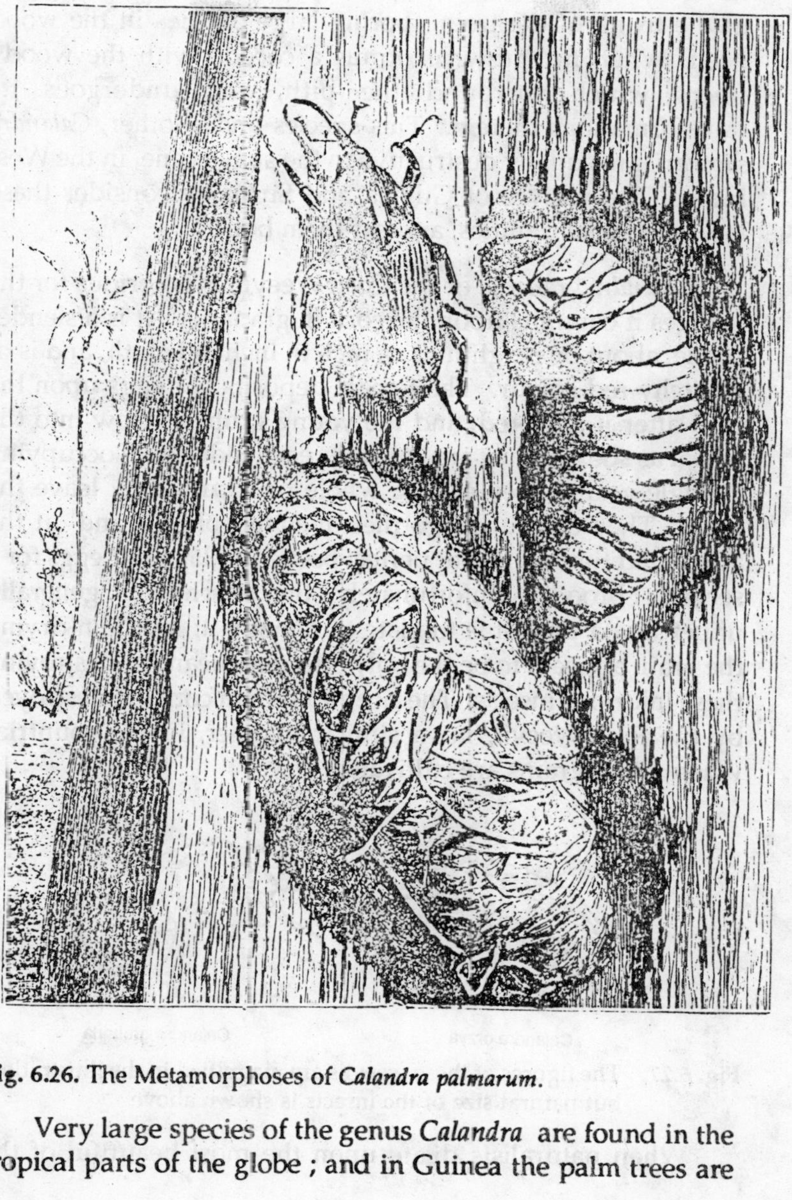

Fig. 6.26. The Metamorphoses of *Calandra palmarum.*

Very large species of the genus *Calandra* are found in the tropical parts of the globe ; and in Guinea the palm trees are more or less riddled by one. The adult *calandra palmarum* is a large velvety black beetle , and is nearly two inches in length; its larva is a great worm-shaped creature, and lives in the pith of the trunk of the trees, boring large cavities in the wood itself. When full-grown it forms a cocoon with the woody fibers which surround the pith, and undergoes its transformations therein. This species and another, *Calandra sacchari*, are equally destructive to the sugar-cane in the West Indies. The natives of Guinea and Surname consider these larvae as great dainties, and eat them boiled.

Calandra granaria, or the Corn Weevil, is notorious for the ravages it commits upon the corn in granaries. It is a slender beetle, about one-eighth of an inch in inch in length, and is of a patchy red colour. The female deposits her eggs upon the corn after it is stored, and the young grubs burrow into the wheat as soon as they are hatched , each individual occupying a single grain : they eat the whole of it, and only leave the husks. They then undergo their metamorphosis, and at the appointed time come out perfect beetles, to lay their eggs for a second brood. Unfortunately, the mischief is generally unperceived until it is too late to remedy it; but kiln-drying the grain appears to be the only effective means of destroying these injurious insects. The rice and Indian corn are destroyed by the Rice Weevil, in a similar manner, in the countries where these grains are cultivated.

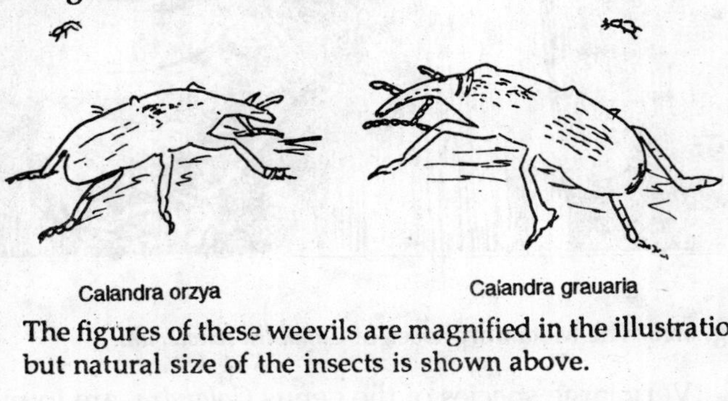

Calandra orzya　　Calandra grauaria

Fig. 6.27. The figures of these weevils are magnified in the illustration, but natural size of the insects is shown above.

When naturalists dilate upon the most beautiful of the

Coleoptera, and upon those which possess the most marvelous richness of colours, and extremely possessed and shining bodies, they refer especially to the *Buprestidae;* it is an immense family, containing no less than forty-two genera and 1,500 species, (Dr. Baird). But those that are found in France and in England have very plain decorations, and are very different in their external appearance to their magnificent brethren in the tropics. They are sometimes called Gold-beaters, but doubtless have many other familiar names.

The brilliant species are not very elegantly shaped; their bodies are long,the prothorax broad, and their legs are short; and they have a character which distinguishes them at once amongst the beetles, for their wings are not longer than the elytra.

The illustration whos one of the *Buprestidae,* with its wings expanded, and it will be observed that they only equal in length the dense elytra. The larvae live in the trunks of trees and in woody tissues, twigs, and stems, and they are somewhat remarkable. They are white insects, without legs, or having only vestiges of them, in the form of tubercles; they have a retractile head, the front part of which along is thick and leathery , but the prothoracic segment is very large and broad, and is clothed with a coriaceous plate that is either granulated or tuberculated. This arrangement enables the larvae to use much force and tearing power when they are perforating trees, to which they do, in the tropics, a great deal of harm. Fortunately, however, our trees do not suffer from them.

The beetles of the genus *Chalcophora* are amongst the largest of the family; and the species *Mariana,* whose metamorphoses are illustrated in the engraving is prettily-sculptured and bronzed beetle, which is common in the pine forests of the south of France, Italy and Germany. Its larva hollows out large galleries and holes in the trunks of the trees.

In the same engraving the larva which has done part of

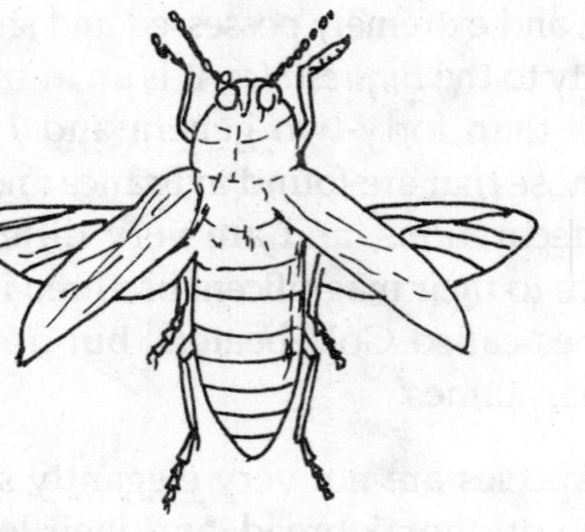

Fig. 6.28. *Chalcophora Mariana,*

this mischief is shown in its gallery, and its small head and enormously developed first thoracic segment may be noticed, the rest of the body being rather slender. It undergoes the metamorphoses in one of the galleries, and in one them a pupa may be observed , whose structures foreshadow those of the adult form very distinctly.

The Longicorn Beetles, or the Goat Beetles, as they are sometimes called, on account of their long, cylindrical antennae which look like horns, belong to the family of the *Cerambycidae.* The adults are made for walking well, and they have usually very strong, and indeed occasionally enormous, mandibles, differently toothed according to the species, and the jaws of the lower lip present many decided modifications of form, according to the peculiar habits of the insects. The beetles are leaf-eaters, and the diversity in the structure of their mouths depends upon the nature of the vegetable matters upon which they feed. The species are world-wide, but their abundance is in distinct relation with the richness of the vegetation of different countries , so that South America, India, Ceylon, and the Islands of Sunda, and the Moluccas contain a great number of the most beautiful and largest Capricorns. It is impossible to confound a beetle belonging to this family with that of any other. There is the greatest resemblance amongst the larvae of the whole family, and they look like stout elongated white

Fig. 6.29. The Metamorphoses of *Chalcophora Mariana*.

The adult insect is represented as crawling upon a branch of a fir-tree which is more or less riddled and hollowed into tunnels or galleries. The larva and chrysalis are also shown.

worms, and the segments of their bodies are very much alike in all. All the segments are a little swollen the first, however, being the largest, and being covered above and below with a leathery plate. They have rudimentary antennae. These larvae live in the trunks and branches of trees, and in the cellular structures of some herbaceous plants. As they never come to the light they are colourless, and have soft integuments, but as they feed upon the wood out of which they form galleries, they have very strong jaws and a very stout head. As they do want to walk much in a narrow gallery, they have no legs, or else they are in a most rudimentary condition, but their swollen segments enable them to climb. This history of the peculiar structures of the larva presents striking analogies with that of the wood-eating larvae of the *Lepidoptera* and *Hymenoptera;* and the existence of similar adaptations in very different insects in order to enable them to live under the same con conditions of existence is very remarkable. But the weak jaws of *Chalcophora Mariana,* which are presumed to do the same kind of work as those of the *Cerambycidae,* must be remembered in considering such generalizations. The strength of the jaws of the larvae of the *Cerambycidae* differs according to the density of the tissue of the plant in which the particular species live. The abdomen of the female beetles of some genera is provided with an ovipositor of considerable length, by means of which they can insert their eggs into the crevices of trees or plants, in the interior of which their larvae live and are hatched. The larvae make a cocoon by joining together fragments of wood and little bits of vegetable tissue with their saliva, and are transformed into nymphs. Some of the beetles are remarkable for the emission of a fragrant odour not unlike that of otto of roses.

The exuberant vegetation of tropical countries can can well afford to be checked by ravages of insects, and some of the largest trees, which require much heat and damp for their growth, are more or less destroyed by some of the largest larvae which are known. thus in Guinea there is an immense beetle called the Titan *(Titanus giganteus)*, whose larva gnaws and makes galleries in the largest trees, and its size almost appears to bear some relation to the magnitude of its food. Again, in India, there are beetles with long mandibles which make them look almost like stag beetles, but the antennae are enormously long and spines at each joint, and they have very

Fig. 6.30. The Metamorphosis of *Acanthophorus serraticornis.*

large larvae. The larva of *Acanthophorus serraticornis* lives under the dark of large trees in the neighbourhood of Pondicherry. It does not come into the light, and, like many other wood-burners, it forms galleries within the wood, and finally makes a cocoon of enemies dimensions out of the vegetable tissue. The engraving exhibits the beetle upon the outside of a tree, the larva being under the bark and a great cocoon in a kind of gallery.

The mimosa trees in the West Indies suffer greatly from a beetle which is called *Lamia amputator.* The larva bores and excavates the branches and delicate saplings of the trees, and undergoes it metamorphosis in them. This mischief having been done, and *coup de grace* is given by the perfect insect, for the beetle has a fancy for gnawing round the branches in a circular line, so as to cut them of completely. An allied species *(Onsiders vomicosa)* does very much the same sort of thing. M. Halite, now the head gardener of the tropical department of the Musee d/Histoire Naturelle in Paris, once lived in the environs of Rib Janeiro, and every night he heard the sound of falling branches of trees belonging to the *Acacia lebbeck.* He found on examination that these branches were sawn all round in circular direction, but their central part, or pith, was not touched, so that the branches broke by their own weight or by the simple force of the wind. The mischief was put down to the evil dispositions of the slaves of the house, but M. Houllet discovered a beetle upon a branch which had been thus cut, and it soon because evident that this insect was the author of the destruction of the tees, for when a branch was carefully opened it was found to contain the living larvae and nymphs of the *Insiders.*

In the engraving a beetle may be noticed with very along horns upon a little twig close to the main branch. Quite low down the fractured end of the branch shows in the center the ragged pith, but between this and the dark the tissues are perfectly smooth, the dark being cut through in a circular manner. The smoothness of the tissues has been produced by

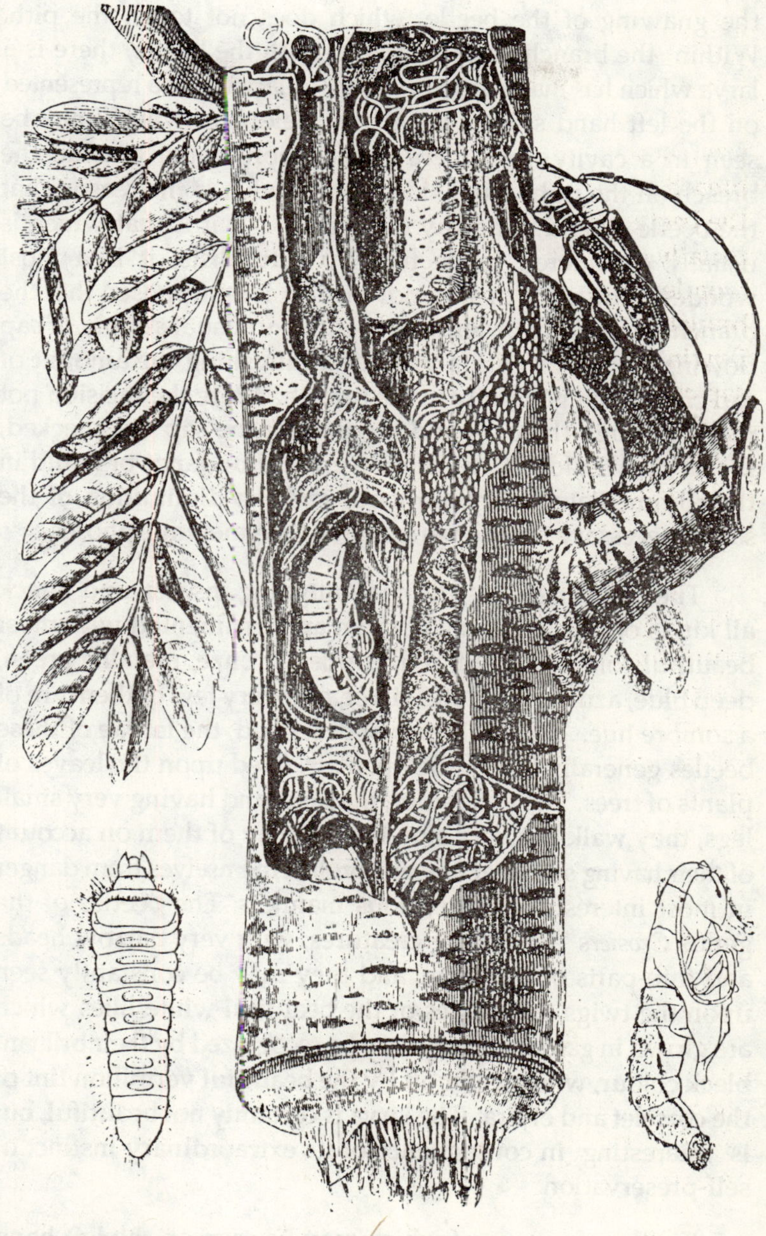

Fig. 6.31. The Mertamorphoses of *onsiders vomicosus*

the gnawing of the beetle, which does not touch the pith. Within the branch, and on a level with the beetle, there is a larva which has made an excavation. A larva is also represented on the left hand side of the plate. Below, a nymph may be seen in a cavity, and there is one represented outside the breach on the right-hand side. Now, what is the reason that the beetle should cut away the branch ? If it is that which is usually considered to be true by naturalists, it is a most wonderful instance of superior instinct It is supposed that the branch is cut off in order to prevent an unusual supply of sap flowing into it, or with a view of diminishing the amount of sap circulating generally in the branch. Were the incision not made through the bark and the flow of the sap not checked, the larvae would suffer from too great abundance of liquid in their immediate neighbourhood; and the diminution of the sap is necessary for the perfection of the metamorphoses.

The *Chrysomelidae* are leaf-outgoing insects, and attack all kinds of plants. Many of them are ornamented with most beautiful colour of gold, red, copper, bronze, metallic green, deep blue, azure violet, and silver, and very few of them are of a sombre hue. Unlike those just described, the larvae of these beetles generally live in the open air, fixed upon the leaves of plants of trees. The y are stout, thick-set and having very small legs, they walk but very slowly, and may of them on account of their having soft integuments, protect themselves from danger in most interesting and curious manners. The beetles of the genus *Crosiers* are oblong creatures, with very narrow heads and fore-parts of the body, and they may be constantly seen upon the twigs and leaves of the beautiful white lilies which are grown in gardens. They may be recognized by their brilliant black colour, which is set off by the beautiful vermilion tint of the corselet and elytra. The larva is certainly not beautiful, but is interesting in consequence of its extraordinary instinct of self-preservation.

The larvae, some of which may be seen on the left hand leaves in the engraving below, have soft integuments, and, therefore, some extraordinary protection must be had against the attacks of other insects and the weather. The vent of the larva is nearly on a level with its back, so that all digested matters which pass from the intestines slide over or collect upon the body. The larva of *Crioceris merdigera* thus lives under most miserable clothing but which is safely and easily replaced, and the insect appears to know this, for if disturbed and cleansed, it sets to work and eats as quickly as it can, knowing that Nature will soon provide another coat. The perfect insects are represented upon the stem of the lily.

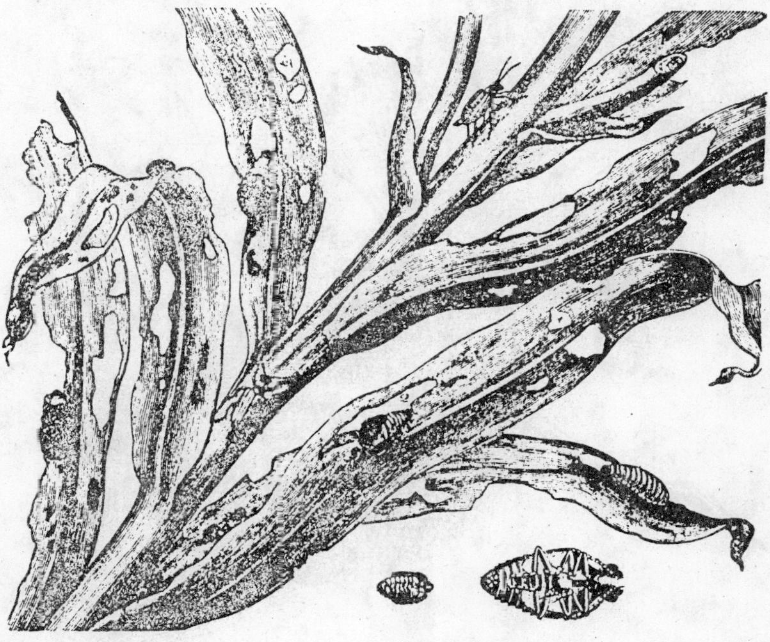

Fig. 6.32. The Metamiorphoses of *Carioceris Merdigrea.*

A nymph seen from above. A nymph (magnified) seen from below.

The larva of another beetle, which is closely allied to the

species just described, has almost as curious a protection. It may be seen on artichokes and large thistles. It is broad, flat, and has long, pointed, or spiny hairs fixed to its sides, and a fork-shaped appendage, which is attached to the end of the body and curved forwards above the insect. The vent is situated at the base of their appendage, and the matters coming from it stick on to the fork which is hoisted up above the body with the lump of dirt upon it like a parasol.

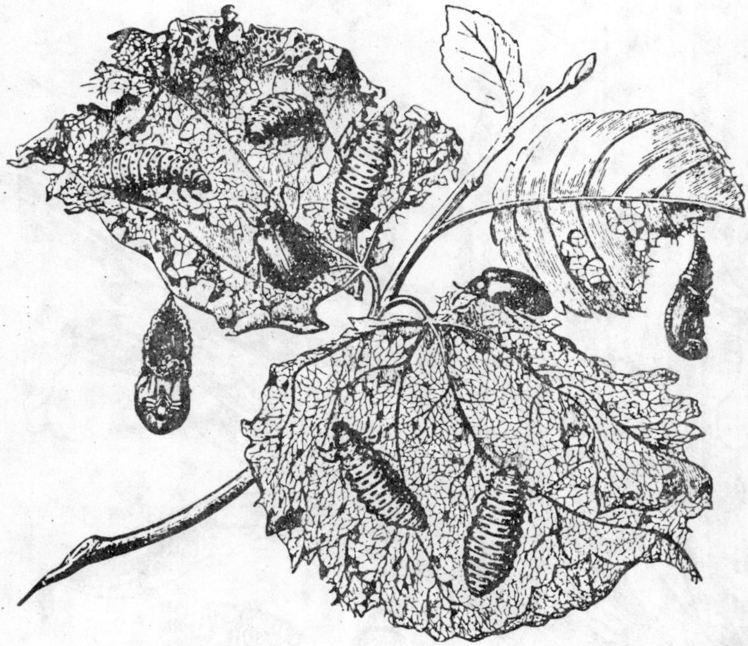

Fig. 6.33. The Metamorphoses of *Lina populi*.

The poplar trees are frequently much damaged by some species of the genus *Lina*. The adult beetle is of a green bronze colour, and has red elytra, but the larva is yellow and is spotted with brilliant black. The larva destroys the foliage, and eats the cellular tissue of the leaves, but does not touch

Fig. 6.34. The Metamorphosis of the Lady Bird (*Coccinella septempunctata*)

the nervures, so that they look very much like lace work. When about to undergo its metamorphosis the larva hangs itself to a leaf by its tail, and after a time the skin cracks and the nmymph may be seen underneath, but it remains thus protected until it is transformed into the beetle.

The nymph is represented hanging from a leaf, the larvae are making lace work of the leaves, and the beetles are crowing upon the plants. This hanging up of the nymph resembles the favourite position of many chrysalids of the *Lepidoptera,* and it is somewhat remarkable that another genus of these beetles has larvae which construct movable tubes, in which they live, and which they carry about after the fashion of many caterpillars.

The last family of the order of the *Coleoptera* that comes under our notice is that of the *Coccinellidae* or the Lady Birds. These are great favourites with children, and even with collectors; and popular sympathy is extended towards them, and certainly very properly, for they do much good by preventing the excessive multiplication of plant lice, and, moreover, they are very pretty things. The lady birds with seven spots (*Coccinella septempunctata*) are very common in Europe, and, indeed, in Asia, Africa, and America. They are small insects, which are so well-known that they need not be described, but it may be often noticed that when they are alarmed of laid hold of they fold up their limbs and eject from the joints a yellow mucilaginous fluid, which has a every disagreeable adoire. Immense swarms of these insects are sometimes observed, especially on the south-eastern coasts of England, and they have been described as extending in dense masses for miles. The adult insects do not eat much,, but the larvae consume enormous quantities of the little insects which are so injurious to vegetation, and they may be seen chasing , or rather creeping up to, the plant lice, and eating one after the other, taking the whole set on a leaf or stem in regular order. The larva is of a lead grey colour, and has a broad yellow spot in front of the head, three little red spots on the sides, besides black spots and little bunches of hairs. When about to undergo its transformation the insect attaches itself to a leaf just like the caterpillar of a butterfly.

In the engraving the lady birds are seen upon the leaves in the center. The re are are two larvae on the right hand and a nymph on a solitary leaf above the others.

7

Metamorphosis in the Orthoptera

There are not many typical forms amongst the *Orthoptera*.oat of them are well-known, and the Earwigs, the Crickets, the Grasshoppers and the Locusts, are the familiar examples.

If we institute a comparison between the *Orthoptera* and the *Coleoptera* some very interesting results may be obtained. so far as their structures are concerned there is a great general likeness between the two groups, but there are enormous differences as regards the progress of their development. The *Orthoptera* are insects that bite and browse like the *Caleoptera*, and the structures of their mouths are not more unlike than they usually are in families belonging to the same order. The organs of flight, however , distinguish the *Orthoptera* from all other insects. The anterior wings differ in their texture and general form from the posterior, they are of almost leathery consistency, and with one exception they cross each other when the insect is quiet. The hind wings are very characteristic. They are membranous, much veined, and their large nervures are almost like straight sticks. These wings fold themselves up longitudinally, like a fan, hence the name of *Orthoptera*.

If we only find moderate differences between the adult *orthoptera* and *Coleptera*, there are nevertheless very great differences between them lid the early part of their existence. The earwigs, the crickets, and the grasshoppers, on leaving the egg have the form the appearance, and almost all the characteristics of the adults whose method of life they also follow. Whilst the *Lepidoptera*, the *Hymenoptera*, and the *Coleoptera* are born in a very embryonic or immature state, the *Orthoptera* spring into life almost as well developed as the adults, and hardly undergo any changes before attaining the perfect form. These changes consist of the development of the organs of flight and reproduction. There are many *Orthoptera*, however, which never obtain their organs of flight, and they become adults without having acquired the same perfection which is common to the rest of the group. They may be said to have had their development arrested , or, perhaps, they are the oldest forms of insect life, metamorphosis and wings having been superadded to the original kinds. Most of the *Orthoptera* change their skin or moult at certain periods of their growth. Some of the insects moults three times successively, and still resemble the creature that first came from the egg; a fourth moult takes place, and then the rudiments of crumpled wings may be observed; then a fifth change of the skin happens, and it leaves the insects with their great wings perfectly finished. The *Orthoptera* which never obtain organs of flight only undergo the first three moultings, and those which only have the rudiments of wings do not change their skin after the fourth time. They never have any period of inactivity; and in consequence of this and of the peculiarities of their growth just mentioned they are called insects whose metamorphoses are incomplete; but it is customary to call those which have no vestiges of wings, larvae, and those which have them in a rudimentary state, nymphs. The larvae, nymphs, and the perfect insects are active and lead the same kind of life.

The species of *Orthoptera* are spread over the whole surface of the world, and are particularly abundant in those tropical climates where vegetation is in excess. There the herbivorous *Orthoptera* which are the most numerous, revel amongst plants of all kinds, which they consume in great quantity. Usually they are beautiful insects, embellished with lively and light colours, the shading of which is very pleasant to the eye; but this is not always the case, as some of them are remarkably ugly and dark.

The *Forficulidae*, which everybody calls the Earwigs, are much more remarkable — especially in their habits—than most of us have any idea of. They have the credit of crawling into the ears and doing some mischief to the brain; but really the only lives they ever endanger are those of the plants and pot-herbs in the garden. Their nippers, as they are familiarly called, and their great activity, appear to have given them a bad character; but they are much to be admired for some things, for they are amongst the few insects that care for their young, and an old earwig may often be seen acting like an old hen.

Fig. 7.1. Earwigs. *(Forficula auricularia)*

The wings of the earwigs are well worth examining. The anterior wings are short, like those of the *Staphylinidae* amongst the beetles , and they do not cross each other as those of the other *Orthoptera* do. The posterior wings are large, and are very pretty when they are extended; the insect folds them first of all like a fan, and then turns them up, and doubles them transversely, so that they fit underneath the front ones.

There is only one genus in this family, and there are several species of it in Europe; but the *Forficula auricularia* is the best known. The earwigs are nocturnal in their habits, and are rarely seen in the day-time, unless disturbed. Sometimes they hide themselves by lying flatly under leaves; but they are more frequently found hidden up under stones, under the dark of old trees which has become rather detached, and in the interior of all sorts of dark hollow places. There a perfect family of them will keep quiet in safety during the day, and it will be found to consist of little ones without wings, those a little advanced in age with rudiments of them, and of the fathers and mothers who know how to fly. They are not industrial in their habits , but the females watch over their eggs and carry them off to a safe place if they are threatened with any danger. These insects attack all sorts of vegetable tissues, and the most beautiful flowers are no more to them than the humbles weeds. Unfortunately, however, they seem to prefer spoiling the flowers whose petals are most elaborate.

The *Blattidae* are generally considered very disagreeable insects. They aren't pleasant to look at; they are offensive to our sense of smell, on account of their very fetid adoure, and they excite our antipathy on account of the waste and destruction they cause. Possessing great agility, and being able to run with extraordinary rapidity, they have something very peculiar in their aspect and habits; generally the body is large, broad, flat, and clothed with very flexible coirasses integuments. The head is almost hidden under the prothorax, and its antennae are very long and as delight as threads. They are common in some places, especially on board ship and in seaport towns;

Fig. 7.2. *Blatta Americana.*

but commerce has carried them more or less over the whole world. They are called by all sorts of names, but we knew them lap black beetles and cock-roaches. Such they are, in fact, as regards their colour, but they are true Orthoptera. They manage to hide themselves up very easily; and from the having flat bodies and very compressible skins, they can squeeze themselves into very narrow places. Being omnivorous they attack everything that comes in their way which can possibly be consumed ; nothing is amiss to them; and they inflict great losses upon merchants, as well as upon householders. They do not lay their eggs one by one in different places, but they shut them up in as sort of capsule, which is rather leathery, land they carry it about with them until the young are about to escape; they then help the little ones to get out of this bag, and at first are very attentive to them.

The *Blatta Americana* is very common and may be frequently noticed in the cargoes of ships. The engaraving represents the the larva and two but its of this kind.

Blatta Oriental, which is not so large as that just noticed, have its wings very doubly developed. It is common in towns, and is particularly found of hiding up in the cracks of chimneys, and of only appearing at night time. The larvae when first born are quite white, and resemble their parents in form, but change their skins six times before they become perfect.

There is one great family of the *Orthoptera* the species of which are all carnivorous, and there are no insects which have such an extraordinary appearance as these *Mantidae.* Their strange attitudes, their great and strong limbs and leaf-like shapes, combined with the great slowness of their movements, do not make them look at all like bloodthirsty creatures. They have a narrow body, a very long prothorax, and a very movable head, which has large eyes and sharp and trenchant mandibles. They have large wings, front legs admirably adapted for seizzing their prey, and the others are slender. The *Mantidae* are generally large and fresh-coloured, land sometimes the posterior wings, which are always more or less transparent, are very prettily ornamented and coloured. They crawl on bushes, where they remain perfectly immobile hour after hour with the front of the body erect and with their front legs folded. This perfect quietude does not raise any suspicions in the insects which are flying about, but if an unfortunate fly comes too close the *Mantis* extends its foot rapidly and too surely.. The *Mantidae* cover their eggs with a kind of capsule, which they hang on plants.

The eggs are laid towards the end of summer, and the young larvae escape and grow like all the other *Orthoptera.* The species of this family inhabit the hot parts of Europe and the Tropics, and wherever they are seen, their positions apparently so meditative have inspired observers with some amount of verneration for them. People are often deceived by appearances, and nothing leads very well disposed persons astray so much as the belief that certain actions represent goodness of disposition. Persons who want to deceive other have only to present to be very sactified, and to pray in the corners of the streets, and act the hypocrite persistently. Success will attend them in an exact relation to their endeavours to deceive. All this holds good for insects, an there is no more respectable and saintly-looking preacher than the *Mantis religiosa,* or the praying *Mantis* unfortunate moths that may admire it on account of its attitudes of supplication soon find out that instead of saying "Let us pray," it says "Let us prey."

The *Empusa* are much handsomer than the insects of the genus Mantis. They have a long, leaf-shaped projection on their heads, short antennae, a very thin thorax, and their legs are ornamented with curious leaf-like expansions. There is a very pretty *Empusa,* which lives in the Maritime Alphs, and which is represented in the engraving. The body of *Empusa* is decorated with a mixture of most delicate grey, green, white, and violet tints, and the wings of the adults are of a clear sea-green colours, the borders and the nervures being lilac.

Fig. 7.3. The eggs and immature and adult forms of *Phyllium siccifolium*.

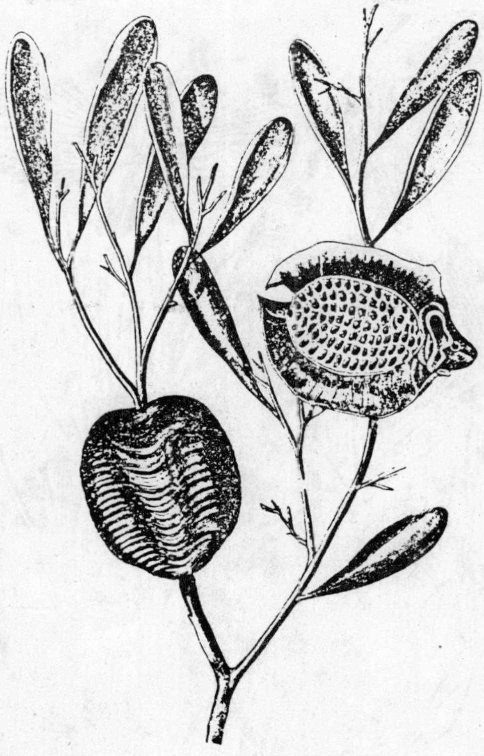

Fig. 7.4. Egg Capsules of *Mantis religiosa.*
A capsule attached to a twig and a section of a capsule.

The *Phasmiaae* are, however, the most curious of all insects. They approach the *Mantidae* in general appearance, but are distinguished by their fore legs being of the ordinary form, and suited, like the rest, for walking. Unfortunately their incomplete metamorphosis has not been studied carefully.

All the *Orthoptera* which we have hitherto noticed may be called walkers or runners; but those which we are now about to describe are termed jumpers or leapers.

Fig. 7.5. Metamorphoses of *Empusa pauperata.*

The grasshoppers are true jumpers, and have their hind legs so formed that they-can lift themselves form the ground and jump considerable distances; the females have a long and stout ovipositor,which has a sabre shape, and is attached to the hind part of the body. Nearly all the "jumpers" have well-developed wings; the front ones, or the elytra, are long, and the posterior are of considerable size; but there are some

Fig. 7.6. The Metamorphoses of the Great Green Grasshopper.

Fig. 7.7. Metamorphosis of the field cricket.

species whose wings are always in a rudimentary state. In every instance the first pair of wings of the male insects are so formed that they can produce musical sounds. At the base of the elytra there is a greater or less space, which is covered by a thing, transparent, and very tense membrane, and it is situated between some twisted and enlarged nervures. When the insects are anxious for a little female society at eventide, they elevate their wings, and rub them lone upon the other, and produce a sound which is rendered intense, sharp, and chirping by the vibration of the membrane. The grasshoppers fly easily, in spite of their heavy bodies; but they walk with difficulty, in consequence of the disproportion which exists between their front and hind legs; so they usually move from place to place by a series of jumps. The jumping is produced by the action of the large muscles of the things acting upon the long legs and feet. These insects use their ovipositor for the purpose of burying in the soils; and it is interesting to observe that several species which live almost in the same manner have different modes of egg-laying and their ovipositors are therefore variously fashioned. The species of the grasshopper genus have a tubercular projection on their foreheads, and their elytra are longer than the hind wings; and the largest kind, *Locusta viridissima,* which people in central and northern Europe very improperly call the *Cigala,* is well-known to everybody.

In the accompanying engraving the female of this green grasshopper is laying her eggs in a hole in the ground, and one is just escaping from the ovipositor. A male is flying about, and it will be observed that he has no ovipositor, but that he has very long front wings, with spaces in them just where they join the body, for his musical apparatus. A male land female are on the ground, and have not yet completed their full growth.

Everybody knows and can recognize a cricket, but naturalists appear to have done their best to make ordinary mortals confound crickets with grasshoppers, and both with locusts. Thus the French call crickets *Grillidae,* and the grasshoppers *Locusts.* The English call the grasshoppers *Grillidae* and *Achetidae.* There are Field Crickets, and House Crickets, and Mole Crickets, and they have great resemblances to the grasshoppers.

The crickets are nocturnal insects, whose colours are dark and of brown and grey tints, but the grasshoppers and

Fig. 7.8. The Metamorphoses of the more Cricket. (*Gryllotalpa vulgaris*)

locusts like the light. They have all long and slender antennae, and legs adapted for jumping, and the elytra of the males have a large musical apparatus, and the females possess an elongated ovipositor.

There are two principal types in this family—the crickets whose front legs and feet are simple in their structure, and the mole crickets, which have flattened fore legs. The musical sound of the crickets has given them their name, and we hear it by night in the fields and in houses hour after hour. These insects live very solitary existences, except when they are disposed for that society which they suppose they can obtain by their everlasting chirp. Each one of them digs a hole, and does not leave it except during the night, and children in the fields, who know this, can often catch a field cricket by poking straws down cracks in the ground. The house cricket, which is of a yellow grey colour, and more or less marked with brown tints, is smaller than the other, and hides itself in holes and cracks in old walls and chimneys. It is a chilly insect, and it seeks its nourishment and delights to live in warm places. In towns it prefers bakeries and the neighbourhood of ovens, and in the country the most humble kitchens are favoured by its marry chirp.

The engraving a cricket emerging from its hole, and a female on the bank above it; the imperfect insects are on the ground in front.

The mole crickets are large insects, and are wonderfully adapted for their particular method of life. They live underground, and very rarely leave their holes. Their body is almost cylindrical, their feet are thick and short; their front legs, which are also short and wonderfully enlarged, and, as it were, fingered, enabled them to grub in the earth very effectually. They are the true moles of the insect class, and they choose made soils, like those of gardens and vineyards, for their dwelling places. They make galleries destroying everything which comes in their way, cutting through roots and eating the fine underground twigs as well as the worms and the grubs which they come across during their excoriations. Some authors have asserted that the mole cricket is truly carnivorous, but this is a mistake. They will eat worms and insects which come in their way, but their nourishment is principally of a vegetable character; and it can be readily imagined that they are very destructive. Their excavations consist of vertical shafts more or less deep, and of long horizontal galleries which lead from them; the insect is, in fact, a regular miner, and the female lays its eggs in the remotest part of the mine.

In the accompanying engraving a mole cricket is seen coming out of its shaft, another is in a gallery, a male is flying, and the immature insects are crawling one the soil.

One tribe of the *Orthoptera* still remains to be noticed, and its members are the true locusts, and belong to the *Acrididae*.

The *Acrididae* have short and thick antennae, but they have not the same kind of musical apparatus as the tribes already described, nor ovipositors; but if they have not this particular musical apparatus they are none the less good musicians—that is to say if the stridulous noise they make can be called music. They produce it is this manner: there are very prominent nervures on the elytra, and the things of the hind legs are furnished with ridges on their internal surface; the insect rubs the thigh sharply against the elytra, and produces on this natural fiddle a sharp vibrating sound. We may not admire it, but it is quite certain that the lady crickets do, and doubtless they constitute a very critical audience. The accompanying engraving exhibits the metamorphosis of the locust (*Acridium peregrinum).*

The locusts deposit their eggs in the earth, and close them up in a kind of tunnel, and the young ones are born without wings, which become developed as they grow older.

Some remarks on the metamorphoses of the *Orthoptera* will be found in this chapter on the *Crustacea*.

Fig. 7.9. The Metamorphoses of Locusts. *(Acridium peregrinum)*.

8

Metamorphosis in the Thysanoptera

The very small black flies which are such a source of annoyance to travellers in the summer-time, and which fly into out eyes and crawl over our faces during the prevalence of warm windy water, principally belong to a kind of insect which is characterized by giving very remarkable wings when in the adult condition. These insects exist by myriads, and there are several species of them; and they are all exceedingly destructive to flowers, and especially to the bloom of cereal plants. The little black insects are to be seen on almost every flower, and they devour the delicate cellular tissue of the petals. *Thrips cerealium* is very destructive when it occur in multitudes upon the wheat, barley, and oats, for it interferes with the proper nutrition of the grain, by nibbling the protecting envelopes and the tissue which connects it to the stalk.

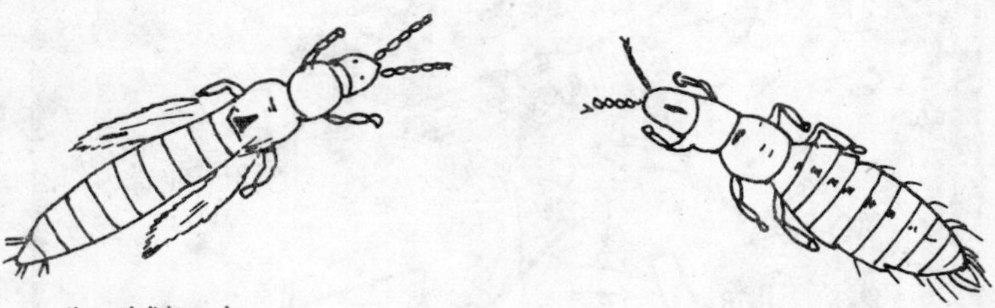

the adult insect

(Magnified)

The larva

Fig. 8.1. The Incomplete Metamorphosis of *Thrips cerealium.)*

All the members of the genus *thrips*—and they alone constitute the order now under consideration—possess four very narrow membranous wings, without any folds or network upon them, but furnished and decorated with beautiful fringes upon the edges. These fringes characterize the order, which in other respects is closely allied to the *Orthoptera*, and they give the name to it. The *Thysanoptera* (fringes; a wing) have filiform antennae and very large eyes, and the different species of the genus *Thrips* have a great diversity of wing fringing. The structure of the wings is somewhat analogous to that observed in the *Lepidoptera*, in the *Pterophorina* and the *Alucitina*.

The metamorphoses of the *Thysanoptera* have not received much attention, but they are known to be of the incomplete kind. The quiet chrysalis condition is not observed, and the larvae are born from the egg, greatly resembling the adults. The absence of wings is that great distinction between the larval and the imago state, as it is in the closely - allied order of the *Orthoptera*. The larva moults several times, and the wings are gradually added, the colour of the insect altering also.

The Neuroptera

All the varieties of metamorphosis may be observed amongst the *Neuroptera*. Some of these insects never have a period of inactivity, and do not undergo greater transformations that the *Orthoptera*. But others have to submit to two metamorphoses before they become perfect, and the transformations, are as sharply defined as they are in the *Lepidoptera*. These last undergo complete, and those just mentioned pass through incomplete, metamorphoses.

Other *Neuroptera* remaining in a condition of inaction for a very short time, and their nymphs lose their activity and undergo a rapid metamorphosis, as it were, at the very moment when the adult insects escape from the pupa case. It is very remarkable that such varied metamorphoses should occur amongst insects which have great structural affinities, and which belong to a very natural order.

The *Neuroptera* are known at once by their peculiar wings. They have four, which are membranous and naked—that is to

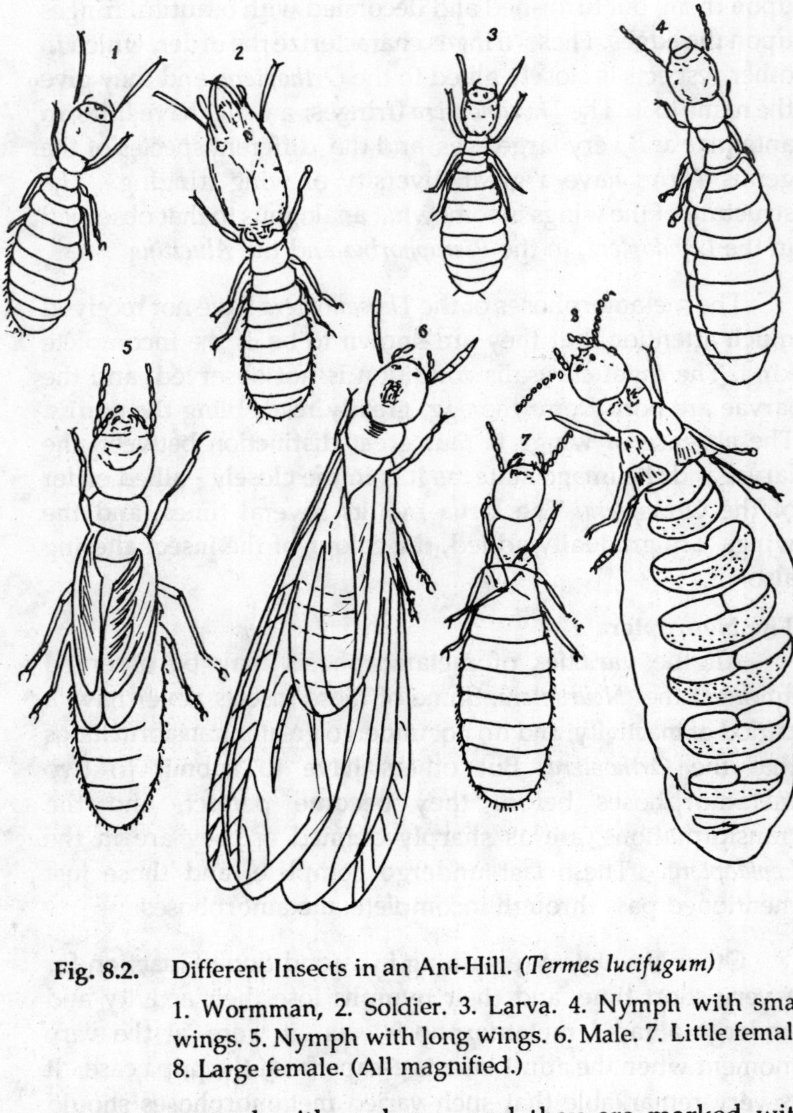

Fig. 8.2. Different Insects in an Ant-Hill. *(Termes lucifugum)*

1. Wormman, 2. Soldier. 3. Larva. 4. Nymph with small wings. 5. Nymph with long wings. 6. Male. 7. Little female. 8. Large female. (All magnified.)

say not covered with scales — and they are marked with membranous nervures so arranged as to look like net-work. They are insects whose wings are therefore much veined and reticulate, their legs are delicate, and their bodies are almost always long and slender. The jaws and the structures of the mouth concerned in feeding are separate, and do not form a sucking apparatus. There are many different forms among the *Neuroptera,* and some have exceptional structures. Thus there are *Neuroptera* whose transparent wings have nervures placed across them and there are others which have hairy wings without any cross-marking, and the nervures branch off. Now, there are many amongst the first kind which greatly resemble the *Orthoptera.* They unite, therefore, these two divisions of the insect world, and at the same time are more or less separable from the characteristic *Neuroptera.*

The *Neuroptera* live in different manners. Some of them are very industrious are found of society, and build large homes, whilst others—and these are the most numerous—do not appear to possess either remarkable talents or instincts, but they are interesting, in consequence of their leading such very different lives in the larva and adult stages. One of the most interesting families—not only of this order but of the whole class of insects—is that of the White Ant, the *Termitidae.* These insects are gregaropis, and form numerous societies, in which may be found males and females, several kinds of neuters, and active larvae and nymphs. They bile very large dwellings, containing multitudes of chambers and galleries and they always work in the dark. If they wish to move from one locality to another, thicken construct first-rate tunnels; and their habits, especially those of construction, have caused them to be compared with the true ants; and they are well-known in all tropical climates, where they commit frighful damage, as white ants. The males and females of *Termes lucifugum* have large wings, the cross or transverse nervures of which can hardly be seen, and they are rudimentary; their heads are large and strong, and there are three ocelli placed between the larger pair of eyes. The neuters, which never have wings, are of two kinds—first, the workers, with a round head and short mandibles, and second, the soldiers, with long heads and large and strong jaws.

M. Lespes has discovered that there exist in the dwellings of *Termes lucifugum* in France, besides the larvae of the neuters and of the males and females , and besides the neuters, workmen, and soldiers, two sorts of nymphs, some very small, with rudiments of very short wings, and other larger, with longer wings and even—and this is a every inexplicable fact—two kinds of males and females, one small, and appearing in May, and the other, much larger, not being observable till the mouth of August. M. Lespes calls the first little kings and little queens; and the second, the great kings and great queens. When the eggs are about to be laid ordinarily a single couple is found in the interior of the nest, and the individuals composing it have lost their wings, and indeed they always appear to fall off at this time. The bodies of the great females are often distended with eggs.

It is of course in hot climates that the *Termites* attain their greatest size and numbers. It is stated that third of the flat country in Ceylon is undermined by them. In Western Africa they resemble the common or ants proper in their habits. Like them they are composed of three kinds of individuals, males, females, and neuters or workers. They erect dwellings in the form of pyramids or cones; and *Termes bellicosus* builds them up to a height of three or four yards, and flanks the edifices with little towers and constructs them so solidly that several men may stand upon them without falling though. The neuters or soldiers which are very large and have great jaws, are always ready to attack anybody molesting their habitations. each house or community consists of a king and queen and laborers and the soldiers, and there are about hundred of the labourers to each soldier. The military *Termes* appear to the labourers in a further state of development. The males and females are possessed of wings, and at the proper time they issue forth from the nest in immense numbers and take their flight. They begin to emigrate at the commencement of the rainy season, and as their wings are not strong, and are only calculated to carry them for a few hours, the number that perish by the heavy rains is enormous. It is said that perhaps not more than one pair out of many thousands reach a place of safety. The fortunate pair which do succeed are generally found by some of the labouring insects—which are continually running over the surface of the ground in the neighbourhood of their covered galleries—and are immediately elected king and queen of a new colony. The queen lays about sixty eggs in a minute, or 80,000 or upwards in the course of twenty-four hours, and they are instantly taken away by the others and carried to a apart of the any hill, where they are hatched and the larvae are attended to and nursed until they are able to shift for themselves. The engraving on the opposite page shows a full-sized queen. The metamorphoses is very incomplete, and the larvae resemble the adults except in not having full-grown wings, but there are other arrests of development, and the limitation of the power of reproduction is very definite. The pupae hardly differ from the larvae and the adult forms.

Fig. 8.3. A Female *Termes* from the African Coast.

There are some small *Neuroptera* which resemble the *Termes* somewhat, but their habits are comparatively uninteresting. They are the *Psocidae.* They have wings which are slightly veined, large heads, with three ocelli on the forehead, setiform antennae, and very slender legs. These insects are found on mosses and upon the trunks of trees and old walls, and a very small species inhabits houses, and is particularly fond of crawling amongst old and damp papers. The wings of the females never become developed, and they therefore remain as it were in a larval condition. It looks something like a little louse, and as it is said to make a noise by clashing its jaws together, it has been called *Psocus pulsatorius.*

The *Perlidae* are very distinct as a family. They are moderately-sized *Neuroptera,* with large hind wings, which are folded near their attachment. They have setiform or thread-shaped antennae and well-developed jaws. Moreover, they present manifest resemblances to the *Orthoptera,* as do the *termites* and the *Psocidae.* Picket of Geneva has paid great attention to the European species of this family. The *Perlidae* undergo incomplete metamorphoses. The larvae are aquatic carnivorous, active, and are often provided with external respiratory organs or gills, which consist of small tufts of filaments that are attached to the lower part of the segments of the thorax. The nymphs or pupae, which are always active, differ from the larvae in having the rudiments of wings.

When the nymphs have attained their full size they leave the water and settle upon the stones or plants on the bank; their skin soon dries and splits along the black, and they the adult or perfect insect scrambles out and takes to flight. The larvae appear to prefer running streams and they pass the winter at the bottom of the water, and shed their skins several times, and undergo their first metamorphosis during the spring. The fly does not make much use of its wings during the day, but is lively in the evening, and the females carry their bag of eggs suspended to their bodies. Some of the species are found in high northern latitudes, and appear over the snow at the every first approach of spring. *Perla marginata,* whose aquatic and perfect conditions are shown in the accompanying engraving has two long articulated processes, something like those of the earwig. It is a common insect, and its larva, which is of a yellow citron colour spotted with black, is found in running streams and swiftly flowing brooks. A closely allied species, the *Nemoura,* lives exactly in the same way as the *Perlae,* but hey are smaller insects. When young they have curious processes sticking out from the body, like those of the

Perlae, but they disappear in the perfect insect. The wings are always perfectly developed in the females, but they are short and rudimentary in the males. This is curious enough, for it is not often that the degraded condition so commonly observable in the wings of female insects is to be found in the males, and it is very interesting that such closely allied insects should have such remarkable differences of structure.

Fig. 8.4. The Metamorphoses of *Perla marginata.*

The *Ephemeridae* exist in enormous numbers, and last of the shortest time imaginable in their perfect state. They are the lightest, the most fragile, and the most delicate creatures in the world. They float away in the air from off the water, in which they have lived during the early stages of their existence, flutter in the sunshine, dance about with every breeze, and they grow faint as the sun gets low, and sink and die with the night. They are truly ephemeral beings. They can be easily distinguished. They have very short antennae, ending with an exteremly delicate silk-like thread; their wings are very fine, the front ones are the largest, the hind ones being very small; the body is terminated by two or three long-jointed bristles; the structures of the mouth red soft, land cannot receive food, and indeed the perfect insects, living as they do only a few hours, never think of wasting their lives in eating and drinking. They all exist for love, and they spends their short day in a instant and active courtship. The eggs are laid in little packets upon the water. As soon as the larvae escape from the egg they take to the water, and grow more or less rapidly, according to the species. The se larvae are essentially aquatic insects, and are organized so that they can respire in water, for there are gills, or delicate folds of skin, which are traversed by numerous tracheae fixed on to their bodies. The larvae of all the species of tea genus *Ephemera* are very similar, and they are believed to live two or three years before they begin their transformations. They reside beneath stones or in burrows at the bottom of running streams, and they undergo an incomplete metamorphosis. They have very strong jaws covered with spines, and mandibles with sharp points to them. They begin to be nymphs when the rudiments of wings appear. When the nymph is full grow, it crawls out of the water on to plants or stones, and its skin cracks own the back and the adult emerges from it into the air. But this perfect insect in covered over and entirely enclosed in a beautiful and most delicate membrane, which it has to escape from before it takes to flight. When thus enclosed it is called the pseud-imago, in contradistinction to the imago or the perfect insect.

The engraving represents the metamorphoses of the May fly In the water, amongst the weeds there are larvae and a large nymph, on a leaf at the side nee is escaping from its covering, and the perfect insects are flying over the lake.

The European *Ephemeridae* have been carefully studied by Pictet, and the kinds which can properly be united together, on account of their having numerous transverse nervures on their wings, and their abdomens being terminated by three

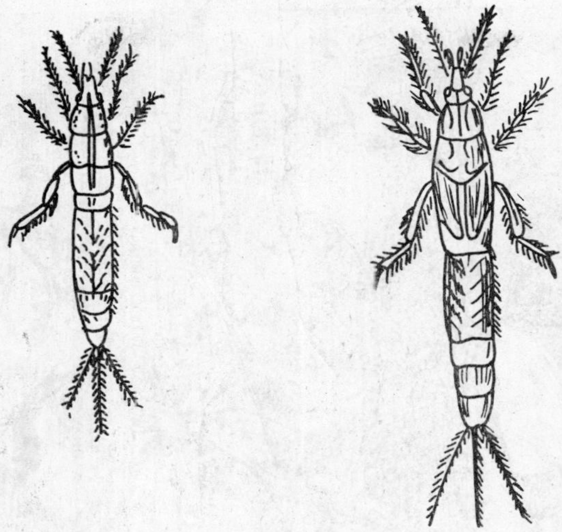

Fig. 8.5. Larva and Pupa of an *Ephemera*.

bristle-shaped hairs, are called *Ephemera*. The commonest are the May flies, which appear in early summer in such enormous quantities. The adults have a short life, which only lasts a few hours; they never take any nourishment; and their beautiful wings which are exquisitely made, and delicately marked with brown spots, appear almost to be wasted, so short is their time of flight. Their larvae and nymphs like those of the species which are allied to them, are long and cylindrical, and have projecting jaws and large claws, which enable them to dig in the mud at the bottom of the streams, where they make galleries, into which they retire out of the way of danger. They breathe with their gills or bronchiae situated on the abdomen, and which are narrow, finely divided , and covered with delicate hire. The nymph is quite as active as the larva.

The insects which are united in the genus *Clop* have very pretty larvae watcher slender, excessively delicate, almost transparent, and provided with leaf-shaped bronchiae, which they vibrate incessantly and with great rapidity. They have tiny legs, and are provided with broad fringes to the body, which act as oars.

Fig. 8.6. The Metamorphoses of the may flies.

These little larvae, so exquisitely elegant, live in the open water, swim with ease, and catch their prey land elude their enemies by the rapidity of their movements. When in the adult state they have their wings vineed, and are furnished with only two bristle-shaped projections on the abdomen. The males have very extraordinary eyes, and each appears to be surmounted by a second, so that this insect with doubt eyes, trasparent wings, and a fawn-coloured thorax, is called *Close bioculata*.

The most important family of the order of the *Neuroptera* is that of the *Libellulidae,* and its members present the characteristics of the group in the highest degree. They are those insects which are constantly flying during the summer by the river side and over ponds and lakes, and which in England are called Dragon Flies; but which, on account of their gauzy wings, brilliant colours, their thin and exquisite shapes, and their elegant motions, are named Demoiselles by the french. this difference in the names is very significant of the national thought. In England we appreciate people for their intrinsic merits; but in France it is too common to the guided by externals. With all their admiration for the beauty of the fair sex, no Frenchman would compare these beautiful and aerial insects with the ladies, if their habits were known; for under this fine exterior there exists a terrible amount of destructive will and power; and they are, perhaps, the most bloodthirsty things in creation. We have not been misguided, for we know that from their very birth, and in every stage of their development, they are worthy of the name of dragon flies. The wings of these insects are nearly equal in length, and they are beautifully reticulated by interlacing nervures; they have a large head and great eyes and small antennae. They have quick sight, and fly with great rapidity backwards, forwards, sideways, upwards, and downwards, and without turning; their mouth is strengthened to the utmost their jaws are strong, and end in sharp point. The mandibles are provided with keen teeth, and the lower lip is very large, and its pulps are short and thick. Thus armed, they chase and pull down every fly, moth, or butterfly which they come across. They rend these delicate creatures, and eat them; but frequently they appear to kill for killing's sake. These *Neuroptera,* especially the males, have curious pincer at the extremity of the body. The lady dragon flies are as sanguinary as the males, from whom they do not appear to receive the most polite treatment; and the courtship is a scene of general bad behaviours and violent. They all undergo incomplete metamorphoses. The egg are laid upon the water, and the larvae take very kindly to that element. As the larvae grow their body becomes larger, their head flatter, and their eyes wider apart then those of the perfect insect. They are not less bloodthirsty than the adults, but they are very slow in their movements, and they do not appear to have the instinct to hide themselves in order to catch their prey. Nevertheless, they have no difficulty in seizing very active insects which come swimming along, for the larvae and nymphs have a most remarkable weapon in their lower lip. The lip is very long, and has pulps formed in the shape of pincers. It is attached to the chin, which is also very long, by a sort of hinge, which allows it to rest against the under part of the body when the insect is quiet. Suppose the larva is looking out for prey; it remains perfectly quiet until some small insect passes by at what might be considered a safe distance, but the victim has not calculated upon the hidden weapon which is suddenly straightened and poked forward, and which clasps the unfortunate with its pincers. The larvae and the nymphs, although they live under water land must respire, have no bronchia or any external organs by which they can breathe. Their method of respiration is unique; they breathe with their intestines. The large intestine is covered with numerous tracheae, and when the animal wishes to breathe it opens the orifice of the intestine and admits a quantity of water. This, of course, contains air mechanically suspended, which is taken up the tracheae just mentioned. Curiously enough, these insects, which walk very slowly, manage to get out of the way of lignums by expelling the water in their intestine with considerable violence, and their projects them to some distance.

Fig. 8.7. The Metamorphoses of the Dragon Fly (*Aeschna - maculatissima.*)

When about to undergo its transformation the nymph leaves the water and crawls upon plants by the side. Then the skin dries, splits, and sets and the perfect insect free. The larvae live about a year before acquiring their full growth, and the nymph is very active, and has the rudiments of wings.

The *Aeschnitae* have a rounded and robust body, and three articulations to their lip. Some have enormous eyes, set close together, so that they nearly touch each lather in the middle line, and others have their eyes set wide apart.

Fig. 8.8. Nymph of the *Libellula* and the perfect insect emerging.

The engraving shows the metamorphosis of *Aeschna maculatissima*. On the right one of them is flying; immediately below, in the water, a nymph may be seen projecting her long under lip with its pincers; by its side is a larva in a quiet condition, and above it, on the leaves, a perfect insect may be seen crawling out of the skin of the nymph preparatory to taking flight.

The *Agrionidae* have slender bodies , and the abdomen is

cylindrical, and in the shape of a rod. They are usually very small dragon flies, and there is a very common one whose males are of a beautiful blue metallic colour, with a greenish-blue band on the wings, and whose females are of a most reliant green colour.

Fig. 8.9. Metamorphoses of the common ant Lion (*Myrmeleon formicarium*).

When we compare some of the members of the family of the *Myrmeleonidae*, or Ant Lions, with the dragon flies, there does not appear to be much difference between them, and there are really very close affinities between the two families, so far as their adults or perfect insects are concerned. But there are very considerable distinction to be noticed in the nature of the metamorphosis and the habits of the larvae. The *Myrmeleonidae* live on dry and in their early conditions of life. They undergo complete metamorphoses, and their larvae, which are short, squat, and always carnivorous, are not in the least like the perfect insects. Moreover, they usually constrict a cocoon when they have attained their full growth, and are transformed into quiet and inactive nymphs. The perfect insects have their wings less reticulate then the dragon flies, their eyes are smaller, and the antennae have many articulations. The mandibles are sharp, and the jaws and lip, which are rather narrowed, have long pulps upon them. There are two divisions in the family.

The ant lions which form the genus *Myrmeleon* have very remarkable habits in the larval condition.

In some sandy places on little hills which have a sunny aspect it is not uncommon to notice small cavities which have a funnel shape. They are made by the larvae of the ant lion, which are insects of a grayish-red colour having on their sides little collections of black hairs; the front and middle sets of legs project forwards, and the hind ones, which are stronger and not so movable, are kept backwards against the body. These larvae can only walk backwards. In burying themselves in the sand they turn round and round, and throw the grains out from the funnel-shaped hole they make by their twisting, using their large and flat heads like a shovel. The sand as it is thrown over the sides of the hole forms a margin to it, and when all is completed, the ant lion sinks itself at the bottom of the trap in the sand, and only shows its long mandibles. The spot which it chooses for its dangerous dwelling is usually in the neighbourhood of the dwellings of ants and other insects. The unfortunate ant which happens to tread on the margin of the hole sets some sand off rolling, and it immediately begins to struggle against falling down, but the larva of the ant lion throws some sand upon it, and it soon comes tumbling down to the bottom of the funnel. The ant is immediately seized between the sharp mandibles of the ant lion which are perforated by tubes, and which allow the bloodthirsty insect to suck the juices of its prey. This perforation of the mandibles is one of the remarkable peculiarities of the organization of these insects. When the larva becomes fully grown it winds a silken cocoon

Fig. 8.10. The Metamorphosis of the Dragon fly ant lion. (*Palpares libelluloides)*

from a silk gland which, unlike those of the *Lepidoptera,* is situated at the extremity of the body; there is not much silk to spare, so a quantity of sand is glued into its walls. The larva lives for about two years, and the chrysalis becomes transformed into the perfect insect in about three weeks. The adults are of a dark grey colour, with yellow spots, and they have large transparent wings, which are ornamented with little black marks. In the tropical regions of the globe there are large ant lions, whose great wings are spotted with brown and black, belonging to the genus *Palpares*. One of these is also found in Provence and is called the Dragon Fly Ant Lion (*Palpares libelluloides,)* and it is represented in the engraving on the previous page. One perfect insect is about to come to the ground, a second is on the ground and is looking at a larva which is nearly completely hidden in the sand, whilst another larva is upon the sand.

Fig. 8.11. *Nemoptera coa* AND *Ascalaphus longicornis.*

It is very remarkable that the larva of this kind should not, like the others, make a funnel-shaped trepan the sand, and that is should dimple bury itself anyhow, leaving, however, its mandibles in the open air.

There is a little group of *Neuroptera* belonging to the tribe of which the ant lions are characteristic forms ; they constitute the genus *Ascalaphus*, and they are very pretty insects, which are to unlike some of the *Lepidoptera*. They have long antennae, which terminate in a large club-shaped end, like those of the genus *Organize;* their wings are very prettily tinted, and are usually of a bright yellow colour, with black markings. Many of them are to be found in Southern Europe, and their larvae, whose habits are unknown, resemble those of the ant lions. In the engraving one of these perfect insects is represented on the right hand of the picture, above the trees.

The *Nemopterae* are remarkable for the great beauty of their forms. They inhabit the shores of the Mediterranean, but unfortunately very little is known about them. A perfect insect of *Nemoptera coa* is represented in the engraving and its silken antennae are very fine and thread-like whilst the hind wings are very long, slender , and oar shaped. The front wings are broad and large, and are of a yellow sulfury colour, spotted and figured with black. It is one of the prettiest of insects.

The *Hemerobinae* are miniature myrmeleon, and have long and slender antennae, and large and well-veined wings. We may often notice these beautiful flies upon rose and elder trees, and they can be recognized easily, on account of their small heads being furnished with prominent, rounded, brilliant eyes, which look like polished gold. The body of the fly is of a delicate green colour, and the large wings, so beautifully reticulated, reflect all the colours of the rainbow.

These pretty insects are sluggish in their motions, and prefer to fly by twilight. They lay their eggs upon plants, attaching them by long and narrow foot-stalk which gives

Fig. 8.12. The Metamorphoses of *Panorpa communis.*

them the look of a fungus. The larvae, more slender than those and twigs, have a peculiar structure on the foot to enable them of the ant lions, and destined to crawl and run over leaves to do this safely. They live a life of constant warfare against the *Aphides*, or plant lice, killing and eating them day after day, so that they are called Aphis Lions. When they have attained their full growth they spin a small cocoon of pure silk, and are transformed into pupae. They live in this condition of fifteen days, and then the perfect insect flies. These pretty *Neuroptera* have a most disgusting scent, as any one that handles them will soon discover.

Fig. 8.13. The Metamorphoses of *Symbols lutaria.*

The Scorpion Flies, *Panorpa communis,* are common in some parts of Europe, in damp places. The adults have a curiously long muzzle, and lather small wings, which are transparent, and spotted with brown colours; the males have the same sort of body-pincers as the dragon flies, and the females have a longish egg tube or ovipositor, which she pokes into the moist ground, and then lays her eggs.

The larvae are almost cylindrical in shape, and have small fore legs, and little tubercles which act as hind legs; they are or a grey red colour, and have a few hairs. They live at a depth of a few lines, or an inch, in moist ground, and make tunnels and galleries amongst the decaying vegetation upon which they feed. They undergo complete metamorphoses, and the nymph remains in one of the underground passages. The engraving represents female imago of *Panorpa communis* laying her eggs in a hole she has made with her ovipositor. The perfect insects are flying over head, and the larvae and nymph are in their galleries.

There are some other *Neuroptera* that undergo complete metamorphoses—the *Semblidae*. The perfect insects of the genus *Symbols* have wings with only a few transverse nervures; their prothorax is rather long, and their antennae are thread-like. The wings are of a dusky colour, and the mandibles are vary short. The perfect insect may often be seen on the borders of ponds and marshes, and it lays its eggs on the tissues of plants. The larvae go down into the water, and spend their time in chasing other water insects, or in hiding up in the mud. They have respiratory membranes joined on to the sides of the body in the form of gills.

When the larvae have attained their full growth they proceed to act differently to those of the other *Neuroptera* for they come out of the water when their time for metamorphosis is at hand, and make a hole in the damp earth, within which they are transformed into nymphs.

The last division of the *Neuroptera* is often separated from the other in a distinct family, for its genera have their wings

Fig. 8.14. The Metamorphoses of the Caddis Flies.
(Phryganea flavicornis, Phryganea rhombica, and *Phryganea fusca.)*

very peculiarly marked and constructed. The Caddis Flies, or *Phryganidae*, undergo complete metamorphoses, and present some resemblance to the *Lepidoptera*. Their wings have no cross reticulations, and are peculiarised by the presence of small hairs, which are stuck on after the manner of the scales

Fig. 8.15. The Metamorphosis of *Cicada fraxini*.

in the butterfly wing. The mouth is useless , and its structures are soft and impervious. The antennae are thick, long, and pointed. The caddis files are numerous in species in temperate climates, and they are usually grayish-brown or yellowish in colour. The adults fly in the evening in the marshes and by the side of rivers, stereos and brooks. The females deposit their eggs in the water, and envelop them in a glutinous mass, so that the bunches stick to stones or aquatic plants.

The caddis flies, or rather the caddis worms, are particularly interesting in their aquatic condition. They are then very carnivorous, but have soft skins; consequently them form tubes and cases our of all kinds of objects, and stick them together with a little silk, like some caterpillars, the *Psychidae,* for instance. They carry their houses, for such are their tubes, about with them, and only project the front part of their bodies. Should any danger threaten them, or should then be alarmed, they retract their bodies within the case immediately. Each species makes its case in a different manner, and employs different materials. Some of the larvae of caddis flies invariably use gravel or very small stones for their cases, and others the small fresh-water snail and other shells, which they glue together. Many use bits of stick or morsels of plants, so that each species may be known by its tube. Some live in fixed habitations, which are made in the same manner as the others, but which are not to be moved.

So numerous were the caddis worms in the olden time that a very thick limestone in the Auvergne is made up of the tubes or indusiae of them.

The larvae always live in water, and they have respiratory filaments fitted for aquatic breathing upon the sides of the abdomen. They have a scaly head, and the thoracic segments are clothed with leathery plates. The legs are long, and the soft skin of the abdomen is terminated by two hooks which anchor the insect in its dwelling.

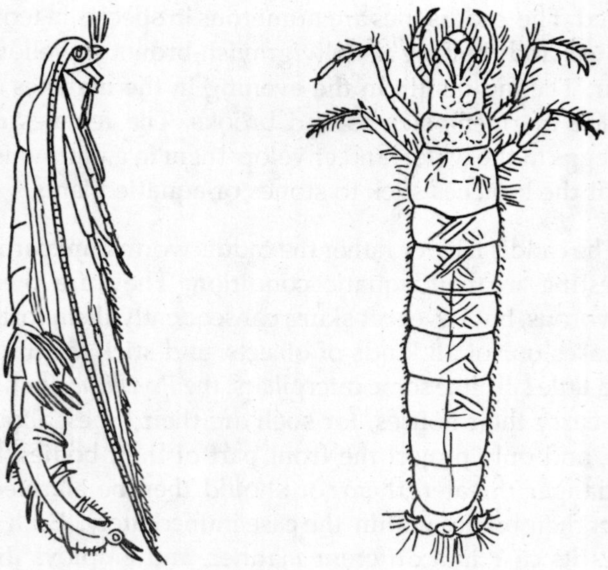

Fig. 8.16. The Nymph and Larva of Species of the Genus. *(Phryganea.* (Magnified.)

When about to undergo their metamorphosis the larvae fix their tube or case upon something and close up the ends. The nymphs have the same kind of respiratory organs as the larvae, and in order tenable its successor to breathe, the larva closes the tube with a sort of open-work fence, which, whilst it prevents any intrusions, allows the water to flow in and out.

The nymph has rudimentary wings, and resembles the larva more than the imago.

The caddis fly, with yellow or fawn-coloured antennae *(Phryganea flavicornis)* when in the larva state, forms its case with large bits of wood. *Phryganea rhombica* chooses the sprigs of aquatic plants, and *Phryganea fusca* makes a central tube of tiny pieces of stone, and adds long straws and shreds to the outside.

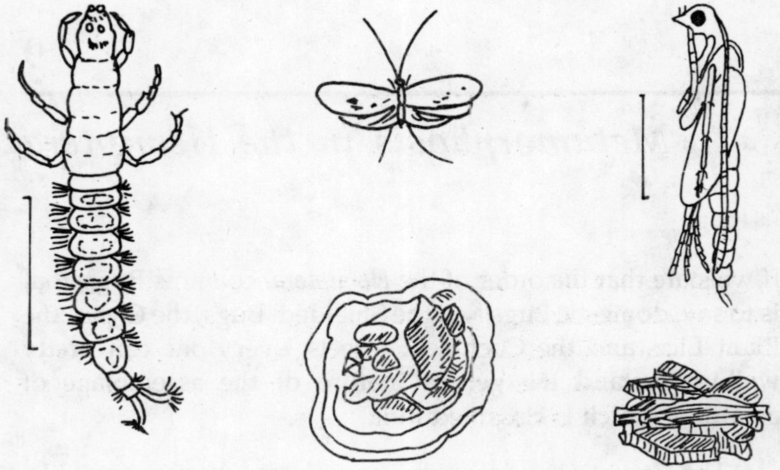

Fig. 10.17. *Hydropsysche atomaria*. Larva, imago, shelter, Pupa and case.

The tubes of these species are represented in the engraving.

Some caddis flies,when in the larval state, form shelters which are not movable, and which afford homes to several individuals. The species of *Hydropsyche* have larvae with bronchial filaments of gills attached to the abdomen, and their metamorphosis into the pupa is complete.

9

Metamorphosis in the Hemiptera

If we state that the order of the *Hemiptera* contains Bugs, that is to say, domestic Bugs and the Shielded Bugs, the *Cigala*, the Plant Lice, and the Cochineal Insects, every one can pretty well-understand the general nature of the assemblage of creatures which is classified in it.

The *Hemiptera* undergo an incomplete metamorphosis like the *Orthoptera*, and are born nearly in the same shape as the perfect insects, and, therefore, they do not suffer any very great external changes during their development. The larvae are, generally speaking only distinguished from the adults by the want of any appearance of wings. They undergo several changes of their skin, and after a particular one they appear with the rundiments of wings. The insects are then considered to be nymphs. After a last change of the skin the wings appear in the perfect state, and then the insect is considered to be adult. Their eggs are often very pretty, being variously ornamented and chamfered; and those of some kinds are in the shape of little barrels of tubs, with a sort of coverlid. Those of the plant-sucking kinds are usually deposited in layers upon the leaves and trees.

Many naturalista have divided the *Hemiptera* into two principal divisions, although all these insects are allied by very evident and natural affinities; but it is perhaps as well to admit that there are two sections of them. In the first—the *Heteroptera*—the wings are cuirasses in that half which is attached to the body, and transparent in the rest, and the suckers spring from the lower part of the head. In the second—the *Homoptera*—the wings are entirely membranous and trasparent, and the sucker is attached to the frontal region of the head.

There are four families in the section of the *Heteroptera*—the *Scutelleridae*, the *Lygeidae*, the *Reduviinae* and the *Nepidae*; and there are also four in the section of *Homoptera*—the *Cicadidae*, the *Fulgoridae*, the *Aphidae*, and the *Coccidae*.

The *Scutelleridae* contain a great number of species, and are commonly called Land Bugs, or shielded Bugs, or Wood Bugs. They are distinguished by having long antennae, and a scutellum that covers the whole of the wings and the abdomen. The body is generally rather short and oval, and the insects are often remarkable for the brilliancy of their colours, and for the red, green, and metallic hues which adorn the greater part of their surface. One set of these insects has the scutellum covering the whole of the body behind the prothorax, another has a triangular which leaves large part uncovered. The first is rarely seen in Europe; but the genus *Tetyra* has some kinds, which are injurious to our wheat crops. The second set is represented by *Pentatoma grisea*, which is found on many common plants and on walls, in groups; and another kind, which frequents the cabbage, is well-known on account of its black and red colouring. Hundreds of these are sometimes found upon cabbages and upon other cruciferous plants, which they injure very much. They have a very characteristic smell, which is similar to that of the ordinary bug.

In the engraving a full-grown adult is flying, and it will be noticed that the front wings are not membranous near the body and the triangular scutellum behind the chest is very readily observed. These insects may be noticed in a state of rest on the leaves; and one on the left-hand side shows how the wings are folded, and are half covered by the scutellum. Several larvae are crawling lover the plants.

The *Lygeidae* have rather elongated bodies, and their scutellum is always very small. One of them, which belongs

Fig. 9.1. The Metamorphosis of *Pentatoma grisea*.

to the genus *Astemma*, whose species, *Astemma apter*, has no ocelli, is found everywhere during summer-time, especially in the fields and gardens; but it often runs about on the ground on roads and paths. This homopterous insect is very peculiarly made about the wings, which are usually imperfectly developed, for their membranous part remains in a rudimentary condition;

however, it happens every now and then that some individuals have their wings full grown and perfect in every respect.

The engraving shows one of these insects with fully developed wings, and others which have them in a rudimentary condition. There are some insects of this set, however, which have ocelli on the forehead; they are very common, and children call them "soldiers and sailors," and they are always scrambling and crawling over each other on the twigs of plants. They are red and black in colour, or they may be of a general lead colour, with red spaces toned down with black spots.

Fig. 9.2. *Astemma opters.*

The name of Bugs *(Cimex)*, which was applied by ancient authors to almost the whole of this division of the *Hemiptera*, is now only given to those species with more or less egg-shaped bodies, antennae very delicate towards the end, and heads slightly narrowed behind. The type of this family is unfortunately well-known to almost every body. and it is the common Beg Bug *(Cimex lectularious).* This insect nourishes itself almost exclusively with the bold of man. Who does not know its habits? During the day-time it remains hides amongst the bed hangings, and in the wood, in cracks in the walls, behind the wall paper, and in any dark and narrow crevices; and its starts forth at night time to look for its victims, and displays a surprising amount of instinct and perseverance. some people remove their beds away from the walls with the notional that they are beyond the reach of such very intimate visitors, but the bugs crawl cup on to the ceiling, and when they come immediately over the sleepers they let themselves fall, and proceed to enjoy themselves forthwith. They lay their eggs in hidden places, and when they are hatched the young ones grow with greater or less rapidity, according to the temperatures of the room and the corpulence of its inhabitants. They change their skins very often, and the husks we see are not dead insect, but those of vigorous ones which are only a little way off. They undergo a very incomplete metamorphosis , and the adult insects never have any wings , and they thus resemble the larvae and nymphs of the other *Hemiptera*. We may thank Providence that they have no wings. They have, as everybody knows, a peculiar smell. The female bug lays her eggs at the beginning of summer, and the larvae, when hatched, are small, white, and semi-transparent, of a slightly different form the parent, and they take eleven weeks to attain their full size.

The *Reduviinae* have rather slender bodies, and are essentially carnivores in their habits. They run swiftly over the ground, and chase other insects, which they kill and suck. Curiously enough, they are especially found of the bed bug, but, unfortunately, they are rather scarce, so that it does not seem possibly to keep them in houses and to train them up to indulge in this very important peculiarity. They have a sharp suctorial organ, which is strong enough to pierce the skins of the animals upon which they live, and its puncture produces much plain. The species are scattered all over the globe, but are much more numerous in warm than in cold countries, only a few being found in Europe. The largest and finest are natives of India and America. The common species, however, is the european *Reduvius personatus*. It is generally found inside houses living amongst the dust and fluff. It is then in the condition of a larva or nymph, and it hides itself by covering its body with those evidences of uncleanly habits. When it becomes an a dult, it obtains its wings and flies about the fields, returning, however, to houses in order to lay its eggs.

A larva may be seen in the engraving on the ground, close to the basket on the right hand side; the perfect insects is entering the house, and immediately beneath it, on the ground, in a nymph.

Fig. 9.3. The Metamorphoses of *Reduvius presonatus*.

The *Hydrometridae* are aquatic *Reduviinae*. They are insects which are commonly observable during the summer upon the surface of running or standing water. Some have long narrow legs, with which the appear to be taking the measure of the top of the water. They are called *Hydrometra stagnorum*. The species of the genus *Gerris*, which are very numerous are also very common and some others belonging to the genus *Halobates* have been found walking over the waves of the tropical seas as easily as the first twirl and twist about on our ponds. In order to enable a *Gerris* to live on the surface of stagnant water, and to skim upon it with great velocity, Nature has given it a body somewhat like that of a wherry in shape. The long middle feet, which are covered with a fine close down like the rest of the body, glide on the water, and even remain motionless without becoming wet. Their food consists of other insects, and they are able in the adult stage to fly from pond to pond. It is very remarkable that the larvae of may species which differ only from the perfect insects by wanting wings, should be able to produce their kind without becoming full grown.

The *Nepidae*, or Water Scorpions, are carnivores, like the *Reduviinae;* they may be known by their very short antennae, which seem to be hidden in cavities below the eyes, and by two long bristly-shaped processes, which project from the end of their tail, and by which they breathe.

There are many groups of them. For instance the *Nepae* or true Water Scorpions, which have slender legs, the front ones being used to seize their prey, the leg being folded on the thigh, as in the *Mantidae*. These insects lay their eggs in packets upon stones or aquatic plants, but some kinds carry them on their back. They swim slowly, often immerse themselves in the mud, and live upon small attack animals. The *Ranetrae* have long, thin, and slender bodies, with immensely long legs; but the *Nepae* we have just motioned have large and flat bodies.

The *Notonectidae* are the true swimmers of this great division of the insects. They have a large head, short fore legs, and very long hind ones, which act as oars. All undergo incomplete metamorphoses. In the engraving on the opposite page *Notonecta glauca*, or the Boatman, may be seen low down. On the right are two water scorpions, *Nepa cinerea*, the smaller one being a larva. On the left hand the great sprawling *Ranatra* is one the top of the water.

Fig. 9.4. The *Nepidae*.

The *Cicada* family introduce us to the Homoptera, and from a well-characterized group, the kinds of which do not differ much amongst themselves, one *Cicada* being very much like another. In Southern Europe the Cigala *(Cicada)* are known to every one, and have been celebrated since the time of the highest antiquity; but they do not live in the center and north of Europe. The perfect insects live upon trees, sucking the juice of the leaves, and the females deposit their eggs,from 500 to 700 in number, in the dead or dried branches, in holes which they bore by means of a very curious ovipositor, which is something like that of the saw flies. They make little slits upon the branches, and drop an egg into each one of them.

The metamorphoses of the *Cigala* are rather more advanced then those of the *Hemiptera,* and are like those of the dragon flies. As soon as the larvae are born—and they look very much like fleas in the first instance—they descend the tree and hide themselves in the earth, and suck the roots in their neighbourhood. They are furnished with spiny legs and large and strong thighs, and these are capital instruments for digging. The nymphs cowl out of the earth and live upon the trunks of trees, or on the plants close by. After remaining perfectly quiet as nymph for a short time only, their skin cracks down the back, and the perfect insect leaves its covering and takes flight. The *Cigala* of the ash and of the elm are common in the south of France, and their wings are transparent and without any colour, but this is not the case with the species of India and southern Africa, for they have their wings magnificently tinted. The song of some of the Brazilian species is said to be heard for the distance of a mile; and one American species which is called the Seventeen Years' Locust' does a great deal of mischief. Ever since the year 1715, when its visit was first recorded at Philadelphia, no causes have affected the regularity of its return at intervals of seventeen years, even to the very month.

It is very common to notice on the leaves of willows and sallows by the river side little spots which look like saliva; if this matter is removed a larva or a nymph may be detected. It is called in England the Cuckoo Spit, or Frog Hopper, and is commonly found frequenting garden plants. The insect is called *Aphrophora spumaria.* The larva and pupa invest themselves with this frothy secretion, in which they lie hid until they complete their transformations. The frothy matter is vulgarly supposed to be really the spittle of the cuckoo. The larvae are quite destitute of any wings, and they are on rudimentary in the nymphs. The adults have legs fitted for leaping, and they luck the juices of plants with the aid of their secrets.

Fig. 9.5. Metamorphoses of the cuckoo spit. (*Aphrophora spumaria.*)

Some of the *Aphidae* of the Plant Lice have wings which are always more transparent than those of the little *Cicadellae* just mentioned; they have also long antennae, which are composed of seven pieces. They are provided with two small tube like projections which stick up from the end of the abdomen, and which communicate with leach other, and with a gland that secrets the sugary liquor of which the ants are so found, and which is said by some naturalists to nourish the newly-born *Aphidae.* There is no doubt about the ants milking the plant lice by tapping them on the end of the tubes, but it is a matter of doubt whether the young *Aphidae* have the sense or the power to avail themselves of what certainly might be very nourishing to them. The males of the plant lice have large wings, but almost all the females are deficient in these organs of flight.

It was formerly great puzzle to understand or to account for the rapid increase of plant lice, and the question has been carefully entertained by some of the most distinguished naturalists and physiologists of the world. It was evident to all, and in fact everybody can observe it, that perfect male and female *Aphidae,* all with wings, fly about in great quantities in the autumn season. A little care will prove that the eggs are laid at this time on plants, and that they are covered up with a sort of mucus which the female places over them to protect them. Now in the spring these eggs are hatched, but they do not produce caterpillars or grubs, but imperfect *Aphidae,* which have no wings, and which are all females. So far as is known, all the males die in the autumn, and the females also as soon as they have laid their eggs, so that in the early spring there are no other *Aphidae* in the world except these imperfect females. Now these, as everybody knows, are the mothers of millions and millions of plant lice, which are exactly like them—that is to say, imperfect because they are wineglass. These children are produced time after time, each set giving birth to others, land sometimes there are nine or ten successive generations. The last generation of *Aphidae* does not produce those like unto itself, but perfect male and female individuals with wings.

Fig. 9.6. The Rose *Aphis*.

All the generations of imperfect females are not born from eggs, but they are born alive, and they may be traced with the microscope like little buds within the old ones. Bonny,Reamer, Owed, Huxley, Carus, Leydig, and Balbiani especially have studied these interesting insects. As soon as they are hatched the imperfect females begin to grow and increase in size, and they attain their full dimenstions in ten or twelve days; and then these curious little inside buds are born the rate of three, four, or seven a day. When they are born they are miniatures of their mother, and in ten days they begin to produce other. The second generation does the same, and so on until the ninth or tenth. Then the imperfect females have inside buds, which, when born, are not miniatures of their mother, but resemble the parents which laid the eggs the year before. These grow rapidly, and soon fly, and lay eggs in their turn. It is very evident that simple arithmetic can show that one plant louse might produce a quintillion during one year. It is easy then to understand the occurrence of enormous multitudes and the sudden appearance of thousands of these insects upon different plants. Fortunately they have a crowd of enemies otherwise they would do a vast amount of harm to vegetation.

They are found on almost every kind of plant, and upwards of three hundred species have been described. The hop *Aphis* sometimes nearly ruins the crop, and the rose *Aphis*, and that wheil is found on the scented geranium also do much mischief. The larvae of the *Coccinella*, or lady birds, and several *Ichneumons*, devour and kill great numbers.

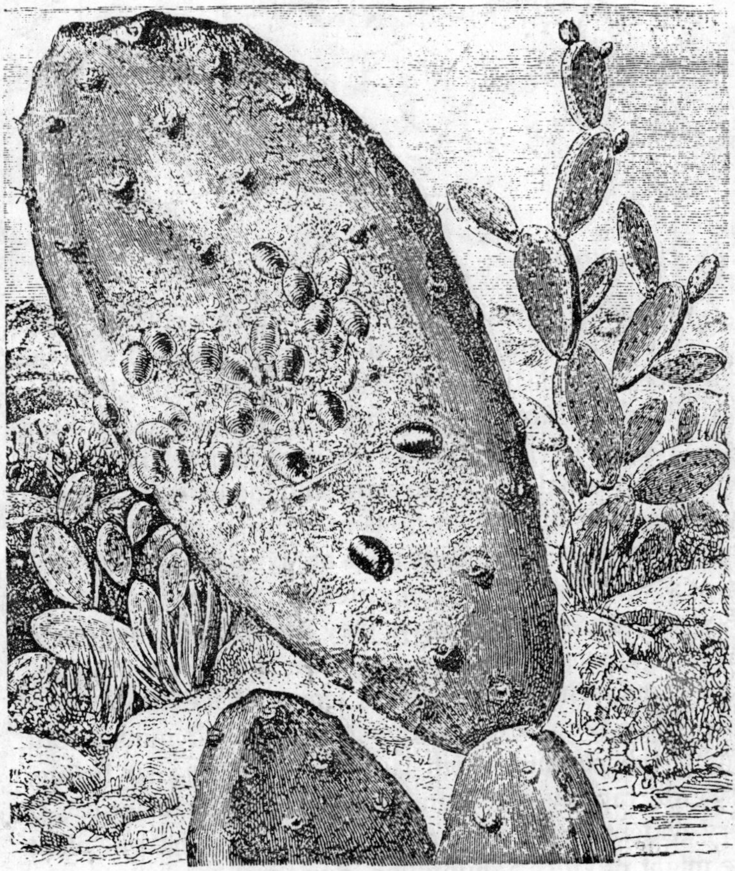

Fig. 9.7. Cochineal Insect. *(Coccus cacti)*.

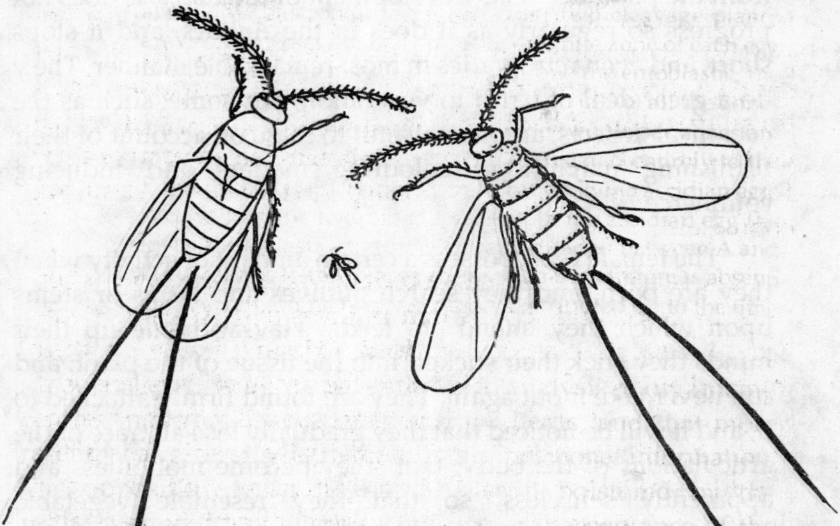

Fig. 9.8. Cochineal Insect. (Magnified.)

The mischief done by these insects is produced by their tapping the delicate tissues of the plants with their suckers, and by their extracting the richest sap. They congregate in groups on the stems and leaves, but in the autumn the last generations emigrate from plant to plant.

Some of the *Aphides* which belong to the genus *Lachunus* do a great deal of harm to apple trees; they are well-known from the curious white cottony stuff which encircles them and protects them from the air. When we crush this white mass the *Aphis* within it bursts, and a red stain is produced like blood. Some others live in groups, which are enclosed in the galls which they produce by irritating the structures of certain plants.

The family of the *Coccidae* are very closely allied to the

Aphidae, and are frequently called scale insects or meals bugs. They are very interesting, and may be easily distinguished from the plant lice. The development of the *Coccidae* does not progress so regularly as it does in the *Aphides*, and it stops short, and even retrogrades in most remarkable manner. The y do a great deal of harm to vegetation; but some, such as the cochineal insects, are most useful to man, on account of their furnishing magnificent colouring matters and inducing commerce.

The female *Cocci* possess a certain amount of activity when they are born, and they search outlives and twigs or stems upon which they intend to feed. Having made up their minds they stick their sucker into the tissue of the plant, and can never take it out again. They are found firmly attached to it, and it will be noticed that they gradually lose all trace of the articulations of the body, that they become motionless and apparently senseless, so that they resemble vegetable excrescences. They are always sucking, and even their courtship is spent whilst they are imbibing the juices of the plant beneath them. They lay a large quantity of eggs, and produce a greater or less abundance of a cottony matter. The egg laying goes on without any movement on the part of the mother, and when she dies her beds dries up and often forms a shelter for the newly born creatures. The larvae, as soon as they are born, spread over the branches and leaves, and remain there during the warm weather. They are usually very small, but they increase to such lan extent that the death of the plant becomes almost certain. It appears that they are reproduced somewhat like the *Aphidae*, but only the males have wings.

The engraving represents the males of the cochineal insect highly magnified.

10

The Metamorphosis in the Diptera—Flies

There are only two large and well-developed wings, which are membranous and without scales, in the *Deptera;* the second pair are quite rudimentary, and form small projections of wingmembrane, which are called balancers or posers. In the common gnat these poisers are like simple foot-stalks surmounted by round knob, and are attached one on each side to that segment of the body which ought to sustain a large hind wing. This is their form in the house fly. They are generally more or less concealed beneath the winglets, which are little appendages to the fore wings. The excessive size of that segment of the body which bears the great proper wings of the *Diptera,* the front pair, appears to render a diminution in the size of the segment which ought to maintain the hind wings inevitable.

The *Diptera* are sucking insects, but their suckers are formed differently to those of the *Hemiptera* or the *Aphaniptera.* Moreover these important organs of the mouth differ greatly in many general of the flies themselves. These insects undergo complete metamorphoses; the larvae are ordinarily vermiform and legles, are known as maggots, and usually possess considerable activity, and can crawl backwards and forwards. They do not required maternal assistance, and are usually placed so that their food is close at hand. The nymphs of the *Diptera* differ greatly; some of them are active, although they take no nourishment; some are inactive, and quite as much so as the chrysalids of the *Lepidoptera* or the nymphs of *Coleoptera*, and others which are perfectly immobile, do not lose the skin belonging to the larvae; these are well shown in the case of the common house fly. The metamorphoses differ, therefore, considerably in this very natural group.

Fig. 10.1. The Metamorphoses of the GNAT.

There are two great divisions in the order of the *Deptera* characterized by distinctions in the antennae; the first section have thread-shaped and the second very short antennae, consisting of three very distinct joints and a slender thread which is called a style. The first section, the *Nemocera,* have long and slender bodies, long wings, and great legs, which are excessively thin and delicate. The Gnats and the Daddy Long Legs are the types.

The natural history of the if full of interest, and even the common gnats, *Culex pipiens,* is well worthy of study; and we may almost wonder at finding such a bloodthirsty and annoying being in so delicate and fragile a body. The gnat has a very elaborately constructed offensive weapon in its mount, the structures of which, although excessively delicate, can do their perforating duty perfectly. All the pieces of the mouth are free, and the mandibles are in the shape of sharp blades, toothed on their margins. Gnats abound in marshy districts and where there is stagnant water, for their larvae are aquatic, but cannot live in running streams. They are very fond of such places as water tubs in gardens, and it does not much matter if the water is not quite fresh.

The male gnats have pretty hairy antennae like little feather, and the females have antennae which are almost plain; it is , therefore, not difficult to distinguish one from the other, and it is rather important, for the females are the flood-suckers. When about to lay their eggs they seek the water, and, with the assistance of their long hind legs, collect and agglutinate them together, and place the little boat-shaped mass upon the surface and leave it to its fate. The larvae are soon hatched, and grow with great rapidity. They swim with great agility, and may be noticed to be constantly ascending and descending, and rushing about in search of prey. They transparence of their tissues, the delicacy of the hairs which cover them here and there, and the extreme mobility of the segments of their bodies, render them very interesting objects for the microscope. They are almost always seen with their heads downwards and their tails towards the surface of the water. There are two antennae on the head, which is also furnished with hairs that direct the food towards the mouth. The first segment of the body is large and broad, and then the others diminish gradually to the last, which carries tufts of hairs, and the last but one is provided with a tube, the extremity of which is perforated for a respiratory orifice. After the larvae have grown to a certain size they undergo a change of their skin, and become nymphs, which, although active and full of life, are, as it were, swathed up and incapable of taking nourishment. They swim with the aid of two large lamellae, which resemble a tail somewhat; and it may be noticed that when the nymphs come up to the surface of the water they do not present their tails, like the larvae, so as to obtain air, but allow their backs to touch the surface just where there are two little respiratory tubes.

When the perfect insects is about to emerge from the nymph this last floats on the surface of the water, perfectly at rest, and then the skin of the back, which is exposed to the air dries and splits open. Then the perfectly-formed gnat begins to come out; first of all lit protrudes its head, then a portion of its body, and, after a short time, lone along leg after another is disengaged from the nymph's skin. But all is not over yet, and the gnat has to take great pains so as not to capsize the frail little boat-like skin which separates it from the water. Very gradually this little creature rises up and extends its wings, which are still moist and limp, but soon, especially if the weather is warm, they become crisp and movable, and then the gnat gives a downward stroke with them and flies off in safety. There is no pseud imago state, as in the *Neuroptera.* When there is no wind this isn't a very difficult operation , but during stormy weather thousands of gnats are drowned, for, if their wings are once wetted, there is an end to the insect's existence within a very short time.

The species of the genus *Tipula,* known commonly as Daddy Long Legs, from their immensely elongated thin legs, frequent grass lands and marshy districts and very frequently commit great havoc upon the turf. The females have a long ovipositor, and penetrate the earth with it and leave their eggs underground, and the larvae eat the fine roots of the all the delicate plants which come in their way. The *Tipula oleracea*, or the Change Crane Fly, is very common, and the perfect insect is of a tawny colour, and its wings are rather, dusky. The larva has a black head, and the rest of its body is of an earthy colour, and the last segment of the body is perforated by the respiratory orifices, and has four tubercles upon it. The nymph is more compact that the larva, and is usually motionless , but it can move with the aid of some spines which are upon the body, so as to approach the surface of the soil when about to under go the second metamorphosis.

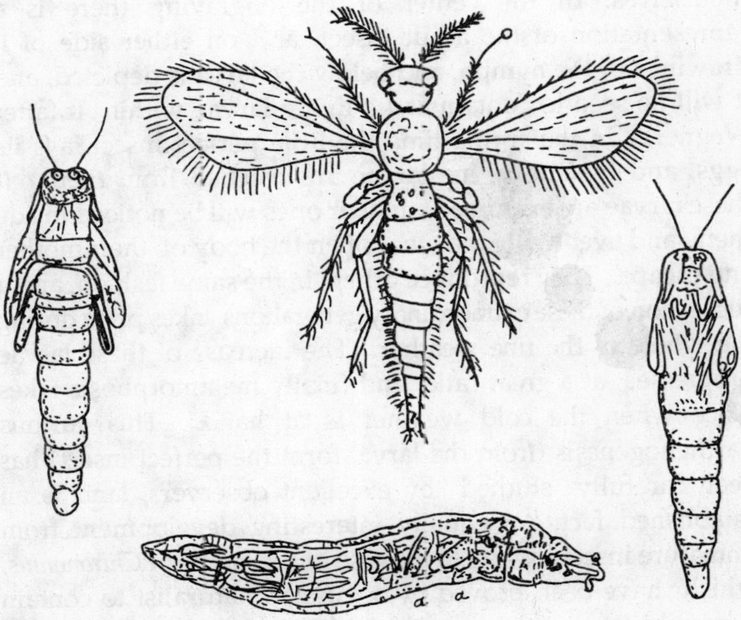

Fig. 10.2. *Cecidomyia* with Viviparous Larvae.

The *Cecidomyidae* are very small *Diptera* which, like the *Cynips*, produce galls on plants in which their larvae live. Never theless, there are exceptions to this gall-making peculiarity, for the wheat fly, *Cecidomyia tritest*, lays its eggs in the center of the flower of the wheat plant, and when its larvae are hatched they eat the pollen, and thus prevent the formation, of the corn. An American species, which is known by the name ofthe Hessian Fly, attacks the lower part of the stem of the wheat plant, and the larva nourishes itself with the sap, which it appears to take by suction, and causes the stem to wither and die.

A very interesting discovery has been made by Nicholas Wagner, a professor at Kazan, to the affect that the larvae of some species of *Cecidomyia* which live underneath the bark of trees have the power of protecting courtiers like unto themselves. In the center of the èngraving there is a representation of the adult insect, and on either side of it drawings of the nymph, and below the larva is depicted, and it will be seen to contain at *a a* living larvae within it (after Wagner). In the spring-time the little perfect insect lays its eggs, and by-and-by the larvae are hatched from them. If these larvae are examined, smaller ones will be noticed inside them, and eventually they tear open the body of their mother and escape. They reproduce others in the same fashion, and a succession of these extraordinary generations takes place during the whole of the fine weather. The increase of these larvae progresses at a great ratio, and finally metamorphosis takes place when the cold weather is at hand. This curious parthenogenesis (from the larva, form the perfect insect) has been carefully studied by excellent observers, land is an established fact. An equally interesting development from immature insects occurs in the case of the larvae of *Chironomus*, which have been proved by a Russian naturalist to contain eggs which are subsequently laid by the pupae. The gall-making species are usually very small as regards their individuals and the females deposit their eggs in the young shoots leaves, or flower, which they pierce by means of their sharp-pointed depositor. There this puncture is made a gall or excrescence takes place, which grows sometimes to a large size, serving often as nest and food to the young larvae.

Fig. 10.3. The Metamorphoses of *Tabanus bovines.*

The *Asilidae* belong to the second division of the *Diptera*, and are very active insects, which had a very formidable sucker, for two mandibles being completely soldered together form a very sharpened piercing blade. The species are very numerous, and are strong predatory insects, living upon live flies, and even upon bees,which they chase and soon kill. Their flight is strong and they make a loud buzzing noise when on the wing. some, which have elongated bodies a short trunks, and a formidable sucked, attack horses and cattle, and sometimes man, and they pierce the skin and suck them blood. One of the commonest European species, *Asilus crabroniformis,* is decorated like some of the *Hymenoptera,* and when it begins to suck produce the same pain as that inflicted by a sharp lancet. It often attacks large animals, but usually it sucks the blood of insects, and especially of caterpillars.

This insect which is so bloodthirsty in adult age, appears to be a vegetarian when in the condition of a larva. The larvae are found underground, where they gnaw roots, and eventually construct cell, in which metamorphosis takes place into the nymphs.

The *Tabanidae,* or Breeze Flies, are generally larger and more powerful than the rest of the *Diptera;* they have a large body, with a depressed head, and their sucker or proboscis is very different to that of the kinds just mentioned. They have large trunk, which is sometimes longer than the whole body, and their piercing mandibles are not joined together, but are separated. Their wings are provided with a greater number of nervures than any others of the order. They live chiefly by sucking the blood of animals,especially of horses, and even of man; but the males are by no means so sanguinary as the females. These pierce the skin of their victims with great facility by means of the lancets of their proboscis, while the former live partially from the juices of flowers. The breeze flies are met with chiefly in woods and pastures, in the middle of summer, and during the heat of the day. When flying they make a loud buzzing noise, and that of *Tabanus bovines* is well-known to cattle and horses, for they become very restive at the sound. This insect is a large pale-brown fly marked on the back by a series of whitish spots, and the larva is a large, dusky yellowish, cylindrical worm, marked by transverse black rings. The larvae know nothing about the love of blood, but live underground, and have no legs, but they manage to feed upon vegetable matters. They are metamorphosed into immobile nymphs, which may be recognized by this having six spines at the end of their bodies.

The perfect insects, one upon a tree and the other in flight, are shown in the engraving on the opposite page, and the protruding proboscis, extended wings, and long legs of these must be compared with the legless condition of the immobile nymph and legless larva which are represented underground in order that the extent of the metamorphosis should be appreciated.

The metamorphoses of *Stratiomys chamaeleo* are, perhaps, as interesting as those of any other of the *Diptera* It is a common fly, and, being, bloodthirsty, it frequents flowers in order to meet with other insects whose juices it can suck. The larva lives in stagnant water, and is a along creature with cuirasses integuments, and its head is small, and is furnished with two hooks which are really its mandibles. The terminal segments of its body are thin, delicate and are so narrowed that they can enter slightly one within the other like the slides of a telescope, or can elongate by being projected backwards; they form a long tube, which is pierced at its extremity by two orifices, which are surrounded by a crown of small hairs. The larva chooses shallow waters to swim about in , and when it requires air and a supply of breath, it sticks up the end of its body land elongates the last segment, and respires through the two holes just mentioned. After a longer or shorter larva life, the insect comes up to the surface of the water and floats; its does not change its shape, but evidently a change has taken place, for the larval condition is at an end. If one of these floating creatures be taken up and cut open, the pupa will be seen within the skin, which acts as the cocoon, and also as a boat for the fly which will spring into life in a short time. The fly, which as two perfect wings, a flat body, long antennae, and which is decorated with beautiful colours, is about as different an insect from the long-tailed, dull-coloured swimming larvae as can well be imagined. In the engraving opposite the perfect insects are seen in flight, and one is upon a rush; two larvae are shown, one on the right hand side has its tail erect, and is breathing in air, and beneath the insect on the leaf, is a nymph.

Fig. 10.4. The Metamorphoses of *Stratiomys chamaleo*.

The *Diptera* with a short and thick trunk, mandibles united to form a single blade, and with broad and very flexible lip, which makes a good sucker, are distinguished as *Syrphidae*.

Their proboscis is thus long and membranous, and they like honey. There are several very interesting genera in this family. The *Volucellae* are large flies, which have black, yellow, and russet colours upon them, and they choose the nests of social bees for their larvae. The female deposits its eggs in the combs, and particular species have their favourite bee-hive or wasp nest, as the case may be. Thus *Volucella bombylans* visits the humble bees, and *Volucella zonaria* attacks the wasps, and curiously enough wears the livery of those dangerous insects, amongst which she leaves her eggs. The larvae of *Volucella* are of a dirty white colour, are covered with small spines, and the last segment of the body is rounded off, and is pierced for the organs of respiration. The membranes legs are furnished with little books. The egg is laid in the cell which contain the wasp larva, and when it is hatched the *Volucellsa* eats its way into its victim, and gradually intrudes the whole of its body except the last segment, which being in contact with the external air, and having the respiratory orifices in it, is the only portion ofthe body by which the parasite can be distinguished, and through which it can take in air to oxygenate its blood.

When the *Volucella* larvae attain their full growth they are transformed into nymphs, which have two small open tubes places upon the prothoracic region, and which communicate internally with the tracheae. This apparatus is like that of the gnat.

The bees are often attacked, according to A.S. Packard, by a fly which belongs to the genus *Phore*. "It is a small insect, about one line and a half long, which is found in Europe during the summer and autumn flying slowly about flowers and windows, and in the vicinity of bee-hives. Its white transparent larva is cylindrical, a little pointed before, but broader behind. The head is small and rounded, and at the posterior end of the body are several slender spines. The pupa case which encloses the delicate chrysalis is oval, consisting of eight segments, flattened above, and with two large spines near the head and four on the extremity of the body. When

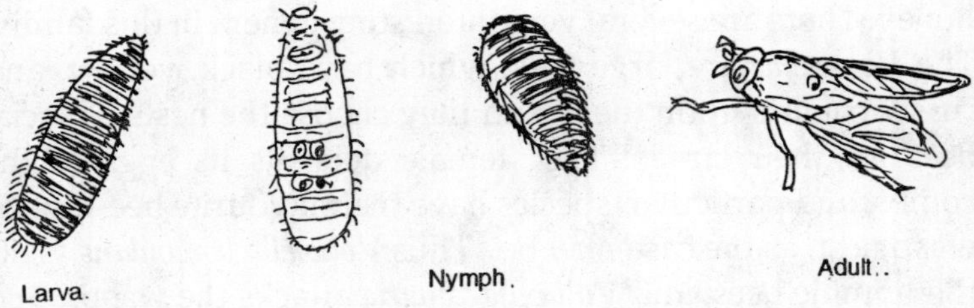

Fig. 10.5. Metamorphoses of *Volucella zonaria.*

impelled by instinct to provide for the continuance of its species, the *Phora* enters the bee-hive, and gains admission to a cell; then it bores with its ovipositor through the skin of the bee larva, lying its long oval egg in a horizontal position just under the skin. The embryo of the *Phora* is already well developed, so that three hours after the egg is inserted in the body of its unsuspecting and helpless host the egg is nearly ready to hatch. In about two hours more the larva actually breaks off the larger end of the egg shell, and at once begins to eat the fatty tissues of its victim, its posterior half remaining in the shell. In an hour more the larva leaves the egg entirely, and buries itself completely in the fatty portion of the young bee. The maggot moults three times. In twelve hours after the last moult it turns round with its head towards the posterior end of the body of its host, and in another twelve hours, having become full fed, it bores through the skins ofthe bee larva, eats its way through the brood curving of the cell, and falls to the bottom of the hive, where it changes to a pupa in the dust and dirt, or else it creeps out of the door, and transforms in the earth. Twelve days after the fly appears."

The habits and metamorphoses of the insects included in the genus *Eristalis* resemble those of *Stratiomys.* The larvae have the same kind of tail-like appendage, which can be elongated when required; they live in shallow water, and move slowly on the bottom by means of very small legs, which are attached to the thorax and to the abdominal segments. The bodies of the large are not supplied with any respiratory

Fig. 10.6. The Metamorphoses of the rat tailed fly (*Eristalis tenax.*)

apparatus on their sides, so that the insects can only breathe by elongating and sticking up their tails, which are perforated, and can act as breathing organs. The larvae are found in very briny waters, and even in the water of salt works, and besides in stinking pools. The perfect insect of this rat-tailed creature has a large blackbody, with russet spots upon the abdomen.

In the engraving the larvae may be seen having two large eyes, a stoutish body, and a long tail.

If one of them be placed in pure water it will live for a short time, and by just covering the insect at first, and adding water gradually, the tail will be noticed to become erect, and to elongate gradually, so as to keep pace with the flow of water, and always to remain above the surface. After having developed its tail to the utmost, should any more water be added the insect is drowned.

The Syrphus Fly, or Aphis Eater, is gaily spotted and banded with yellow, so that it resembles a small wasp. They may be seen upon flowers in fine weather, and they lay Their eggs upon bushes and plants. The larvae are very small creatures, and are very slender in front, and they walk, or rather crawl over the foliage, principally with the aid of their mandibles.

They are of a pale green colour, and may often be observed upon rose trees where they feed upon the plant lice, which they suck and destroy with great rapidity. The larvae of other *Syrphidae* even attack caterpillars, and manage to eat their way into the bodies of their victims, but they always leave the last segment of the body in the open air, so as to breathe, for the respiratory orifices are situated at the end.

The true flies, or *Muscidae,* are a great puzzle to those naturalists who indulge in classification. They have a distinct proboscis, which is thick, membranous, and terminated by two large lobes. It is formed by the united mouth-pieces, and especially by the large lower up, and the mandibles, which are partly united together, form a sharp blade. These *Deptera* which differ so much in the adult stage, resemble each other in the larva condition. The white maggots are so well-known that it is not necessary to describe them, but it must be observed that they do not moults or change their skin during their growth, and that when the time for metamorphosis is at hand, the body shortens, the skin becomes hard, and of a brown colour, and the chrysalis or pupa forms within. Many of the larva live as parasites within other insects.

In the accompanying engraving the larva of *Gymnosoma rotundata* is shown within the body of one of the *Hemiptera*. It has a long hooked-shaped tube, which is attached to its last segment, and this tube is protruded through one of the spiracles or respiratory orifices of its victim so that it eats the juices and fat, and breathes through the body of the creature or "host" upon which it is parasitic. The perfect insect and the pupa are also depicted.

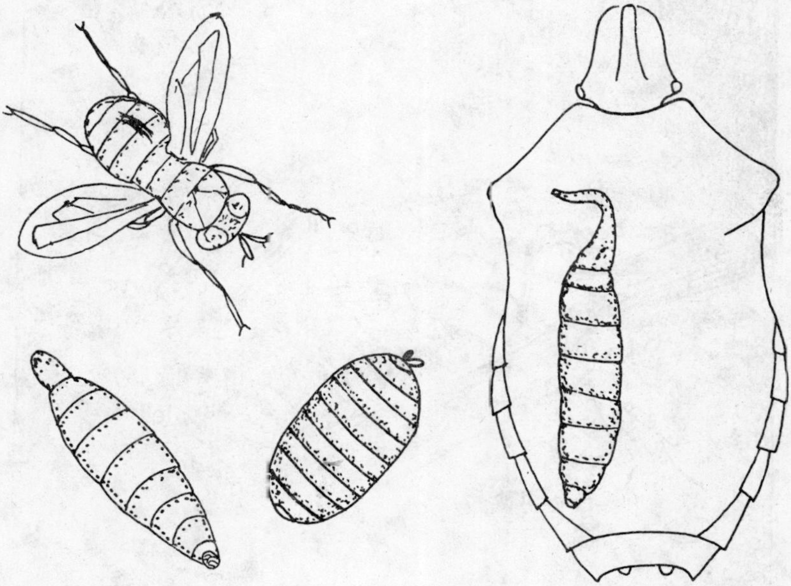

Fig. 10.7. The Metamorphoses of *Gymnosoma rotundata.*

The Flesh Fry and the Blue Bottle are very useful as larvae, or, as they are called, Gentles, for they devour dead bodies of all kinds, and many species, belonging to the genus *Sarcophaga*, live upon stercoraceous matter.

The curious legless larvae are very well-known, and so are the egg-shaped, dark-brown pupae. The larvae, which at first are called "fly blows," increase in size with very great rapidity if they have sufficient food; and there in some truth in the saying that a pair of maggots will eat up a carcas as soon as lion, for the reproductive powers of the female are enormous.

Fig. 10.8. The Metamorphoses of the Meat Flies. (*Calliphora wmitoria* and *Sarcophaga carnaria.*)

The Bot Flies are very extraordinary creatures. They are large velvety flies, which have very small antennae and a rudimentary trunk. When in the imago or perfect condition they do not take any nourishment; but inasmuch as they chase horses oxen and sheep, and all kinds of ruminants, and frighten them with their buzzing, they are generally supposed to sting the animals, but really they are incapable of doing anything of the kind.

The Gad Flies, or *Tabani*, chase the animals in order to suck their blood, but the bot flies seek them in order to lay their eggs, for their larvae live as parasites upon the domestic animals, and ruminating *Mamamalia* generally. The adult fly lays it eggs on the skin of some particular animal, and in some instances they are licked off the skin and are conveyed into the stomach, where they hatch, and the larvae live by sucking the juices of the animal or the matter which their presence produces. Some a larvae are hatched on the surface of the skin, and they then bore into the body underneath, and feed there. The larvae are thick, lf fleshy bots without feet, and they are covered with rows of spines and tubercles, by which they move about, and they breathe through spiracles, which open at the end of the body. They moults twice and just before attaining the pupa state they leave the bodies of the creatures they have lived in, and reach the ground, where they undergo their metamorphoses, the pupa being protected by the dried skin of the larva. The bot fly of the horse buzzes about horses and lays its eggs rapidly, and attaches them with a sticky substance to the hair about those parts of the body which the animal is in the habit of licking. The eggs are hatched, and the horse swallows the larvae and they pass down into the stomach, to which they become fixed by means of their hooked mandibles. After a certain time they are passed out with the excrement and undergo their metamorphoses in cracks under the ground the bot which affects sheep is an object of great tirade to them. It lives in its larval state up the passages which lead from the nose amongst the front bones of the head, and produces much suffering.

The larvae of some bots, especially those which attack oxen, deer, and goats, dig themselves under the skin of those animals, and produce little tumours; and this is well-known to some birds, for the sterlings may be constantly seen at certain seasons of the year looking out for them on the backs and flanks of deer.

There are several well-authenticated instances of bot flies attacking men, but it is probable that this is an accidental occurrence, and that they are parasites of the monkey or dog.

There are some very extraordinary flies which inhabit the sea, and Puckered describes the larva of a species living on floating sea grass. He noticed that they were transformed into pupae, and that in nineteen days the second metamorphosis into the fly was perfected. He also describes an *Ephydra* whose larva lives in the strong salt works of Illinois and under the sea-weed on the shores of Narragansett Bay.

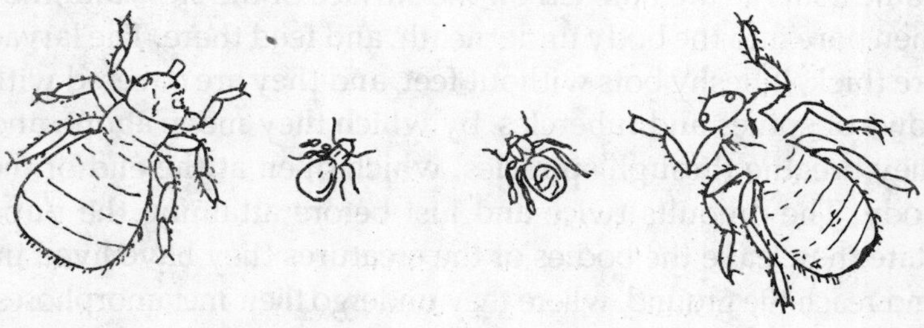

Fig. 10.9. *Melophagus avis,* *Leptatana curve.*

The Ticks are *Diptera* without wings, or which have them in a rudimentary condition. They have a sucker, and live parasitically upon all sorts of creatures.

The Sheep Tick, *Melophagus avis,* has a wingless and hairy body and a long proboscis. The young are developed within the body of the parent until they attain to pupa state, and then the pupa in its case, which is nearly as large as the whole insect, is cast forth. Packard notices the Bat Tick, which has a small body and enormous legs, and it is either blind or provided with four simple eyes. These insects are of small size, being only a line or two in length.

The Stag Tick, *Leptotoena survey*, like the Sheep Tick, has no wings, and remain attached to the animal; but some Bird Ticks have very small wings, which enable them to leave one animal for another. These degraded *Diptera* are the connecting links between the true insects and the spider and mite families.

The Bee Louse or Tick may be compared to a spider fly, and is about two-thirds of a line long. It is pupiparous, like the sheep tick, and the very day it is hatred it sheds its skin and changes to an oval pupa of dark brown colour. Packard states that its habits resemble those of the flea, and that it is evidently a connecting link between the flea and the two winged flies. Like the former, it lives and brings forth its young on the body of its victim or host, and draws its food by plunging its stout beak into the skin of the bee; and sometimes one of these industrious insects may be seen to be weighted down by as many as a hundred of these bloodthirsty creatures.

The varieties of metamorphosis in the *Diptera* are thus very considerable and it must strike every observer that the most decided transformations are those of the parasitic flies. This is one of the proofs that the metamorphoses has been gradually added to the evolution of the insect, and that it is preservative of the species.

THE ANOPLURA—THE THYSANURA

These are the most degraded forms of insects, and they never obtain, during any time of their lives, those elaborate structures which are usually considered to be characteristic of the class. The members of the species of *Anoplura* and *Thysanura* are invariably winglass, and their young are hatched almost as fully developed as the adults. They do not undergo metamorphosis but their growth and development are accompanied by the phenomena of skin changing.

It appears that when the parasitic lice or *Pediculi*—which constitute an important portion of the class *Anoplura*—are hatched from the egg, they have not much subsequent alteration to submit to, as they are nearly perfect. Moreover, the young louse is not more perfect than the adult, and, therefore, no retrograde metamorphosis takes place. The caterpillars of *Psyche* and *Orgyia* turn either into male moths with beautiful wings or into females with the most rudimentary organs of flight, and this occurs because these last submit to a retrograde metamorphosis. The adult female *Orgyia* is absolutely lower in organization than the caterpillar. But, although there is some resemblance between these female moths and the *Anoplura* it must be remembered that the lice retain their larval condition through life. There are some proofs however, that there is a regrograde development in the louse before it's hatched. Until late in its left within the legs the louse grows, and is formed like the embryos of other insects, and then, just before hatching, takes on its peculiar degraded aspect and essential structural detail. Then the parasite is born closely resembling its parents.

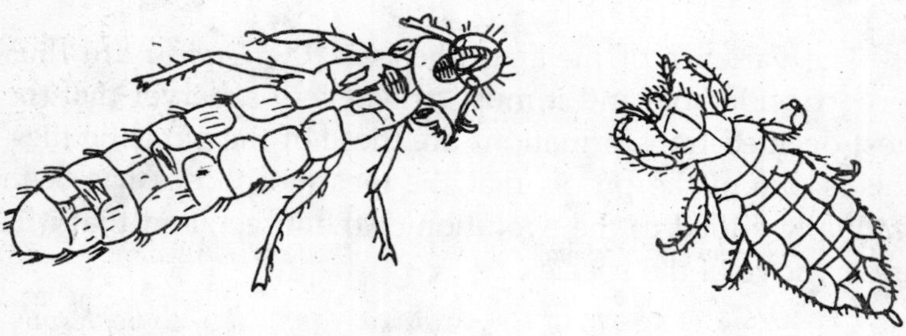

Fig. 10.10. *Philopterus selcifrons* *Pediculus capitis.*

There are two great divisions of the *Anoplura;* in one, the true lice, the insects have suckers; and in the other, the bird lice, there are jaws. The *Pediculi* which are parasitic upon the hair and clothes of man are, of course, disgusting on account of their association with dirt and disease, but nevertheless they are interesting creatures to the naturalist, on account f their wingless and degraded condition, as well as from their not undergoing metamorphosis. Moreover, there is much interest accompanying their birth, for the females preponderate, and they can lay fertile eggs time after time without the help of the male, and generations of females succeed each other and produce fertile eggs under the same exceptional circumstances.

The Head Louse, drawn above, is not every unlike a long bedbug *(Cimex)* in its early stage.

The lice without suckers, and with jaws that can bite through the skin, afflict most domestic mammals and all sorts of birds. In tracing their development within the egg it becomes evident that some of the earlier characters ofthe embryo are retained persistently; thus the permanently enlarged forehead of the bird louse can be recognized during very early egg life, but this structure desappears very soon in the embryos of the true lice with suckers.

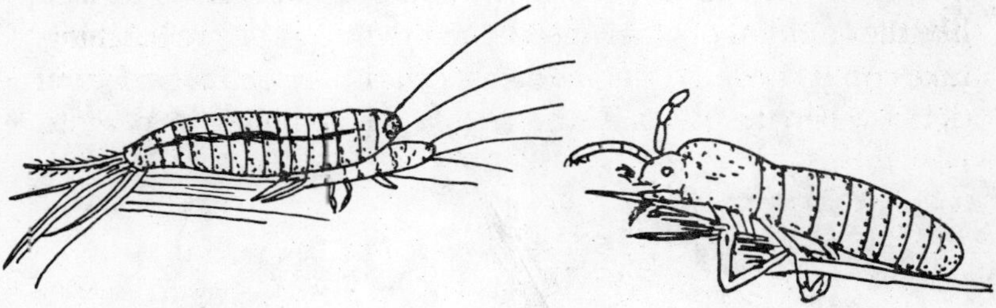

Fig. 10.11. The Fish Scale (*Lepisma saccharina*) The Skip Tail (*Podura plumbea*)

The legs and claws of the different species of *Philopterous* or bird lice vary much is structure, on account of the peculiar conditions in which the parasites are placed and live. One genus, which has a species parasitic on the porpoise—the only marine parasitic true insect—is vary degraded in its structures and development; and the forms classified as *Goniodes* are also remarkable for their defective organization, as well as for the presence of a claw upon the antennae of the males. The philosophical naturalists may ponder with interest upon the absence of the metamorphoses in these parasitic creatures, and consider their evident anatomical, and therefore classificatory relation with those spiders which have tracheae as their respiratory organs; and he must be struck with the anatomical relations of these groups with the other insects that undergo incomplete or imperfect metamorphoses.

The *Thysanura* or skip tail tribe are small insects which are better known on account of the beauty of their microscopic body scales than for any interesting habits or instincts. They do not undergo metamorphosis, and are wingless.

Sir John Lubbock's careful researches have resulted in his being satisfied that in the genus *Smynthurus* there are tracheae for respiratory purposes, and that they communicate with the external air by means of spiracles, which are not situated along the margins of the thorax, but in the head, where it is attached to the body.